AF389482

Nanocellulose and Nanocarbons Based Hybrid Materials

Nanocellulose and Nanocarbons Based Hybrid Materials: Synthesis, Characterization and Applications

Editors

Djalal Trache

Vijay Kumar Thakur

MDPI • Basel • Beijing • Wuhan • Barcelona • Belgrade • Manchester • Tokyo • Cluj • Tianjin

Editors
Djalal Trache
Teaching and Research Unit of Energetic
Processes, Ecole Militaire Polytechnique
Algeria

Vijay Kumar Thakur
Biorefining and Advanced Materials Research
Center, Scotland's Rural College (SRUC)
UK

Editorial Office
MDPI
St. Alban-Anlage 66
4052 Basel, Switzerland

This is a reprint of articles from the Special Issue published online in the open access journal *Nanomaterials* (ISSN 2079-4991) (available at: https://www.mdpi.com/journal/nanomaterials/special_issues/cellulose_carbons_hybrid).

For citation purposes, cite each article independently as indicated on the article page online and as indicated below:

LastName, A.A.; LastName, B.B.; LastName, C.C. Article Title. *Journal Name* **Year**, *Article Number*, Page Range.

ISBN 978-3-03943-374-2 (Hbk)
ISBN 978-3-03943-375-9 (PDF)

Contents

About the Editors

Djalal Trache has been working as an Associate Professor at Ecole Militaire Polytechnique (EMP), Algeria, since 2016. He received his Engineer degree in chemical engineering, Magister in Applied Chemistry and Doctor of Sciences in chemistry at EMP. He has made several presentations at national and international conferences, published over 80 scientific papers in international peer-reviewed journals in the field of chemical sciences/materials science, eight book chapters, and one book. He is a reviewer of more than 50 international reputed journals. Prof. Trache has particular expertise in energetic materials, bio-based materials, polymer composites and their characterization. He also has interests in nanomaterials and their applications, phase equilibria and kinetics. Besides, he has successfully supervised many engineers, MScs and doctoral students.

Vijay Kumar Thakur is currently a Professor of Biomass in the Biorefining and Advanced Materials Research Centre at SRUC, Edinburgh, U.K., and also holds an Adjunct Professor position in the Research School of Polymeric Materials, Jiangsu University, China, and is a Visiting Professor at Shiv Nadar University, India, and a Visitor at Cranfield University, U.K. He has previously held faculty positions at Cranfield University, U.K., Washington State University, U.S.A., and Nanyang Technological University, Singapore. His research activities span the disciplines of Biorefining, Chemistry, Chemical Engineering, Manufacturing, Materials Science, and Nanotechnology, as well as all aspects of Sustainable and Advanced Materials. He has been a PI/Co-I on several projects sponsored by BAE Systems, EPSRC (EP/T024607/1), Royal Academy of Engineering (IAPP-33-24/01/2017; IAPP18-19 295), UKIERI (DST/INT/UK/P-164/2017), Innovate UK, and others. He has published over 200 SCI journal articles, 2 patents, 50 books, and 37 book chapters in areas concerning polymers, nanotechnology, and materials science. He sits on the editorial board of several SCI journals (e.g., *Nature Scientific Reports, Industrial Crops & Products, Journal of Renewable Materials, Advances in Polymer Technology, International Journal of Polymer Analysis and Characterization, Polymers for Advanced Technologies, Biomolecules, Nanomaterials, Surfaces and Interfaces, Sustainable Chemistry and Pharmacy, Current Opinion in Green and Sustainable Chemistry, and Nano-Structures & Nano-Objects*) as an editor or Editorial Advisory Board member.

Preface to "Nanocellulose and Nanocarbons Based Hybrid Materials: Synthesis, Characterization and Applications"

Since the emergence of nanotechnology in recent decades, the development and design of hybrid bio-nanomaterials has become an important field of research. Looking at the growing concern about the environment and sustainability, such nanomaterials find many applications in a wide range of domains that influence our society and our way of life. The improvement in properties and the discovery of new functionalities are key goals that cannot be reached without a well controlled and better understanding of the preparation, characterization, and manufacturing of new hybrid nanomaterials, which constitute the starting points of the design of specific and adequate systems. The investigation of nanocellulose/nanocarbons hybrid materials has demonstrated both academic and technological importance and offered great research opportunities within cross-disciplinary areas. Nanocellulose, which refers to the cellulose with a nanoscale dimension, encompasses cellulose nanocrystals, cellulose nanofibrils, and bacterial cellulose. It displays outstanding features such as biocompatibility, eco-friendliness, renewability, interesting reinforcing potential, hydrogen-bonding capacity, low thermal expansion coefficient, dimensional stability, high elastic modulus, high specific surface area, and low density. Various nanocellulose-based composites, which found application in different fields, have been developed during the last two decades. However, numerous challenges need to be addressed to achieve the full requirement of advanced materials. Accordingly, several research activities continue to be performed by different research groups worldwide to develop the next generation of nanomaterials and fully explore the potential of nanocellulose. Nanocarbon is another exciting type of potential nanomaterials, which has received tremendous attention since the discovery of fullerenes over thirty years ago. They include carbon nanotubes, graphene and carbon dots. They exhibit outstanding thermal, electrical, optical and mechanical features, and they can be employed in a wide range of applications such as composites, microelectronics, biomedical, conductive films, sensors, adsorbents, catalysis, energy storage, and coatings, among others. Based on these premises, the present book aims to further contribute to the momentum of research and development in nanocellulose/nanocarbon hybrids, by featuring several different chapters authored and reviewed by experts in the field. This book targets a broad readership of materials scientists, chemists, physicists, and nanotechnologists, among others. Most of the chapters highlight theoretical concepts and practical approaches of interest for real-world applications related to nanocellulose and nanocarbons.

In summary, this book advances not only our understanding of the emerging and significant role of nanocellulose and nanocarbons in several fields, but also of the challenges and future research directions needed to fully explore their outstanding features in practical ways. It is also expected that this book will encourage multidisciplinary research activities on hybrid bio-nanomaterials, extending the range of potential practical applications taking into account the scaling-up of the systems, the economic viability, the impact on the environment and human health as well as the long-term stability and recyclability.

Djalal Trache, Vijay Kumar Thakur

Editors

 *nanomaterials*

Editorial

Nanocellulose and Nanocarbons Based Hybrid Materials: Synthesis, Characterization and Applications

Djalal Trache [1],* and Vijay Kumar Thakur [2,3]

[1] Energetic Materials Laboratory, Teaching and Research Unit of Energetic Processes, Ecole Militaire Polytechnique, BP 17, Bordj El-Bahri, Algiers 16046, Algeria

[2] Biorefining and Advanced Materials Research Center, Scotland's Rural College (SRUC), Kings Buildings, Edinburgh EH9 3JG, UK; vijay.kumar@sruc.ac.uk

[3] Department of Mechanical Engineering, School of Engineering, Shiv Nadar University, Uttar Pradesh 201314, India

* Correspondence: djalaltrache@gmail.com

Received: 2 September 2020; Accepted: 7 September 2020; Published: 10 September 2020

Since the emergence of nanotechnology in recent decades, the development and design of hybrid bio-nanomaterials has become an important field of research. Looking at the growing concern about the environment and sustainability, such nanomaterials find many applications in a wide range of domains that influence our society and our way of life [1]. The improvement in properties and the discovery of new functionalities are key goals that cannot be reached without a well-controlled and better understanding of the preparation, characterization, and manufacturing of new hybrid nanomaterials, which constitute the starting points of the design of specific and adequate systems [2,3]. Investigation of nanocellulose/nanocarbons hybrid materials has demonstrated both academic and technological importance and offered great research opportunities within cross-disciplinary areas.

Nanocellulose, which refers to the cellulose with a nanoscale dimension, encompasses cellulose nanocrystals, cellulose nanofibrils, and bacterial cellulose [4,5]. It displays outstanding features such as biocompatibility, eco-friendliness, renewability, interesting reinforcing potential, hydrogen-bonding capacity, low thermal expansion coefficient, dimensional stability, high elastic modulus, high specific surface area, and low density [6–9]. Various nanocellulose-based composites, which found application in different fields, have been developed during the last two decades. However, numerous challenges need to be addressed to achieve the full requirement of advanced materials [10–14]. Accordingly, several research activities continue to be performed by different research groups worldwide to develop the next generation of nanomaterials and fully explore the potential of nanocellulose.

Nanocarbon is another exciting type of potential nanomaterials, which has received tremendous attention since the discovery of fullerenes over thirty years ago [15–18]. They include carbon nanotubes, graphene and carbon dots. They exhibit outstanding thermal, electrical, optical and mechanical features, and they can be employed in a wide range of applications such as composites, microelectronics, biomedical, conductive films, sensors, adsorbents, catalysis, energy storage, and coating, among others.

Based on these premises, the present Special Issue in *Nanomaterials* aims to further contribute to the momentum of research and development in nanocellulose/nanocarbon hybrids, by featuring ten (10) original research articles as well as two (2) review articles, authored and reviewed by experts in the field. It targets a broad readership of materials scientists, chemists, physicists, and nanotechnologists, among others. Most of the research papers highlight theoretical concepts and practical approaches of interest for real-world applications related to nanocellulose and nanocarbons.

Some interesting research works dealing with nanocellulose-based materials have been published in the present Special Issue. Imtiaz et al. prepared wood-based cationic cellulose nanocrystals (CNCs) and assessed their cytotoxicity as potential immunomodulators [19]. They firstly transformed anionic

CNCs to cationic CNCs with a rod-like shape through the grafting of cationic polymers containing pendant $^+NMe_3$ and $^+NH_3$. The authors proved that such an interesting cellulosic derivative, which exhibited very low toxicity, is considered as a good candidate as an immunomodulator and as vaccine nano-adjuvants. In another work, Han et al. investigated the physicochemical properties of cationic nanocellulose (cationic NC) and starch nanocomposites [20]. The authors prepared cationic NC from cationic-modified microcrystalline cellulose (MCC) using different approaches, i.e., acid hydrolysis, high-pressure homogenization, and high-intensity ultrasonication. They demonstrated that the nanocomposite prepared by cationic-NC produced by high-pressure homogenization and starch displayed a good dispersion, better thermal stability and the best water vapor barrier properties compared to other nanocomposites, which suggest that such nanocomposite can find potential application as a biodegradable food-packaging material. The study by Sharma et al. assessed the potential of carboxy cellulose nanofibers (CNF) extracted from untreated jute biomass using nitro-oxidation method as the reinforcing agent of natural rubber latex (NRL) [21]. It was revealed that the incorporation of CNF into NRL enhanced the mechanical characteristics even in the non-vulcanized state for which the optimal amount of CNF is around 0.2 wt.%. Another paper by Hai et al. describes the preparation and characterization of cellulose nanofibers and chitosan nanofiber blends, which are expected to be applied for active food packaging [22]. The obtained nanocomposite displayed a remarkable improvement in the mechanical features. The higher content of chitosan nanofibers exhibited higher antioxidant activity, better UV-light protection, increased water vapor transmission rate, and better mechanical properties. They claimed that such a nanocomposite could be useful for active food packaging application. The report by Lu et al. describes the preparation of a nanocellulose/metal-organic-framework-derived carbon-doped CuO/Fe_3O_4 nanocomposite via direct calcination [23]. Its catalytic efficiency through the reduction in 4-nitrophenol, methylene blue, and methyl orange has been demonstrated.

Similarly, interesting nanocarbon-based materials have been also reported in the present Special Issue. One research article focused on the reinforcement of high-molecular-weight, multimodal, high-density polyethylene by microwave-induced plasma graphene platelets using melt intercalation [24]. It was found that graphene platelets were homogenously distributed and dispersed within the polymer matrix through a strong interfacial bonding. The developed nanocomposites exhibited better thermal stability and interesting mechanical features. In another study, Sharma et al. developed a now-reduced graphene oxide incorporated gun tragacanth-cl-*N*, *N*-dimethylacrylamide hydrogel composite as a reusable adsorbent for Hg^{2+} and Cr^{6+} ions [25]. The authors optimized the experimental conditions of the synthesis based on the microwave-assisted method. It is proved that the adsorption effectiveness of 99% and 82% were obtained for Hg^{2+} and Cr^{6+} ions, respectively, confirming that the developed adsorbents are highly efficient and can be employed for environmental remediation applications. A pure nanocarbon hybrid based on multi-walled carbon nanotube and graphene nanoplatelet (GNP) for flexible strain sensors has been developed by Huang et al. [26]. The production of such hybrid was assisted with surfactant Triton X-100 and the sonication through a vacuum filtration process. The authors revealed that the increase in the content of GNP from 0 to 5% decreased the tensile strength of the hybrid film, whereas the electrical conductivity increased. It is also found that the increase of the GNP content leads to an increase in the strain sensitivity with good stability and repeatability.

Considered as a promising class of nanomaterials, the combination of nanocellulose and nanocarbon has received tremendous interest in recent years. This interest is promoted by the exceptional properties and outstanding synergetic effects that these powerful nanomaterials offer, which allow developing new opportunities and ways for the preparation and employment of novel materials in nanotechnology spanning from water treatment, energy and environment, optics and photonics, medical, biosensing and optoelectronics. For such specific nanohybrids, two review papers and two research articles have been published. The first review article compiles five years of recent research findings on new development on the emerging generation of cellulose nanocrystals/graphene-based nanomaterials [27]. The methodologies of their production, their properties and applications have been thoroughly discussed. It is demonstrated that such nano-hybrids exhibited prominent innovative features due to synergetic effects, which are unachievable by taking cellulose nanocrystals and graphene-based nanomaterials separately. The authors reported that the real use of such hybrids as the next generation of materials necessitates more efforts and further improvements in functionality and performance, in addition, the decrease in the production costs and environmental impacts. The second review by Bacakova et al. is dedicated to the application of nanocellulose (nanofibrils and nanocrystals)/nanocarbon (fullerenes, graphene, nanotubes and nanodiamonds) composites with a special focus on biotechnology and medicine [28]. Such promising hybrids displayed unique features such as high mechanical strength coupled with flexibility and stretchability, shape memory effect, photodynamic and photothermal activity, electrical conductivity, semiconductivity, thermal conductivity, tunable optical transparency, intrinsic fluorescence and luminescence, and high adsorption and filtration capacity. They can be used in several applications such as wound dressing, tissue engineering, electrical stimulation of damaged tissues, isolation of different biomolecules, drug delivery, among others. However, the authors claimed that the risk of their potential cytotoxicity and immulogenicity needs to be deeply investigated.

The first research paper dealing with nanocellulose/nanocarbon hybrids is studied by Yu et al. [29]. These authors prepared cellulose nanofibrils (CNFs)/carbon nanomaterial hybrid aerogels for adsorption removal of cationic and anionic organic dyes. They employed two classes of nanocarbons, i.e., carbon nanotubes (CNT) and graphene nanoplates (GnP), which are combined with different amounts of CNFs. They reported that the combination CNFs: GnP 3:1 provided the best performance toward day adsorption for which the pseudo-second-adored kinetic and monolayer Langmuir adsorption isotherm are followed. The authors claimed that such a hybrid could find promising application as an adsorption material for wastewater treatment. The last work presented in the Special Issue studied by Wang et al. describes the preparation and characterization of self-healable electro-conductive hydrogels based on core-shell structured nanocellulose/carbon nanotubes hybrids, expected to be used as flexible supercapacitors [30]. The developed hydrogel exhibited interesting mechanical features, high water-content, excellent flexibility, self-healing capability, higher electrical conductivity and specific capacity, rendering it a potential candidate for advanced personalized wearable electronic devices.

In summary, the present Special Issue advances not only our understanding of the emerging and significant role of nanocellulose and nanocarbons in several fields, but also of the challenges and future research directions needed to fully explore their outstanding features in practical ways. It is also expected that this Special Issue will encourage multidisciplinary research activities on hybrid bio-nanomaterials, extending the range of potential practical applications taking into account the scaling-up of the systems, the economic viability, the impact on the environment and human health as well as the long-term stability and recyclability.

Author Contributions: All authors contributed to the editorial. All authors have read and agreed to the published version of the manuscript.

Funding: This research received no external funding.

Acknowledgments: We want to thank all authors for their outstanding contributions. The time and effort devoted by reviewers of the articles and their constructive comments are also highly appreciated. Furthermore, we would like to acknowledge the members of the Editorial Office of Nanomaterials for their help, promptness, administrative and editorial support during all this long period from the point of designing the issue and throughout its implementation and completion.

Conflicts of Interest: The authors declare no conflict of interest.

References

1. Ates, B.; Koytepe, S.; Ulu, A.; Gurses, C.; Thakur, V.K. Chemistry, structures, and advanced applications of nanocomposites from biorenewable resources. *Chem. Rev.* **2020**, *120*, 9304–9362. [CrossRef] [PubMed]
2. Chen, S.; Skordos, A.; Thakur, V. Functional nanocomposites for energy storage: Chemistry and new horizons. *Mater. Today Chem.* **2020**, *17*, 100304. [CrossRef]
3. Neibolts, N.; Platnieks, O.; Gaidukovs, S.; Barkane, A.; Thakur, V.; Filipova, I.; Mihai, G.; Zelca, Z.; Yamaguchi, K.; Enachescu, M. Needle-free electrospinning of nanofibrillated cellulose and graphene nanoplatelets based sustainable poly (butylene succinate) nanofibers. *Mater. Today Chem.* **2020**, *17*, 100301. [CrossRef]
4. Miyashiro, D.; Hamano, R.; Umemura, K. A review of applications using mixed materials of cellulose, nanocellulose and carbon nanotubes. *Nanomaterials* **2020**, *10*, 186. [CrossRef]
5. Trache, D.; Hussin, M.H.; Haafiz, M.M.; Thakur, V.K. Recent progress in cellulose nanocrystals: Sources and production. *Nanoscale* **2017**, *9*, 1763–1786. [CrossRef]
6. Trache, D.; Tarchoun, A.F.; Derradji, M.; Mehelli, O.; Hussin, M.H.; Bessa, W. Cellulose fibers and nanocrystals: Preparation, characterization and surface modification. In *Functionalized Nanomaterials I: Fabrication*; Kumar, V., Guleria, P., Dasgupta, N., Ranjan, S., Eds.; Taylor & Francis: Boca Raton, FI, USA, 2020; pp. 171–190.
7. Trache, D.; Tarchoun, A.F.; Derradji, M.; Hamidon, T.S.; Masruchin, N.; Brosse, N.; Hussin, M.H. Nanocellulose: From fundamentals to advanced applications. *Front. Chem.* **2020**, *8*, 392. [CrossRef]
8. Hussin, M.H.; Trache, D.; Chuin, C.T.H.; Fazita, M.N.; Haafiz, M.M.; Hossain, M.S. Extraction of cellulose nanofibers and their eco-friendly polymer composites. In *Sustainable Polymer Composites and Nanocomposites*; Springer: Berlin/Heidelberg, Germany, 2019; pp. 653–691.
9. Trache, D. Nanocellulose as a promising sustainable material for biomedical applications. *AIMS Mater. Sci.* **2018**, *5*, 201–205. [CrossRef]
10. Lin, N.; Tang, J.; Dufresne, A.; Tam, M.K. *Advanced Functional Materials from Nanopolysaccharides*; Springer: Berlin/Heidelberg, Germany, 2019.
11. Dufresne, A. Nanocellulose processing properties and potential applications. *Curr. For. Rep.* **2019**, *5*, 76–89. [CrossRef]
12. Nascimento, D.M.; Nunes, Y.L.; Figueirêdo, M.C.; de Azeredo, H.M.; Aouada, F.A.; Feitosa, J.P.; Rosa, M.F.; Dufresne, A. Nanocellulose nanocomposite hydrogels: Technological and environmental issues. *Green Chem.* **2018**, *20*, 2428–2448. [CrossRef]
13. Kargarzadeh, H.; Mariano, M.; Gopakumar, D.; Ahmad, I.; Thomas, S.; Dufresne, A.; Huang, J.; Lin, N. Advances in cellulose nanomaterials. *Cellulose* **2018**, *25*, 2151–2189. [CrossRef]
14. Kargarzadeh, H.; Ahmad, I.; Thomas, S.; Dufresne, A. *Handbook of Nanocellulose and Cellulose Nanocomposites*; Wiley Online Library: Hoboken, NJ, USA, 2017.
15. Boudjellal, A.; Trache, D.; Khimeche, K.; Hafsaoui, S.L.; Bougamra, A.; Tcharkhtchi, A.; Durastanti, J.-F. Stimulation and reinforcement of shape-memory polymers and their composites: A review. *J. Thermoplast. Compos. Mater.* **2020**. [CrossRef]
16. Benhammada, A.; Trache, D.; Kesraoui, M.; Chelouche, S. Hydrothermal synthesis of hematite nanoparticles decorated on carbon mesospheres and their synergetic action on the thermal decomposition of nitrocellulose. *Nanomaterials* **2020**, *10*, 968. [CrossRef] [PubMed]
17. Tiwari, S.K.; Sahoo, S.; Wang, N.; Huczko, A. Graphene research and their outputs: Status and prospect. *J. Sci. Adv. Mater. Devices* **2020**, *5*, 10–29. [CrossRef]
18. Xie, F.; Yang, M.; Jiang, M.; Huang, X.-J.; Liu, W.-Q.; Xie, P.-H. Carbon-based nanomaterials–a promising electrochemical sensor toward persistent toxic substance. *TrAC Trends Anal. Chem.* **2019**, *119*, 115624. [CrossRef]
19. Imtiaz, Y.; Tuga, B.; Smith, C.W.; Rabideau, A.; Nguyen, L.; Liu, Y.; Hrapovic, S.; Ckless, K.; Sunasee, R. Synthesis and cytotoxicity studies of wood-based cationic cellulose nanocrystals as potential immunomodulators. *Nanomaterials* **2020**, *10*, 1603. [CrossRef]
20. Han, L.; Wang, W.; Zhang, R.; Dong, H.; Liu, J.; Kong, L.; Hou, H. Effects of preparation method on the physicochemical properties of cationic nanocellulose and starch nanocomposites. *Nanomaterials* **2019**, *9*, 1702. [CrossRef]

21. Sharma, S.K.; Sharma, P.R.; Lin, S.; Chen, H.; Johnson, K.; Wang, R.; Borges, W.; Zhan, C.; Hsiao, B.S. Reinforcement of natural rubber latex using jute carboxycellulose nanofibers extracted using nitro-oxidation method. *Nanomaterials* **2020**, *10*, 706. [CrossRef]

22. Hai, L.V.; Zhai, L.; Kim, H.C.; Panicker, P.; Pham, D.; Kim, J. Chitosan nanofiber and cellulose nanofiber blended composite applicable for active food packaging. *Nanomaterials* **2020**, *10*, 1752. [CrossRef]

23. Lu, H.; Zhang, L.; Ma, J.; Alam, N.; Zhou, X.; Ni, Y. Nano-cellulose/mof derived carbon doped CuO/Fe_3O_4 nanocomposite as high efficient catalyst for organic pollutant remedy. *Nanomaterials* **2019**, *9*, 277. [CrossRef]

24. Ahmad, I.A.; Koziol, K.K.; Deveci, S.; Kim, H.-K.; Kumar, R.V. Advancing the use of high-performance graphene-based multimodal polymer nanocomposite at scale. *Nanomaterials* **2018**, *8*, 947. [CrossRef]

25. Sharma, B.; Thakur, S.; Trache, D.; Yazdani Nezhad, H.; Thakur, V.K. Microwave-assisted rapid synthesis of reduced graphene oxide-based gum tragacanth hydrogel nanocomposite for heavy metal ions adsorption. *Nanomaterials* **2020**, *10*, 1616. [CrossRef] [PubMed]

26. Huang, J.; Her, S.-C.; Yang, X.; Zhi, M. Synthesis and characterization of multi-walled carbon nanotube/graphene nanoplatelet hybrid film for flexible strain sensors. *Nanomaterials* **2018**, *8*, 786. [CrossRef] [PubMed]

27. Trache, D.; Thakur, V.K.; Boukherroub, R. Cellulose nanocrystals/graphene hybrids—a promising new class of materials for advanced applications. *Nanomaterials* **2020**, *10*, 1523. [CrossRef]

28. Bacakova, L.; Pajorova, J.; Tomkova, M.; Matejka, R.; Broz, A.; Stepanovska, J.; Prazak, S.; Skogberg, A.; Siljander, S.; Kallio, P. Applications of nanocellulose/nanocarbon composites: Focus on biotechnology and medicine. *Nanomaterials* **2020**, *10*, 196. [CrossRef] [PubMed]

29. Yu, Z.; Hu, C.; Dichiara, A.B.; Jiang, W.; Gu, J. Cellulose nanofibril/carbon nanomaterial hybrid aerogels for adsorption removal of cationic and anionic organic dyes. *Nanomaterials* **2020**, *10*, 169. [CrossRef]

30. Wang, H.; Biswas, S.K.; Zhu, S.; Lu, Y.; Yue, Y.; Han, J.; Xu, X.; Wu, Q.; Xiao, H. Self-healable electro-conductive hydrogels based on core-shell structured nanocellulose/carbon nanotubes hybrids for use as flexible supercapacitors. *Nanomaterials* **2020**, *10*, 112. [CrossRef]

Article

Synthesis and Cytotoxicity Studies of Wood-Based Cationic Cellulose Nanocrystals as Potential Immunomodulators

Yusha Imtiaz [1], Beza Tuga [1], Christopher W. Smith [1], Alexander Rabideau [1], Long Nguyen [1], Yali Liu [2], Sabahudin Hrapovic [2], Karina Ckless [1,*] and Rajesh Sunasee [1,*]

[1] Department of Chemistry, State University of New York at Plattsburgh, Plattsburgh, New York, NY 12901, USA; yimti001@plattsburgh.edu (Y.I.); btuga001@plattsburgh.edu (B.T.); c.w.smith022@gmail.com (C.W.S.); arabi005@plattsburgh.edu (A.R.); lnguy007@plattsburgh.edu (L.N.)

[2] Aquatic and Crop Resource Development Research Centre, National Research Council Canada, Montreal, QC H4P 2R2, Canada; yali.liu@cnrc-nrc.gc.ca (Y.L.); sabahudin.hrapovic@cnrc-nrc.gc.ca (S.H.)

* Correspondence: kckle001@plattsburgh.edu (K.C.); rajesh.sunasee@plattsburgh.edu (R.S.)

Received: 30 July 2020; Accepted: 12 August 2020; Published: 15 August 2020

Abstract: Polysaccharides have been shown to have immunomodulatory properties. Modulation of the immune system plays a crucial role in physiological processes as well as in the treatment and/or prevention of autoimmune and infectious diseases. Cellulose nanocrystals (CNCs) are derived from cellulose, the most abundant polysaccharide on the earth. CNCs are an emerging class of crystalline nanomaterials with exceptional physico-chemical properties for high-end applications and commercialization prospects. The aim of this study was to design, synthesize, and evaluate the cytotoxicity of a series of biocompatible, wood-based, cationic CNCs as potential immunomodulators. The anionic CNCs were rendered cationic by grafting with cationic polymers having pendant $^+NMe_3$ and $^+NH_3$ moieties. The success of the synthesis of the cationic CNCs was evidenced by Fourier transform infrared spectroscopy, dynamic light scattering, zeta potential, and elemental analysis. No modification in the nanocrystals rod-like shape was observed in transmission electron microscopy and atomic force microscopy analyses. Cytotoxicity studies using three different cell-based assays (MTT, Neutral Red, and LIVE/DEAD®) and three relevant mouse and human immune cells indicated very low cytotoxicity of the cationic CNCs in all tested experimental conditions. Overall, our results showed that cationic CNCs are suitable to be further investigated as immunomodulators and potential vaccine nanoadjuvants.

Keywords: cellulose nanocrystals; immunomodulator; synthesis; polymerization; characterization; cytotoxicity

---

1. Introduction

Polysaccharides, such as 2,3-O-acetylated-1,4-β-D-glucomannan, have been shown to have immunomodulatory properties, including the stimulation of cytokines, interleukin 1-beta (IL–1β), and tumor necrosis alpha (TNF-α) in human cell lines [1,2]. The immune system is a complex network of different cell types and mediators responsible for the defense of an organism against pathogens as well as the maintenance of tolerance to innocuous antigens [3]. Immunomodulation plays a crucial role in physiological processes as well as in the treatment and/or prevention of several diseases, including infectious diseases. Therefore, studies focusing on polysaccharide-based nanomaterials with immunomodulatory activities can have an important impact on development of novel vaccine adjuvants. Adjuvants, in the context of vaccines, are defined as components capable of enhancing and/or shaping antigen-specific immune responses [4]. Advances in particle engineering allow the design of new nanomaterials with desired physico-chemical properties, such as composition, size, shape, surface characteristics, and degradability [5], which ultimately will impact their effect as immunomodulators. It is well known that cationic substances, especially particulate materials, can act as immunomodulators [6,7]. In fact, the immunostimulatory activity of cationic cellulose nanocrystals (CNCs) was first described in our previous work in which we found that they induced the secretion of the inflammatory cytokine, IL-1β, in mouse and human macrophage cells [8,9]. CNCs are elongated crystalline rod-like (or needle-like) nanoparticles derived from the most abundant natural biopolymer on earth, cellulose [10–12]. CNCs are green and biocompatible materials with unique physico-chemical properties, including high aspect ratio, low density, large specific surface area, tunable surface chemistry (presence of abundant hydroxyl groups), non-toxic, colloidal stability, and optical and mechanical properties. These favorable features make CNCs as attractive nanoscale materials for applications in various fields such as biomedical, pharmaceutical, nanocomposites, electronics, among many others [13–15]. CNCs can be manufactured on a large scale and a recent study showed that the current standing of industrially produced CNCs was positive, indicating that the evolution of commercial-scale applications would not be hindered by CNCs production [16]. The emerging interest in CNCs for biomedical purposes led us to synthesize a novel functionalized series of wood-based CNCs possessing surface positive charges with potential immunomodulatory activities that could hopefully be further developed in newly engineered vaccine nanoadjuvants. For applications that required a positively charged CNCs surface, synthetic routes for cationization of CNCs were developed such as nucleophilic ring-opening of epoxide, esterification, copper(I)-catalyzed 1,3-cycloaddition and surface initiated-atom transfer radical polymerization [17–22]. In this work, we reported the design, synthesis, characterization and cytotoxicity of wood-based cationic CNCs as potential immunomodulators. A series of colloidally stable cationic CNCs was prepared via surface-initiated single electron transfer living radical polymerization (SI-SET-LRP) technique with different sizes and surface charges. The choice of [2-(methacryloyloxy)ethyl] trimethylammonium chloride (METAC) and aminoethyl methacrylate hydrochloride (AEM) monomers with pendant cationic groups (+NMe$_3$ and + NH$_3$ respectively) was guided in part due to a recent study highlighting the use of NH$_2$-functionalized aluminum oxyhydroxide nanorods for an enhanced immune adjuvant activity [7]. In addition to an effective synthetic route and sound physico-chemical characterization, another important aspect for the potential application of these cationic CNCs as vaccine nanoadjuvant is safety of these nanomaterials. In general, vaccine adjuvant should not induce an overwhelming immune response, otherwise it would have a toxic effect. Therefore, the cytotoxicity of these cationic CNCs was evaluated using three different cell-based assays (MTT, Neutral Red, and LIVE/DEAD®) and relevant immune cells, including two mouse cell lines (J774A.1 and BV2) as well as human peripheral blood mononuclear cells (PBMCs). Overall, we successfully synthesized and characterized a series of cationic CNCs that showed very low cytotoxicity in all tested experimental conditions indicating that these cationic cellulose-based nanomaterials are suitable to be further investigated as immunomodulators and potential vaccine nanoadjuvants.

2. Materials and Methods

2.1. Materials and Reagents

Spray-dried sulfated CNCs (prepared by sulfuric acid mediated hydrolysis of hardwood pulp) was kindly supplied by InnoTech Alberta Inc (Edmonton, AB, Canada). The spray-dried CNC powder was stored at 4 °C and used as obtained. Triethylamine (TEA), 4-dimethyl-aminopyridine (DMAP), 2-bromoisobutyryl bromide, *N,N,N',N'',N''*-pentamethyldiethylenetriamine (PMDETA), copper (I) bromide, [2-(methacryloyloxy)ethyl] trimethylammonium chloride (METAC) (75 wt% in H$_2$O), 2-aminoethyl methacrylate hydrochloride (AEM), 2,2'-bipyridyl (bpy), methanol, ethanol, tetrahydrofuran (THF), acetone, and potassium bromide (KBr) were purchased from Sigma-Aldrich (St Louis, MO, USA).

2.2. Synthesis of Wood-Based Cationic CNCs

The initiator modified CNC materials (CNC-BriB) were first synthesized according to our previously reported procedure with slight modifications [23]. Pristine spray-dried CNCs were reacted with 2-bromoisobutyryl bromide (BriB bromide) at two different ratios with respect to anhydroglucose units (AGU) in CNCs ([Br]/[AGU]), 5:3 for CNC-BriB-1 and 5:12 for CNC-BriB-2 (for a detailed experimental procedure of CNC-BriB, refer to SI & Supplementary Figure S1). CNC-BriB was then grafted with either poly (METAC) or poly (AEM) via SI-SET-LRP at different ratios of monomer to AGU such as [METAC/AGU] = 50:3 or 60:3 and [AEM/AGU] = 50:3 (Table 1). A general procedure for the preparation of cationic CNCs is as follows: CNC-BriB (350 mg) was dispersed in a MeOH:H_2O solvent mixture (100 mL, 1:1 v/v) in a 250 mL Schlenk flask. The reaction mixture was degassed under N_2 gas for 45 min prior to the addition of the monomer METAC or AEM (50 and/or 60 mmol) and copper (I) bromide (0.5 mmol). The suspension was degassed again before addition of bpy or PMDETA (1 mmol) and the reaction mixture was vigorously stirred at room temperature for 24 h. The resulting crude cationic CNC was then centrifuged (3 × 12,000 rpm at 10 °C for 30 min) with a 1:3 ratio of H_2O:MeOH. The residual solid was resuspended in water and extensively dialyzed (MWCO 3500 Da) against deionized water for 1 week with daily constant change of water. The suspension was then freeze-dried to afford purified CNCs as white flaky material. The syntheses of other cationic CNCs are described in detail in the supporting information and Supplementary Table S1.

Table 1. Different molar ratios of [Br]/[AGU] and [monomer]/[AGU] for the fabrication of cationic CNCs.

Sample	[Br]/[AGU]	[Monomer]/[AGU]
CNC-METAC-1A	5:3	50:3
CNC-METAC-1B	5:3	60:3
CNC-METAC-2A	5:12	50:3
CNC-METAC-2B	5:12	60:3
CNC-AEM-1A	5:3	50:3
CNC-AEM-2A	5:12	50:3

2.3. Characterization of Pristine and Modified CNCs

2.3.1. Fourier Transform Infrared Spectroscopy (FTIR)

FTIR spectra of pristine spray dried sulfated CNCs and lyophilized esterified and cationic CNCs were recorded on a PerkinElmer FTIR spectrophotometer (Spectrum Two) (Norwalk, CT, USA) at room temperature. KBr pellets were prepared by grinding in a mortar and compressing about 2% of the CNC samples in KBr (previously well-dried in an oven). Background measurement using a neat KBr pellet was first obtained to correct for light scattering losses in the pellet and any water absorbed by KBr. Spectra in the range of 4000–400 cm^{-1} were obtained with a resolution of 4 cm^{-1} by cumulating 32 scans.

2.3.2. Zeta Potential and Dynamic Light Scattering

Zeta (ζ-) potential and dynamic light scattering (DLS) measurements were conducted using a Malvern Zetasizer Nano ZS instrument (model: ZEN3600; Malvern Instruments Inc., Westborough, MA, USA). This instrument is equipped with a 4.0 mW helium-neon laser (λ=633 nm) and an avalanche photodiode detector and works at a 173° scattering angle. A 0.25 wt% CNC dispersion in Milli-Q water was used for zeta potential while for DLS, the hydrodynamic apparent particle size was measured for a 0.05 wt% CNC dispersion in Milli-Q water. Prior measurements, the suspensions were sonicated and filtered through a 0.45 μm filter membrane and analyses were performed immediately at 25 °C. For DLS results, triplicate samples were measured 12 times each and the average particle size distribution was obtained. The standard deviation was reported as the error for the measurements. Results for zeta potential measurements were recorded in triplicate and the averages were reported.

2.3.3. Elemental Analysis

Carbon, hydrogen, nitrogen, and sulfur content by mass for freeze-dried pristine and cationic CNC samples were determined using Thermo Scientific Flash 2000 Organic Elemental Analyzer. Analysis (Edmonton, AB, Canada) was performed in duplicate and the averages were reported.

2.3.4. Transmission Electron Microscopy

Transmission electron microscopy (TEM) images of pristine and cationic CNC samples were obtained using Hitachi H-7500 operating at 80 kV in HR mode. All samples were prepared at 0.1 mg/mL in double distilled water and bath sonicated for 15 min at room temperature prior to the sample immobilization on TEM grids. 6 µL of sample were deposited on the TEM grid (Cu-300CN, Pacific Grid-Tech, San Francisco, CA, USA) with excess solution removed by using the edge of a wet filter paper. Samples were stained in 1 wt% phosphotungstic acid (PTA) in water (pH 7) for 60 s and air-dried in the dark prior to the TEM analysis. The samples were observed in the magnification range of 40,000–600,000 Å in order to provide enough information about sample purity and morphology.

2.3.5. Atomic Force Microscopy

Atomic force microscopy (AFM) micrographs of pristine and cationic CNCs were obtained using a Nanoscope IV (Digital Instruments Veeco, Santa Barbara, CA, USA) with a silicon tip operated in tapping mode. Particle analysis of the micrographs was performed using the software Scanning Probe Image Processor™ (5.0.8.0; Image Metrology, DK-2800 Lyngby, Denmark)

2.4. Cytotoxicity Studies

2.4.1. Preparation of the Colloidal Suspension of CNCs for Cell-Based Assays

For cytotoxicity assays, the CNC suspensions were prepared at 1 mg/mL in ultrapure water. The CNC suspensions were vortexed for 15 s, sonicated for 2 min at 70 output, followed by filtration with 0.45 µm polytetrafluoroethylene (PTFE) filter and autoclaved at 121 °C, 15 psi for 30 min. The suspensions were aliquoted and kept at −20 °C for future use.

2.4.2. Cell Culture and Experimental Conditions

Mouse macrophage-like cell line (J774A.1, Sigma) and BV2 cell line (murine microglia, kindly provided by Dr. Amy Ryan, SUNY Plattsburgh) were seeded at 3–5×10^5 cells/mL in 96 well plate (3–5×10^4 cells/well) using RPMI 1640 and DMEM medium (GIBCO), respectively. Both media were supplemented with 10% fetal bovine serum (FBS), penicillin, streptomycin, and L-glutamine. Both cell lines were cultured at 37 °C in a 5% CO_2-supplemented atmosphere for at least overnight before the treatment with 10, 25, 50, and 100 µg/mL of CNCs. The peripheral blood mononuclear cells (PBMCs) were extracted from Leukotrap blood filters from healthy blood donors. The Leukotrap filters were obtained from UVM Health Network-CVPH North Country Regional Blood Center, Plattsburgh, NY. To retrieve blood cells, the filter was reverse flushed with 3×50 mL of calcium and magnesium-free PBS, followed by PBMCs isolation using a separation medium (LSM). Briefly, 4 mL of LSM was added to a 15 mL conical centrifuge tube and 6 mL of 1:1 diluted blood in calcium and magnesium-free PBS was carefully layered on LSM, creating a sharp blood-LSM interface. The conical tubes were centrifuged at $400\times g$ at room temperature for 25 min. The top layer containing plasma was discarded and the mononuclear cell layer was carefully transferred to a clean tube and an equal volume of calcium and magnesium-free PBS was added. The cell suspension was centrifuged at $250\times g$ to pellet cells and the cells were washed once in calcium and magnesium-free PBS and finally resuspended in RPMI medium for seeding. The PBMCs were seeded in 48 well plate at 1×10^6 cells/mL and cultured at 37 °C in a 5% CO_2-supplemented atmosphere for at least overnight before the treatment with 10, 25, 50, and 100 µg/mL of CNC derivatives.

2.4.3. Cell Viability Assays

After 24 h of treatment, the medium from J774A.1 and BV2 treated and non-treated (control) cells was removed, and 100 µL of fresh culture medium containing 500 µg/mL of 3-(4,5-dimethylthiazol-2-yl)-2,5-diphenyltetrazolium bromide (MTT) or 50 µg/mL of Neutral Red (NR) was added to each well to access cell viability. The MTT assay involves the conversion of the water soluble MTT (yellow) by mitochondrial dehydrogenases in viable cells to a water-insoluble formazan (purple/blue) [24] while the NR assay is based on the ability of viable cells to incorporate and bind the dye neutral red in the lysosomes [25]. In both assays, the intensity of the color is directly proportional to cell viability. The cells with MTT or NR loading medium were incubated for at 37 °C in a 5% CO_2-supplemented atmosphere. After 30 min, the respective loading medium was removed, and the attached cells were gently washed once with PBS. To solubilize the formazan crystals, 100 µL/well of dimethyl sulfoxide (DMSO) was added and the absorbance was measured at 570 nm using a microplate reader (Synergy H1 Hybrid Multi-Mode, BioTek/Agilent, Winoosky, VT, USA). To extract NR dye from lysosomes, 100 µL of acidified ethanol (1% glacial acetic acid, 50% aqueous ethanol) was added and the plate was placed on the plate shaker for ~10–15 min with protection from light. The absorbance at 540 nm and at 690 nm was measured in the microplate reader within 60 min of adding NR Desorb solution. For calculations, 690 nm was subtracted from 540 nm. The non-treated cells (control) were considered 100% of viable cells in both MTT and NR assays. For statistical significance, both cytotoxicity assays were repeated at least 3 times in triplicates.

The LIVE/DEAD® viability assay is a quick and easy two-color assay to determine the viability of cells in a population. Typical results will distinguish live cells which are stained with green-fluorescent calcein-AM to indicate intracellular esterase activity, from dead cells that are stained with red-fluorescent ethidium homodimer-1 to indicate loss of plasma membrane integrity. The assay was performed according to the manufacturer's instructions. Briefly, after 24 h treatment with respective CNCs, the cell culture medium was gently removed and 250 µL PBS, 2 µL of Calcein (40 µM) and 2 µL of ethidium homodimer-1 2 mM were added, mixed, and the cell suspension was incubated in the dark RT for 15 min. The loading solution was discarded and 250 µL of cold PBS + EDTA was added to each well and the plate was placed on ice for 30 min to facilitate the release of PBMCs from the plate. The bottom of the wells was scraped with a pipette tip and the cells were removed by gentle pipetting up/down, 100 µL of the cell suspension was utilized to assess live and dead cells by flow cytometry (BD Accuri™ C6 cytometer, BD, Franklin Lake, NJ, USA) using FL-1 filter Calcein (green, live) and FL-2 filter Ethidium homodimer (red, dead). The percentage of labeled cells was calculated by BD Accuri software using H_2O_2 treated cells as positive control for gating dead cells. The calculated data was expressed using the average of fold of change in the (%) of dead cells (ethidium bromide staining) corrected by (%) dead cells in the control (non-treated cells), from at least 4 experiments.

2.4.4. Statistical Analysis

The data were statistically analyzed by using the two-way analysis of variance test followed by Dunnett's multiple comparison test using GraphPad Prism version 8.2 software (San Diego, CA, USA). Statistical significance was defined as $p < 0.01$.

3. Results and Discussion

3.1. Synthesis and Characterization of Cationic CNCs

CNCs with cationic polymer brushes were synthesized by grafting poly (METAC) or poly(AEM) via SI-SET-LRP from the surface of pristine CNCs (Scheme 1). The first step required the covalent attachment of an initiator on the surface of CNCs using a well-known esterification method of the hydroxyl groups of CNCs with an acyl bromide [23,26,27]. CNC-initiator (CNC-BriB) were prepared at two different initiator loading capacity with respect to anhydroglucose units (AGU) in CNCs ([Br]/[AGU]), 5:3 for CNC-BriB-1 and 5:12 for CNC-BriB-2). The next key step involved a grafting from approach to introduce cationic poly(METAC) or poly(AEM) brushes by SI-SET-LRP on CNCs under mild reaction conditions (Scheme 1B). SET-LRP is a robust and versatile polymerization technique for materials synthesis [28] and had been previously used to graft thermoresponsive and methacrylamide-based polymers from the surface of CNCs [23,26,27,29]. A series of cationic CNCs was synthesized by varying the ratio between the immobilized initiator and the monomers (Table 1). After the polymerization, the resulting cationic CNCs were extensively purified by several centrifugations and dialysis over a week to ensure the removal of unreacted materials or any trace impurities.

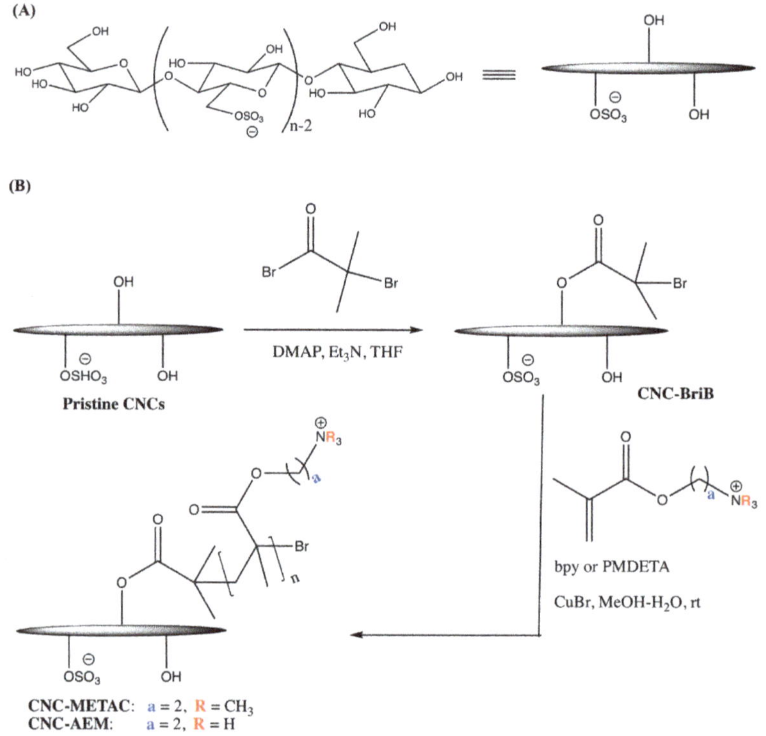

Scheme 1. (**A**) chemical structure and schematic representation of pristine CNCs; (**B**) synthetic strategy for the fabrication of a series of cationic CNCs.

The success of the grafting of cationic brushes on the surface of CNCs was confirmed by FTIR spectroscopy, DLS, zeta potential and elemental analysis. Pristine CNCs displayed typical IR peaks characteristic of cellulosic functional groups at 3000–3600 cm^{-1}, 1645 cm^{-1} and 900–1150 cm^{-1} corresponding to the -OH, -OH bending of water and -C-O-C vibrations, respectively [30,31] (Figure 1, spectrum A; Supplementary Figure S2). The spectra of CNCs after the polymerization showed a notable change with the appearance of new IR peaks at 1730 cm^{-1} and 1728 cm^{-1} for CNC-METAC-1A (Figure 1, spectrum B) and CNC-AEM-2A (Figure 1, spectrum C), respectively. These peaks were attributed to the typical carbonyl stretching vibration (C = O) of ester moieties confirming the presence of poly (METAC) and poly (AEM) on the surface of CNCs [27].

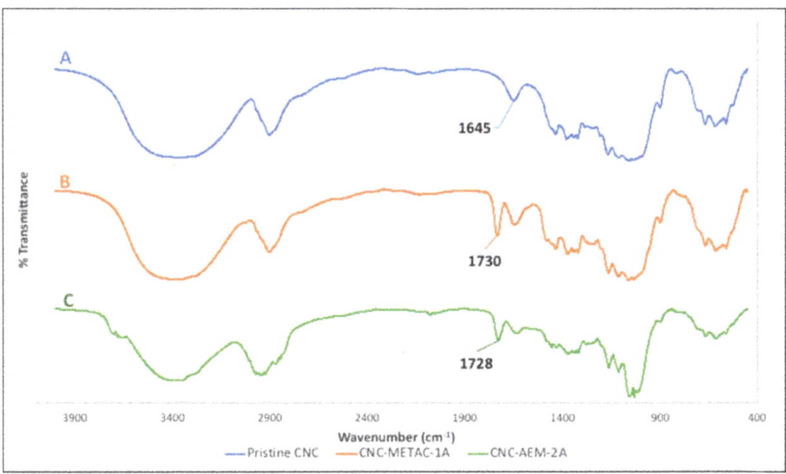

Figure 1. FTIR spectra of cationic CNCs: (A) pristine CNC, (B) CNC-METAC-1A, (C) CNC-AEM-2A.

Dynamic light scattering (DLS) and zeta potential measurements were conducted to explore the surface properties and stability of CNC suspensions before and after polymer grafting. Although the DLS technique is used for spherical particles, the measurements have been commonly employed to compare the relative sizes of rod-like CNCs before and after a chemical modification as well as the state of dispersion of the particles [16,21]. However, particle size measurements by DLS are not absolute and should not be considered as the exact dimensions of pristine and modified CNCs. The apparent hydrodynamic diameter of CNC particles was obtained by DLS in water at a 0.05 wt% concentration. Pristine CNCs showed an apparent particle size of ~101 nm which is in close agreement with the data obtained from AFM analysis (Figure 2D). An increase in apparent particle size was observed for all cationic CNCs (Table 2, Supplementary Figure S3) which indicated the presence of the polymer brushes on the surface of CNCs and as expected, the cationic CNCs would be highly hydrated. Furthermore, DLS analysis indicated no apparent aggregation of CNC particles at low concentrations. The nature of the surface charges on pristine and cationic CNCs was examined by measuring the zeta potential of the particles in water at 0.25 wt% concentration. Pristine CNCs had a negative zeta potential (-34.8 mV) due to the presence of anionic sulfate half-ester groups. On the other hand, cationic CNCs displayed positive zeta potential values in the range of +31.8 to +45.0 mV (Table 2). This further confirmed a successful polymerization reaction and the grafted cationic polymer brushes shielded the anionic sulfate half-ester groups. An increase in zeta potential (becomes more positive) was observed with an increase in monomer concentration ([monomer/AGU]: 50:3 v/s 60:3, Table 1). For instance, CNC-METAC-1B which had a higher monomer concentration compared to CNC-METAC-1A showed a more positive zeta potential value (+44.9 mV v/s +31.8 mV). The same trend was observed for CNC-METAC-2B (+38.2 mV) compared to CNC-METAC-2A (+32.0 mV). Overall, the cationic nature and colloidal stability of CNC-METAC and CNC-AEM materials are crucial for consistent results in biological studies.

The atomic composition of CNC samples before and after polymer grafting was measured by elemental analysis (Table 3). As expected, a small percentage of sulfur was detected for pristine CNCs due to the presence of the sulfate half-ester groups. While pristine CNCs displayed the absence of nitrogen, all the prepared cationic CNCs indicated the appearance of nitrogen deriving from the cationic monomers (METAC or AEM). This further supported the success of the polymerization reaction. The cationic CNC with a fixed initiator content and increased monomer concentration (CNC-METAC-1B v/s CNC-METAC-1A, Table 1) had a higher nitrogen content (~7.68%).

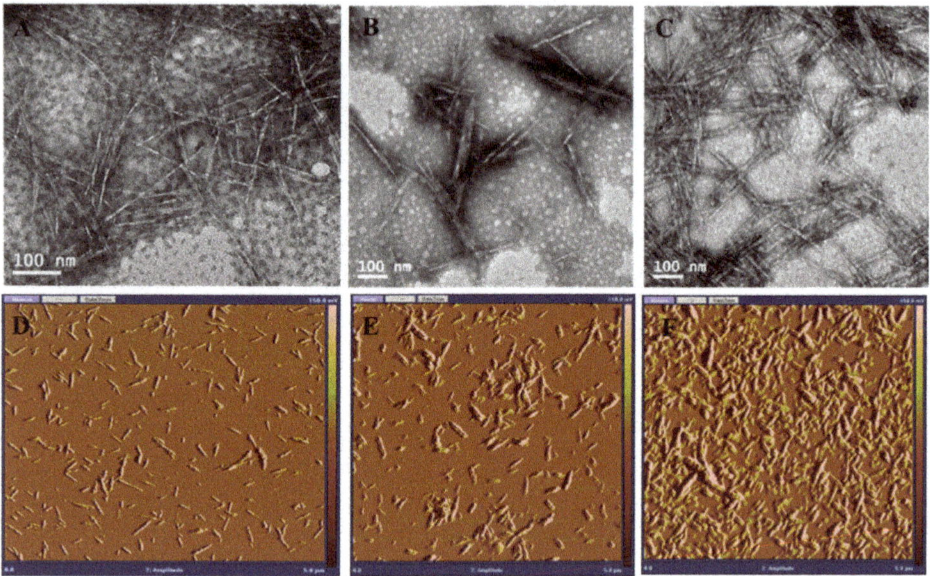

Figure 2. Morphological analyses of pristine and cationic CNCs by TEM and AFM: Representative TEM images of (**A**) pristine CNCs; (**B**) CNC-METAC-1A; (**C**) CNC-METAC-1B and AFM phase images of (**D**) pristine CNCs, (**E**) CNC- METAC-2A; (**F**) CNC-METAC-2B. TEM and AFM images of other cationic CNCs are depicted in Figures S4 and S5.

Table 2. Apparent particle size (DLS) and zeta potential of pristine and modified CNC samples.

Sample	Apparent Particle Size (nm)	Polydispersity Index (PdI)	Zeta Potential (mV)
Pristine CNCs	101.6 ± 0.72	0.23	−34.8 ± 2.16
CNC-BriB-1	96.3 ± 1.65	0.18	−28.9 ± 1.47
CNC-BriB-2	99.9 ± 1.80	0.22	−30.2 ± 1.69
CNC-METAC-1A	123.4 ± 1.32	0.20	+31.8 ± 2.89
CNC-METAC-1B	136.2 ± 1.50	0.22	+44.9 ± 3.93
CNC-METAC-2A	178.1 ± 2.53	0.24	+32.0 ± 0.94
CNC-METAC-2B	203.2 ± 2.66	0.31	+38.2 ± 0.94
CNC-AEM-1A	172.0 ± 9.08	0.38	+45.0 ± 1.44
CNC-AEM-2A	215.3 ± 2.86	0.31	+41.4 ± 3.15

Table 3. Elemental analysis of pristine and cationic CNCs.

Sample	% Carbon	% Hydrogen	% Nitrogen	% Sulfur
Pristine CNCs	40.92	6.07	0.00	0.30
CNC-METAC-1A	44.32	6.94	5.90	<0.20
CNC-METAC-1B	44.38	7.47	7.68	<0.20
CNC-METAC-2A	42.58	6.56	1.70	<0.20
CNC-METAC-2B	43.10	6.59	1.82	<0.20
CNC-AEM-1A	41.97	6.67	5.80	0.22
CNC-AEM-2A	40.45	6.35	4.83	<0.10

The morphological features of the cationic CNCs were analyzed by both transmission electron microscopy (TEM) and atomic force microscopy (AFM) to verify whether the characteristic rod-like crystalline structure of pristine CNCs (Figure 2A,D) was maintained after the surface modification reactions. Figure 2B–F, Figures S4 and S5 indicated that the rod-like morphology of the cationic CNCs was retained and the samples did not degrade into more simple carbohydrates [32]. Accurate size measurement of the rod-like particles by TEM was difficult as the edges of the grafted CNCs were less defined as well as a tendency to aggregate during air-drying or staining [33]. AFM analysis of pristine CNCs depicted a good distribution of the nanorods possessing an average length of 104 ± 68 nm and diameter of 4.0 ± 0.5 nm with a resultant aspect ratio of 26 (Figure 2D). The size of the cationic CNCs was found to be in the range of 112–164 nm in average length and 4.9–5.3 nm in cross-section. Moreover, AFM phase images of the cationic CNCs (Figure 2D,E) showed a halo that was indicative of the presence of grafted polymer brushes [26]. Aggregation of nanoparticles was observed for CNC-METAC-2B (Figure 2F) as it was prepared with a high monomer concentration (compared to CNC-METAC-2A (Figure 2E)), and hence a high degree of polymerization was expected [34].

3.2. Cytotoxicity Studies of Cationic CNCs

Assessment of toxicity in cell-based assays is a rapid, simple, and affordable approach to address the initial safety of nanomaterials with potential for biomedical application. These cytotoxic assays can be considered as a crucial part of the characterization of nanomaterials [35]. To evaluate the cytotoxicity of the cationic CNCs, we chose to perform MTT and NR assays using the firmly attached J774A.1 and BV2 cell lines because (i) these assays are largely used for screening of toxic and harmless compounds, (ii) quick analysis and low manipulation of attached cells after treatment and (iii) satisfactory reproducibility. In addition to cell lines, we tested the toxicity in a more relevant cell system, such as human cells PBMCs since the biological application of these nanomaterials can involve systemic circulation throughout the body. To assess the CNCs impact on cell viability in PBMCs, which remain loosely attached to the plate when not primed, we chose LIVE/DEAD® assay followed by flow cytometry analysis since it is an easy-to-use and sensitivity assay and PBMCs are not the most suitable cell type to access cytotoxicity by MTT or NR using a plate reader. Overall, our results indicated that CNCs have none or very low negative impact (<20%) on the cell viability assessed by all three cytotoxicity assays in all the cell types tested. However, results differed among CNCs and some cell types are more sensitive than others, depending on the assay used. Using MTT assay, which assesses cell viability by mainly mitochondria dehydrogenases, our data showed that in BV2 cells, the pristine CNC and CNC-AEM-1 did not decrease cell viability in any tested condition (Figure 3A). However, the METAC-modified CNCs did show a small decrease in cell viability, statistically significant, but not in a dose-response manner (Figure 3B). Using the same assay in J774A.1 cells, all the AEM-(Figure 3C) and METAC-modified CNCs (Figure 3D) did not impact negatively the cell viability in all tested concentrations. Interestingly, at the lowest concentration of CNC-METAC-1B (10 µg/mL), an increase in the conversion of tetrazolium to formazan was observed (Figure 3D). This effect does not indicate that there was an increase in cell viability above the control but rather an increase in the mitochondria dehydrogenase activity. This effect is not unknown since we have previously observed similar increases with other functionalized CNCs in other cell types, including MCF-7, a breast cancer cell line [23]. One reason for such an effect could be attributed to more tetrazolium dye entering into the cells, and therefore being more available to the dehydrogenases. A plausible explanation for potentially more dye entering the cells is that engineered nanomaterials, particularly high aspect ratio materials, can cause plasma membrane perturbation among other effects [36].

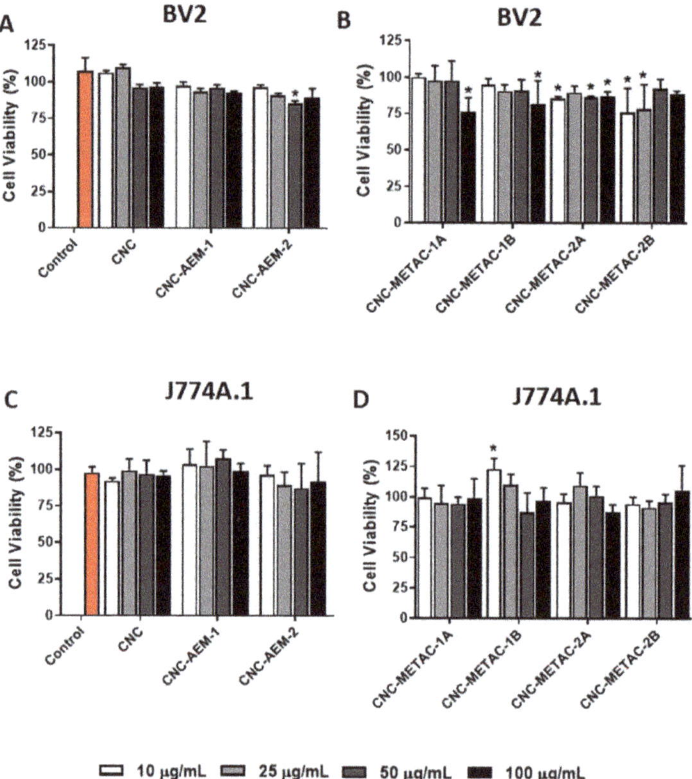

Figure 3. Cytotoxicity of unmodified and modified CNCs assessed by MTT assay. After 24 h of treatment, cell viability in BV2 and J774A.1 cell lines was determined spectrophotometrically. (**A,C**) pristine CNC and AEM modified and (**B,D**) METAC modified CNCs. Representative of 3 independent experiments, in triplicates. * $p < 0.01$ compared to control, non-treated cells.

Data from NR assay indicated that in BV2 cells, CNC-AEM-2 appeared to cause an augmentation of NR dye uptake mainly in the highest concentrations (Figure 4A). A marginal negative effect on cell viability was observed in the cells treated with CNC-METAC-1B at the highest dose, 100 µg/mL (Figure 4B). Similar to what was observed with MTT assay, the J774A.1 cell line appeared to be less sensitive than BV2 cells. The AEM-modified CNC did not affect J774A.1 cells (Figure 4C) and only CNC-METAC-1B had a negative effect on the viability of these cells, although not a dose-response effect (Figure 4D). These differences could be attributed to potential physico-chemical changes on the nanomaterials since some nanomaterials might tend to form large agglomerates, sometimes larger than 100 nm size due to particle interaction with serum proteins [37], which was observed to have a significant effect on particle dispersion for certain materials. Furthermore, the cell lines used in this study were grown in a different medium, for instance, while J774A.1 cells were cultivated in RPMI, BV2 cells were grown in DMEM. Another aspect to consider is the differences in cellular responses was the biological differences among cell lines per se.

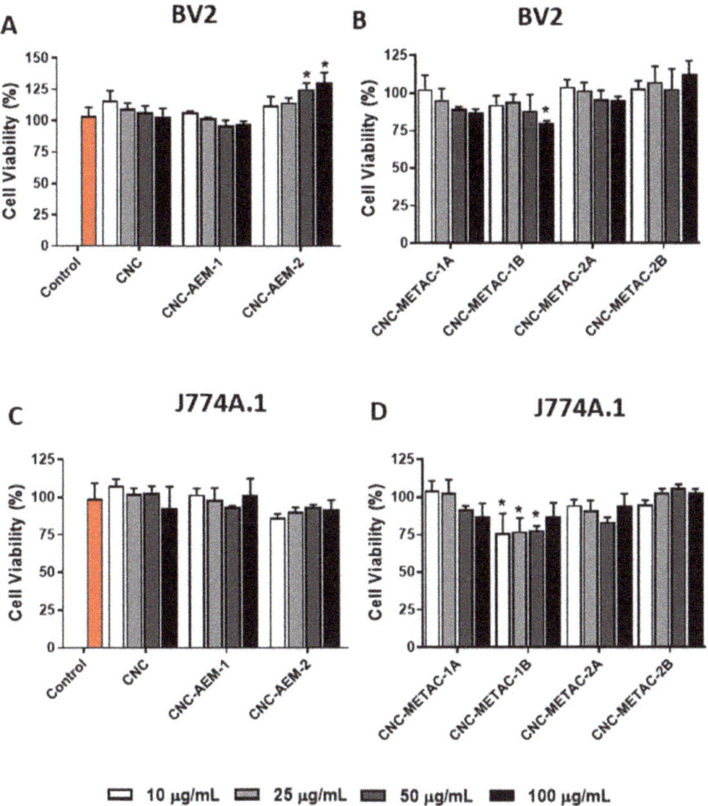

Figure 4. Cytotoxicity of unmodified and modified CNCs assessed by Neutral Red assay. After 24 h of treatment, cell viability in BV2 and J774A.1 cell lines was determined spectrophotometrically. (**A,C**) pristine CNC and AEM modified and (**B,D**) METAC modified CNCs. Representative of 3 independent experiments, in triplicates. * $p < 0.01$ compared to control, non-treated cells.

We also evaluated the cytotoxicity of these modified CNCs using PBMCs and LIVE/DEAD® assay. Corroborating with what was observed with cell lines, the overall cell viability in primary cells PBMCs was not affected significantly and in a dose-response manner. Despite the variation of the data, the percentage of ethidium bromide-labeled cells did not change significantly in the presence of different concentrations of AEM- or METAC-modified CNCs, in comparison to the non-treated cells (Figures 5 and 6, respectively). As expected, H_2O_2 (positive control) increased the dead cell population as observed in the significant increases in the percentage of ethidium bromide labeling (Figure 5D,F). Although we observed that CNC-METAC-1B (Figure 6B) and CNC-METAC-2B (Figure 6D) displayed different flow cytometer spectrum from other CNCs, this difference appeared not to be related with an increase in cell death since it neither showed a characteristic peak in the H_2O_2 intensity range (Figure 5,D) nor a dose-response effect. We also did not observe differences in Calcein-AM staining between treatments and control. The SSC vs. FSC plot of non-stained cells was used for gating cells and excluding debris (Figure S6). A typical LIVE DEAD® flow cytometer SSC vs. FSC plot of staining cells (Ethidium bromide and Calcein-AM) treated of 100 μg/mL is displayed in the supplemental materials (Figure S7).

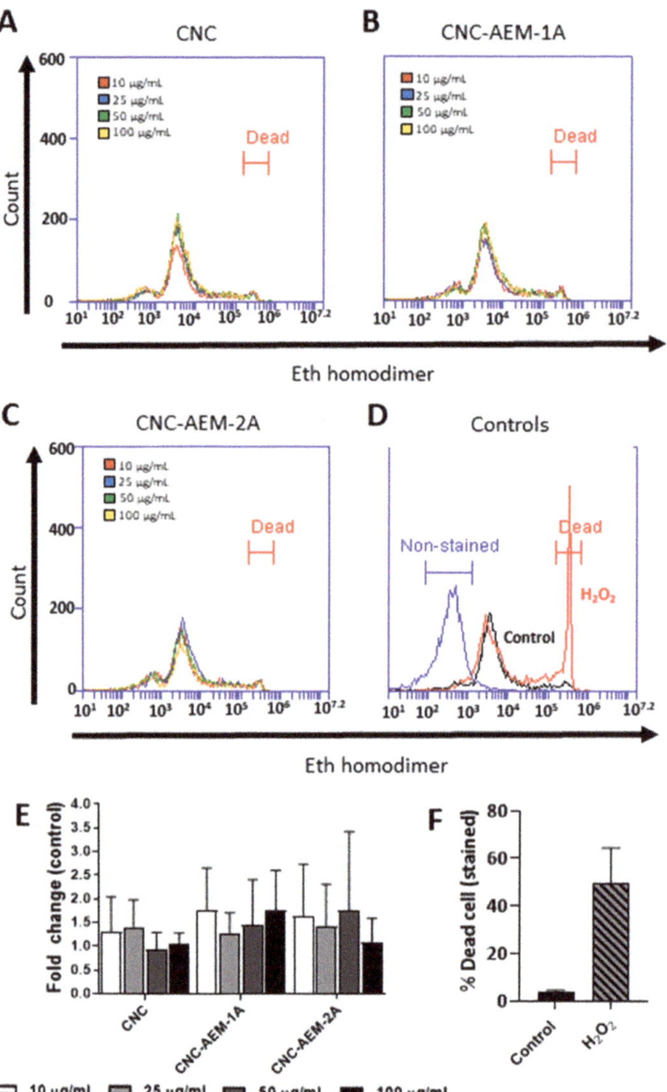

Figure 5. Flow cytometer plots from a representative experiment displaying the cytotoxic effects of unmodified (**A**) and AEM-modified CNCs (**B**,**C**) on PBMCs, as well as controls for gating (**D**). After 24 h of treatment, cell viability was determined by LIVE/DEAD® assay. The graph (**E**) displayed the average of fold of change in the (%) of dead cells (ethidium bromide staining) corrected by (%) dead cells in the control (non-treated cells), from at least 4 experiments. H_2O_2 500 µM, 4h was used as positive control for gating dead cells for the flow cytometer analysis (**D**,**F**).

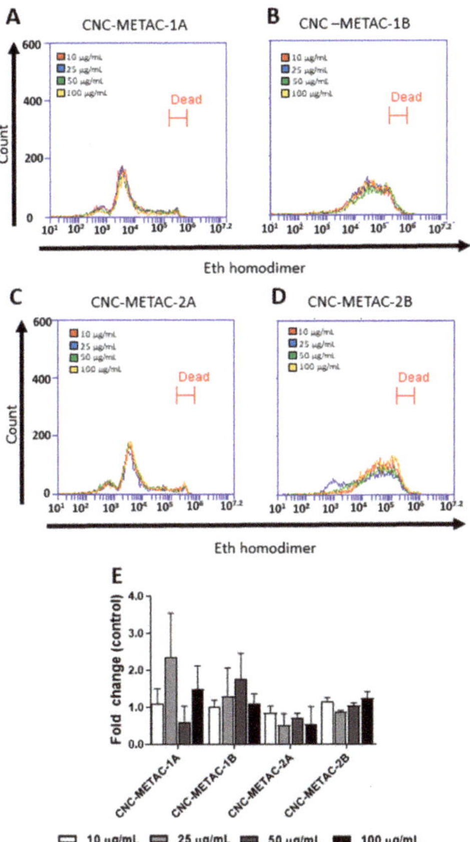

Figure 6. Flow cytometer plots from a representative experiment displaying the cytotoxic effects of METAC-modified CNCs (**A–D**) on PBMCs. After 24 h of treatment, cell viability was determined by LIVE/DEAD® assay. The graph (**E**) displayed the average of fold of change in the (%) of dead cells (ethidium bromide staining) corrected by (%) dead cells in the control (non-treated cells), from at least 4 experiments.

4. Conclusions

In conclusion, a series of novel cationic CNCs were successfully synthesized via SI-SET-LRP at room temperature. The cationic CNCs displayed rod-like structures with different sizes and positive surface charges. The morphological features and structural integrity of the cationic CNCs were retained as evidenced by both TEM and AFM analyses. In general, the cationic CNCs showed low toxicity and the slight decreases in cell viability were cell-type dependent and did not indicate any correlation with physico-chemical characteristics. This initial analysis is crucial for further biological applications, suggesting that these cellulose-based nanomaterials would be good candidates to be investigated as immunomodulators and further developed as potential vaccine nanoadjuvants.

Supplementary Materials: The following are available online at http://www.mdpi.com/2079-4991/10/8/1603/s1: Preparation of CNC-initiators and cationic CNCs, Figure S1: FTIR spectra of pristine CNCs, CNC-BriB-1 and CNC-BriB-2, Table S1: Amounts of reactants and reagents used for the preparation of cationic CNCs, Figure S2: FTIR spectra of cationic CNCs, Figure S3: Intensity-averaged size distribution profiles for pristine CNCs and CNC-METAC-1B in water, Figure S4: TEM images of cationic CNCs, Figure S5: AFM height and phase images of cationic CNCs, Figure S6: controls and gating for flow cytometry, Figure S7: Typical representative flow cytometry SSC vs. FSC plots.

Author Contributions: Conceptualization, R.S. and K.C.; investigation, Y.I., B.T., C.W.S., and A.R., (synthesis and characterization), Y.L. and S.H. (AFM and TEM analysis), and L.N. and K.C. (cytotoxicity assays); data curation, Y.I., B.T., Y.L., and K.C.; writing—review and editing, R.S. and K.C.; supervision, R.S. and K.C.; project administration, R.S. and K.C.; funding acquisition, R.S. and K.C. All authors have read and agreed to the published version of the manuscript.

Funding: This material is based upon work supported by the National Science Foundation under Grant No 1703890.

Acknowledgments: The authors would like to thank InnoTech Alberta Inc., Edmonton, AB, Canada for generously providing spray-dried sulfated CNC for this study. Dante Quinones and Angela Pacherille are acknowledged for scaling up the preparation of CNC-BriB samples.

Conflicts of Interest: The authors declare no conflict of interest.

References

1. He, T.B.; Huang, Y.P.; Yang, L.; Liu, T.T.; Gong, W.Y.; Wang, X.J.; Sheng, J.; Hu, J.M. Structural Characterization and Immunomodulating Activity of Polysaccharide from Dendrobium officinale. *Int. J. Biol. Macromol.* **2016**, *83*, 34–41. [CrossRef] [PubMed]

2. Huang, Y.P.; He, T.B.; Cuan, X.D.; Wang, X.J.; Hu, J.M.; Sheng, J. 1,4-Beta-d-Glucomannan from Dendrobium officinale Activates NF-small ka, CyrillicB via TLR4 to Regulate the Immune Response. *Molecules* **2018**, *23*, 2658. [CrossRef] [PubMed]

3. Delves, P.J.; Roitt, I.M. The Immune System. First of Two Parts. *N. Engl. J. Med.* **2000**, *343*, 37–49. [CrossRef] [PubMed]

4. Reed, S.G.; Orr, M.T.; Fox, C.B. Key Roles of Adjuvants in Modern Vaccines. *Nat. Med.* **2013**, *19*, 1597–1608. [CrossRef] [PubMed]

5. Gause, K.T.; Wheatley, A.K.; Cui, J.; Yan, Y.; Kent, S.J.; Caruso, F. Immunological Principles Guiding the Rational Design of Particles for Vaccine Delivery. *ACS Nano* **2017**, *11*, 54–68. [CrossRef] [PubMed]

6. Nakanishi, T.; Kunisawa, J.; Hayashi, A.; Tsutsumi, Y.; Kubo, K.; Nakagawa, S.; Nakanishi, M.; Tanaka, K.; Mayumi, T. Positively Charged Liposome Functions as an Efficient Immunoadjuvant in Inducing Cell-Mediated Immune Response to Soluble Proteins. *J. Control. Release* **1999**, *61*, 233–240. [CrossRef]

7. Sun, B.; Ji, Z.; Liao, Y.P.; Chang, C.H.; Wang, X.; Ku, J.; Xue, C.; Mirshafiee, V.; Xia, T. Enhanced Immune Adjuvant Activity of Aluminum Oxyhydroxide Nanorods through Cationic Surface Functionalization. *ACS Appl. Mater. Interfaces* **2017**, *9*, 21697–21705. [CrossRef]

8. Sunasee, R.; Araoye, E.; Pyram, D.; Hemraz, U.D.; Boluk, Y.; Ckless, K. Cellulose Nanocrystal Cationic Derivative Induces NLRP3 Inflammasome-Dependent IL-1β Secretion Associated with Mitochondrial ROS Production. *Biochem. Biophys. Rep.* **2015**, *4*, 1–9. [CrossRef]

9. Guglielmo, A.; Sabra, A.; Elbery, M.; Cerveira, M.M.; Ghenov, F.; Sunasee, R.; Ckless, K. A Mechanistic Insight into Curcumin Modulation of the IL-1beta Secretion and NLRP3 S-glutathionylation Induced by Needle-like Cationic Cellulose Nanocrystals in Myeloid Cells. *Chem. Biol. Interact.* **2017**, *274*, 1–12. [CrossRef]

10. Habibi, Y.; Lucia, L.A.; Rojas, O.J. Cellulose Nanocrystals: Chemistry, Self-Assembly, and Applications. *Chem. Rev.* **2010**, *110*, 3479–3500. [CrossRef]

11. Klemm, D.; Kramer, F.; Moritz, S.; Lindstroem, T.; Ankerfors, M.; Gray, D.; Dorris, A. Nanocelluloses: A New Family of Nature-Based Materials. *Angew. Chem. Int. Ed.* **2011**, *50*, 5438–5466. [CrossRef] [PubMed]

12. Moon, R.J.; Martini, A.; Nairn, J.; Simonsen, J.; Youngblood, J. Cellulose Nanomaterials Review: Structure, Properties and Nanocomposites. *Chem. Soc. Rev.* **2011**, *40*, 3941–3994. [CrossRef] [PubMed]

13. Sunasee, R.; Hemraz, U.D.; Ckless, K. Cellulose Nanocrystals: A Versatile Nanoplatform for Emerging Biomedical Applications. *Expert Opin. Drug Deliv.* **2016**, *13*, 1243–1256. [CrossRef] [PubMed]

14. Thomas, B.; Raj, M.C.; Joy, J.; Moores, A.; Drisko, G.L.; Sanchez, C. Nanocellulose, a Versatile Green Platform: From Biosources to Materials and Their Applications. *Chem. Rev.* **2018**, *118*, 11575–11625. [CrossRef]

15. Sunasee, R. Nanocellulose: Preparation, Functionalization and Applications. In *Reference Module in Chemistry, Molecular Sciences and Chemical Engineering*; Elsevier: Amsterdam, The Netherlands, 2020. [CrossRef]

16. Reid, M.S.; Villalobos, M.; Cranston, E.D. Benchmarking Cellulose Nanocrystals: From the Laboratory to Industrial Production. *Langmuir* **2017**, *33*, 1583–1598. [CrossRef]

17. Hasani, M.; Cranston, E.D.; Westman, G.; Gray, D.G. Cationic Surface Functionalization of Cellulose Nanocrystals. *Soft Matter* **2008**, *4*, 2238–2244. [CrossRef]

18. Jasmani, L.; Eyley, S.; Wallbridge, R.; Thielemans, W. A Facile One-pot Route to Cationic Cellulose Nanocrystals. *Nanoscale* **2013**, *5*, 10207–10211. [CrossRef]

19. Feese, E.; Sadeghifar, H.; Gracz, H.S.; Argyropoulos, D.S.; Ghiladi, R.A. Photobactericidal Porphyrin-Cellulose Nanocrystals: Synthesis, Characterization, and Antimicrobial properties. *Biomacromolecules* **2011**, *12*, 3528–3539. [CrossRef]

20. Sunasee, R.; Hemraz, U.D. Synthetic Strategies for the Fabrication of Cationic Surface-Modified Cellulose Nanocrystals. *Fibers* **2018**, *6*, 15. [CrossRef]

21. Rosilo, H.; McKee, J.R.; Kontturi, E.; Koho, T.; Hytönen, V.P.; Ikkala, O.; Kostiainen, M.A. Cationic Polymer Brush-modified Cellulose Nanocrystals for High-Affinity Virus Binding. *Nanoscale* **2014**, *6*, 11871–11881. [CrossRef]

22. Zoppe, J.O.; Dupire, A.V.M.; Lachat, T.G.G.; Lemal, P.; Rodriguez-Lorenzo, L.; Petri-Fink, A.; Weder, C.; Klok, H.A. Cellulose Nanocrystals with Tethered Polymer Chains: Chemically Patchy versus Uniform Decoration. *ACS Macro Lett.* **2017**, *6*, 892–897. [CrossRef]

23. Hemraz, U.D.; Campbell, K.A.; Burdick, J.S.; Ckless, K.; Boluk, Y.; Sunasee, R. Cationic Poly(2-aminoethylmethacrylate) and Poly(N-(2-aminoethylmethacrylamide) Modified Cellulose Nanocrystals: Synthesis, Characterization, and Cytotoxicity. *Biomacromolecules* **2015**, *16*, 319–325. [CrossRef] [PubMed]

24. van Meerloo, J.; Kaspers, G.J.; Cloos, J. Cell Sensitivity Assays: The MTT Assay. *Methods Mol. Biol.* **2011**, *731*, 237–245. [PubMed]

25. Repetto, G.; del Peso, A.; Zurita, J.L. Neutral Red Uptake Assay for the Estimation of Cell Viability/Cytotoxicity. *Nat. Protoc.* **2008**, *3*, 1125–1131. [CrossRef]

26. Zoppe, J.O.; Habibi, Y.; Rojas, O.J.; Venditti, R.A.; Johansson, L.S.; Efimenko, K.; Osterberg, M.; Laine, J. Poly(N-isopropylacrylamide) Brushes Grafted from Cellulose Nanocrystals via Surface-Initiated Single-Electron Transfer Living Radical Polymerization. *Biomacromolecules* **2010**, *11*, 2683–2691. [CrossRef]

27. Xu, Q.; Yi, J.; Zhang, X.; Zhang, H. A Novel Amphotropic Polymer Based on Cellulose Nanocrystals Grafted with Azo Polymers. *Eur. Polym. J.* **2008**, *44*, 2830–2837. [CrossRef]

28. Anastasaki, A.; Nikolaou, V.; Nurumbetov, G.; Wilson, P.; Kempe, K.; Quinn, J.F.; Davis, T.P.; Whittaker, M.R.; Haddleton, D.M. Cu(0)-Mediated Living Radical Polymerization: A Versatile Tool for Materials Synthesis. *Chem. Rev.* **2016**, *116*, 835–877. [CrossRef]

29. Jimenez, A.S.; Jaramillo, F.; Hemraz, U.D.; Boluk, Y.; Ckless, K.; Sunasee, R. Effect of Surface Organic Coatings of Cellulose Nanocrystals on the Viability of Mammalian Cell Lines. *Nanotechnol. Sci. Appl.* **2017**, *10*, 123–136. [CrossRef]

30. Morandi, G.; Heath, L.; Thielemans, W. Cellulose Nanocrystals Grafted with Polystyrene Chains through Surface-Initiated Atom Transfer Radical Polymerization (Si-Atrp). *Langmuir* **2009**, *25*, 8280–8286. [CrossRef]

31. Hu, Z.; Berry, R.M.; Pelton, R.; Cranston, E.D. One-Pot Water-Based Hydrophobic Surface Modification of Cellulose Nanocrystals Using Plant Polyphenols. *ACS Sustain. Chem. Eng.* **2017**, *5*, 5018–5026. [CrossRef]

32. Gallagher, Z.J.; Fleetwood, S.; Kirley, T.L.; Shaw, M.A.; Mullins, E.S.; Ayres, N.; Foster, E.J. Heparin Mimic Material Derived from Cellulose Nanocrystals. *Biomacromolecules* **2020**, *21*, 1103–1111. [CrossRef] [PubMed]

33. Gauche, C.; Felisberti, M.I. Colloidal Behavior of Cellulose Nanocrystals Grafted with Poly(2-alkyl-2-oxazoline)s. *ACS Omega* **2019**, *4*, 11893–11905. [CrossRef] [PubMed]

34. Zoppe, J.O.; Osterberg, M.; Venditti, R.A.; Laine, J.; Rojas, O.J. Surface Interaction Forces of Cellulose Nanocrystals Grafted with Thermoresponsive Polymer Brushes. *Biomacromolecules* **2011**, *12*, 2788–2796. [CrossRef]

35. Nel, A.; Xia, T.; Meng, H.; Wang, X.; Lin, S.; Ji, Z.; Zhang, H. Nanomaterial Toxicity Testing in the 21st Century: Use of a Predictive Toxicological Approach and High-Throughput Screening. *Acc. Chem. Res.* **2013**, *46*, 604–621. [CrossRef] [PubMed]

36. Wang, X.; Sun, B.; Liu, S.; Xia, T. Structure Activity Relationships of Engineered Nanomaterials in inducing NLRP3 Inflammasome Activation and Chronic Lung Fibrosis. *NanoImpact* **2017**, *6*, 99–108. [CrossRef] [PubMed]

37. Murdock, R.C.; Braydich-Stolle, L.; Schrand, A.M.; Schlager, J.J.; Hussain, S.M. Characterization of Nanomaterial Dispersion in Solution Prior to In Vitro Exposure Using Dynamic Light Scattering Technique. *Toxicol. Sci.* **2008**, *101*, 239–253. [CrossRef]

 *nanomaterials* 

Article

Effects of Preparation Method on the Physicochemical Properties of Cationic Nanocellulose and Starch Nanocomposites

Lina Han [1,†], Wentao Wang [1,2,†], Rui Zhang [1], Haizhou Dong [1], Jingyuan Liu [1], Lingrang Kong [3,*] and Hanxue Hou [1,*]

[1] College of Food Science and Engineering, Shandong Agricultural University, Tai'an 271018, China; hanlina1994@126.com (L.H.); wwtlxm@126.com (W.W.); xuyuyisha@163.com (R.Z.); hzhdong@sdau.edu.cn (H.D.); LJY1664961793@126.com (J.L.)

[2] State Key Laboratory of Biobased Material and Green Papermaking, Qilu University of Technology, Shandong Academy of Sciences, Jinan 250000, China

[3] College of Agronomy, Shandong Agricultural University, Tai'an 271018, China

* Correspondence: lkong@sdau.edu.cn (L.K.); hhx@sdau.edu.cn (H.H.)

† These authors contributed equally to this work.

Received: 31 October 2019; Accepted: 19 November 2019; Published: 28 November 2019

Abstract: Nanocellulose (NC) has attracted attention in recent years for the advantages offered by its unique characteristics. In this study, the effects of the preparation method on the properties of starch films were investigated by preparing NC from cationic-modified microcrystalline cellulose (MD-MCC) using three methods: Acid hydrolysis (AH), high-pressure homogenization (HH), and high-intensity ultrasonication (US). When MD-MCC was used as the starting material, the yield of NC dramatically increased compared to the NC yield obtained from unmodified MCC and the increased zeta potential improved its suspension stability in water. The NC prepared by the different methods had a range of particle sizes and exhibited needle-like structures with high aspect ratios. Fourier transform infrared (FTIR) spectra indicated that trimethyl quaternary ammonium salt groups were introduced to the cellulose backbone during etherification. AH-NC had a much lower maximum decomposition temperature (T_{max}) than HH-NC or US-NC. The starch/HH-NC film exhibited the best water vapor barrier properties because the HH-NC particles were well-dispersed in the starch matrix, as demonstrated by the surface morphology of the film. Our results suggest that cationic NC is a promising reinforcing agent for the development of starch-based biodegradable food-packaging materials.

Keywords: nanocellulose; cationic microcrystalline cellulose; high-intensity ultrasonication; high-pressure homogenization; acid hydrolysis; starch nanocomposite films

---

1. Introduction

Cellulose is the most abundant renewable natural resource and thus has attracted the interest of researchers around the world [1]. Native cellulose consists of amorphous and crystalline regions; when subjected to appropriate treatments such as mechanical, chemical, or enzymatic methods; nanocellulose (NC) can be obtained by breaking down amorphous regions [2]. In recent years; NC has become one of the most promising nanomaterials and has attracted increasing attention in the field of nanocomposites because of its appealing intrinsic properties which include its high specific surface area; high aspect ratio; low density; high chemical reactivity; high tensile strength; and high Young's modulus [3–6].

Multiple technologies have been developed to prepare NC from cellulosic materials, including ultrasonication [7], high-pressure homogenization [8], steam explosion treatment [9], high-speed grinding [10], acid hydrolysis [11], TEMPO (2,2,6,6-tetramethylpiperidine-1-oxyl radical) oxidation [12], and combined treatments [13]. The morphology and properties of NC particles determine their application properties and are affected by preparation methods [14]. Recently, there has been increasing interest in the production of NC from microcrystalline cellulose (MCC) because MCC has a high cellulose content, which is affected by its pectin, hemicellulose, lignin, and other lignocellulosic components [15]. The main drawback of acid hydrolysis, ultrasonication, and high-pressure homogenization methods using MCC is their low NC yield [16–18]. Several studies have attempted to increase the NC yield from MCC; however, their methods were limited by complicated pretreatment steps, high-energy consumption, long treatment times, and high water consumption [18–20].

Nanocellulose has been widely used as a reinforcing agent for various advanced composite applications because of its unique advantages [21], which include improving the mechanical and water-vapor barrier properties of starch films [22]. However, NC particles tend to agglomerate via van der Waals forces and hydrogen bonding during film preparation because a large number of hydroxyl groups are present on the NC surface, which restricts the advantages they offer. The dispersion of NC in starch matrices and the interfacial adhesion between NC and starch are widely accepted as the critical factors determining the reinforcement effect of NC in starch [14]; the advantages offered by NC—i.e., large aspect ratio, high modulus, and large surface area to interact with the starch matrix—can only be fully realized when NC is homogeneously distributed in the starch matrix.

Surface modification of NC to improve its dispersion and compatibility with polymer matrices has been widely studied and different surface modification methods have been reported, such as etherification [23], esterification [24], ionic interaction [25], silylation [26], and oxidation [27]. Most of the previous studies modified the NC directly. However, surface modification of NC is complicated and has limited effectiveness because of the highly aggregated structure of NC [28]. Therefore, it was difficult to prepare NC with a modified group. Alkaline solutions are known to swell cellulose samples. Alkali-swelling can disrupt hydrogen bonding between microfibrils in pulp fibers and could facilitate nanofibrillation [29]. This suggests that pre-swelling can be conducive to nanofibrillation of the MCC. This has positive significance for resource conservation and environmental protection. Thus, the preparation of NC from previously swelling and modified MCC is a new strategy that warrants study.

In order to efficiently obtain modified NC with high dispersibility in starch film, a new preparation process has been developed. Microcrystalline cellulose was swelling by sodium hydroxide solution and first modified with 3-chloro-2-hydroxypropyl trimethylammonium chloride. Then, cationic NCs were prepared from previously modified MCC by acid hydrolysis, high-pressure homogenization, and high-intensity ultrasonication. The yield, zeta potential, average particle size, dispersion stability, morphology, crystallinity, chemical structure, and thermal stability of the NC particles were studied. Finally, the effects of the NC preparation methods on the physico-chemical properties of starch nanocomposites were investigated.

2. Materials and Methods

2.1. Materials

Cotton microcrystalline cellulose (MCC) was purchased from Huzhou City Linghu Xinwang Chemical Co., Ltd. (Huzhou, China), (3-chloro-2-hydroxypropyl)trimethylammonium chloride (CHPTA) was obtained from Chengdu Aikeda Chemical Reagent Co., Ltd. (Chengdu, China), and starch was purchased from Hangzhou Starpro Co., Ltd. (Hangzhou, China).

2.2. Cationic Modification of Microcrystalline Cellulose (MCC)

Microcrystalline cellulose was soaked in a 10% sodium hydroxide solution with a solid to liquid ratio of 1 to 10 (1 g/10 mL) at 25 °C for 24 h, whose purpose was to swell cellulose so that CHPTA could enter the cellulose. Then the MCC suspension was centrifuged at 3000 rpm for 10 min and neutralized with diluted hydrochloric acid. After centrifugation, the MCC was dried in an oven at 60 °C for 48 h and subsequently ground into powder.

A 30 g sample of the MCC powder was dispersed in 600 mL of deionized water and 9.87 g sodium hydroxide was added while stirring at room temperature for 30 min. The cationic etherifying agent was then gradually added with continuous stirring and the molar ratio of sodium hydroxide to CHPTA was fixed at 1.2:1. The reaction mixture was stirred for 5 h at 65 °C and the resultant suspension was centrifuged at 5000 rpm for 10 min to obtain the precipitate. The precipitate was suspended in deionized water to remove CHPTA, then dried and ground into powder. The modified MCC is referred to as MD-MCC.

2.3. Preparation of Nanocellulose (NC)

For acid hydrolysis, 10 g of the MCC/MD-MCC was dispersed in 100 mL of 60% (*v/v*) sulfuric acid. Hydrolysis was conducted at 55 °C with constant stirring for 1 h, after which the reaction was stopped by adding cold water (10-fold dilution). The suspension was centrifuged at 5000 rpm for 15 min and dialyzed with distilled water for several days until the dialysate became neutral; the neutral suspension was centrifuged at 10,000 rpm for 10 min to recover the colloidal suspension. The resultant NC is referred to as AH-NC.

For high-pressure homogenization, 1 g of MCC/MD-MCC was added to 100 mL of deionized water and homogenized 4 times at 800 bar using a SCIENTZ-150 high pressure homogenizer (Ningbo Xinzhi Biotechnology Co., Ltd., Ningbo, China). The resultant cellulose suspension was centrifuged at 10,000 rpm for 5 min to recover the precipitate, which was then suspended in deionized water. This process was repeated 3 times. The NC colloidal suspension obtained as the supernatant is referred to as HH-NC.

For high-intensity ultrasonication, MCC/MD-MCC was soaked in deionized water for 24 h with a solid to liquid ratio of 1:100. The suspension was homogenized using the TJS-3000 ultrasonicator (1750 W for 30 min) and subsequently centrifuged at 10,000 rpm for 5 min to recover the precipitate, which was suspended in deionized water. This process was repeated 3 times. The NC colloidal suspension obtained as the supernatant is referred to as US-NC.

The yield of NC was calculated gravimetrically according to Equation (1):

$$\text{Yield (\%)} = \left(\frac{\text{Weight of NC}}{\text{Weight of MCC}}\right) \times 100\% \tag{1}$$

2.4. Preparation of Starch/NC Composite Films

Starch/NC composite films were prepared using a solution casting method. A certain amount of NC (5% w/w of starch) was dispersed in 100 mL of deionized water by ultrasonication at 600 W for 5 min; 3 g of starch and 0.9 g of glycerol were subsequently added and the suspension was stirred at 85 °C for 1 h. The sample solution was poured onto a PTFE glass plate (24 cm × 12 cm), dried at 60 °C for 6 h, and then peeled off and kept at 23 °C and 53% relative humidity for at least 7 d prior to testing.

2.5. Characterization of MCC and NC

2.5.1. Particle Size and Zeta Potential

The average particle size and zeta potential of the NC in aqueous suspensions were determined using a Nanobrook ZetaPlus Potential Analyzer (Brookhaven Instruments Corporation, Holtsville, NY, USA) under the following conditions: 1.3328 water refractive index, 90° angle, and 25 °C.

2.5.2. Scanning Electron Microscopy (SEM)

The morphology of the MCC and MD-MCC were analyzed by scanning electron microscope (QUANTA FEG 250, FEI, Hillsboro, OR, USA) at a voltage of 5.0 kV. The specimens were placed on a bronze stub and sputter-coated with gold before testing.

The morphological characteristics of NC were studied by transmission electron microscopy (TEM) with a TECNAI 20 U-TWIN microscope (PHIA, Eindhoven, Netherlands) using an acceleration voltage of 100 kV. The prepared suspension was spotted on to a carbon coated copper grid. The grid was dried before TEM analysis. The length and diameter of NC were measured using image analysis (Nano Measure software) at least 100 randomly selected NC fibrils in certain TEM images.

2.5.3. X-ray Diffraction (XRD)

X-ray diffraction (XRD) analysis of the NC samples were determined with a D8 X-ray diffractometer (Bruker-AXS, Karlsruhe, Germany) equipped with a copper target (λ = 0.15406 nm) at 30 mA and 40 kV. Data were recorded in the range (2θ) of 5–40° at a scan rate of $0.02°\cdot s^{-1}$. The crystallinity index was calculated with Equation (2):

$$\text{CrI} = \frac{I_{200} - I_{am}}{I_{200}} \tag{2}$$

where *CrI* is the crystallinity index, I_{200} is the maximum intensity of the diffraction from the 200 plane, and I_{am} is the intensity of particles scattered by the amorphous part of the sample.

2.5.4. Fourier Transform Infrared (FTIR) Spectroscopy

The fourier transform infrared (FTIR) spectra of MCC, MD-MCC and NC were recorded using a Nicolet iS5 spectrometer with iD5 ATR sampling accessory (Thermo Fisher Scientific, Waltham, MA, USA). All spectra were collected from an accumulation of 32 scans at wavelengths ranging between 4000 and 600 cm^{-1}.

2.5.5. Thermogravimetric Analysis (TGA)

Thermogravimetric analysis (TGA) was performed to analyze the thermal properties of NC samples using a TA-60 thermogravimetric analyzer (SHIMADZU, Japan) under nitrogen flow of 50 mL/min. Heating temperature ranged from 25 °C to 500 °C at a heating rate of 10 $°C\cdot min^{-1}$.

2.6. Characterization of Starch/NC Composite Films

2.6.1. Mechanical Properties

Tensile strength (TS, MPa), elongation at break (EAB, %) and elastic modulus (EM, GPa) of the films were measured with an XLW auto tensile tester (Labthink Instruments Co. Ltd., Jinan, China) according to ASTM (American Society of Testing Materials) D882-12 (2012). The samples were cut into strips (dimensions, 150 mm × 15 mm). The initial distance between the grips was 100 mm and the test speed was set at 100 mm/min. Each test was repeated at least six times.

2.6.2. Water Vapor Permeability (WVP)

Water vapor permeability (WVP) was measured according to ASTM E96/E96M-16 (2016) using a PERME™ W3/030 automatic water vapor permeability tester (Labthink Instruments Co., Ltd., Jinan, China). Films were cut into round specimens (80 mm in diameter) with a special sampler. The test was performed at 38.0 °C and 90% RH (relative humidity) with a preheating time of 4 h and a weighing interval of 120 min. The WVP of each sample was obtained from the average of three measurements.

2.6.3. Morphology of Starch Nanocomposite Films

The morphology of starch composite films was analyzed by scanning electron microscope (QUANTA FEG 250, FEI, Hillsborough, OR, USA) at a voltage of 5.0 kV. The specimens were placed on a bronze stub and sputter-coated with gold before testing.

2.7. Statistical Analysis

Statistical differences between the properties of MCC, MD-MCC, NC, and starch films were determined by analysis of variance (ANOVA) via SPSS 21 (IBM Co., Armonk, NY, USA). Comparisons of mean values were performed using Duncan's multiple range tests ($p < 0.05$).

3. Results and Discussion

3.1. Effect of Preparation Method on the Yield of NC Prepared from MCC and Modified Microcrystalline Cellulose (MD-MCC)

Low NC yield is a limiting factor in industrial production and commercial applications. The yields of NC prepared from MCC and MD-MCC acid hydrolysis, high-pressure homogenization, and high-intensity ultrasonication are shown in Table 1. The yield of AH-NC prepared from MD-MCC was 30.63% which increased by 48.1%, compared with that prepared from MCC. Moreover, the yields of HH-NC and US-NC prepared from MCC were only 2.04% and 3.57%, respectively, whereas the yields from MD-MCC were 33.08% and 14.18%; these correspond to increases of approximately 16.2 and 3.97 times, respectively. These results demonstrate that cationic etherification of MCC could significantly increase the NC yield of different preparation methods, especially high-pressure homogenization and high-intensity ultrasonication. The NC yields of AH-NC and HH-NC prepared from MD-MCC are comparable to other reports of NC obtained under the similar conditions [30,31], and it can be increased with the optimization of the treatment conditions; however, there is minimal information on the effect of separate ultrasound processing on NC yield as ultrasonication has primarily been used as an auxiliary processing method.

Table 1. Effect of preparation method on the yield, zeta potential, and average particle size of NC prepared from MCC and MD-MCC.

	NC Samples	Raw Materials	
		MCC	**MD-MCC**
Yield (%)	AH-NC	20.68 ± 0.04	30.63 ± 0.08
	HH-NC	2.04 ± 0.14	33.08 ± 0.34
	US-NC	3.57 ± 0.21	14.18 ± 0.10
Zeta potential (mV)	AH-NC	−5.82 ± 0.09	−5.17 ± 1.18
	HH-NC	−0.71 ± 1.02	15.66 ± 0.44
	US-NC	−1.41 ± 0.85	16.01 ± 0.66
Average particle size (nm)	AH-NC	69.5 ± 0.3	69.0 ± 0.3
	HH-NC	253.3 ± 0.4	161.4 ± 1.4
	US-NC	347.9 ± 5.6	255.0 ± 2.9

Natural cellulose is insoluble and tends to agglomerate in water as it forms an extensive network of intermolecular and intramolecular hydrogen bonds, which can clog valves in homogenizers. Thus, the HH yield of NC prepared from MCC is typically low [32]. During the preparation of US-NC, ultrasonication affected the surface to the inner amorphous regions of MCC and caused MCC to break into submicron fragments instead of directly forming NC. However, the small size of the MCC fragments impeded the ultrasonication process and resulted in the low NC yield [12,18]. Conversely, acid molecules can rapidly penetrate into the inner amorphous regions of the cellulose fibrils where they disintegrate amorphous regions, reduce the size of the cellulose fibers, and ultimately release cellulose nanofibrils [33]; thus, AH-NC had a higher yield than HH-NC or US-NC. Considering these processes, cationic modification of MCC can substantially affect NC preparation by acid hydrolysis, high-pressure homogenization, and high-intensity ultrasonication methods.

3.2. Effect of Preparation Method on the Suspension Stability of NC Prepared from MCC and MD-MCC

The suspension states of NC prepared from MCC and MD-MCC by acid hydrolysis, high-pressure homogenization, and high-intensity ultrasonication are shown in Figure 1. All NC samples were well-dispersed in water and their suspensions were stable and uniform without any stratification when fresh. After 3 days at 20 °C, flocculation and precipitation occurred in the HH-NC and US-NC samples prepared from MCC whereas the other NC samples remained homogeneous, which indicates that the NC particles prepared by MD-MCC were relatively stable in water.

Figure 1. The suspension stability of NC prepared from MCC (AH-1, HH-1, US-1) and MD-MCC (AH-2, HH-2, US-2) t = 0 day and t = 3 days.

Net charge is a critical characteristic that affects the stability of NC particles. In general, higher absolute values of zeta potential correspond to better dispersion and stability [2]. The results from this study reveal that the introduction of trimethyl quaternary ammonium groups by cationic modification of MCC (confirmed by the FTIR results) helped to increase the NC zeta potential and further improved the stability and dispersion of the NC particles. Based on the measured yield and application properties, the NC samples prepared from MD-MCC were selected for further analysis.

3.3. NC Morphology

The SEM micrographs of MCC and MD-MCC shown in Figure 2A reveal that MCC particles had irregular shapes with different dimensions and MD-MCC particles were swollen and porous with a rough surface. Moreover, cationic modification damaged MCC particles and eroded their surface such that the outer layer of the fibers was disrupted and cracked along the inner structure, exposing the fibril strand. These changes to the MCC granular structures facilitated the substantial increase in NC yield as discussed above.

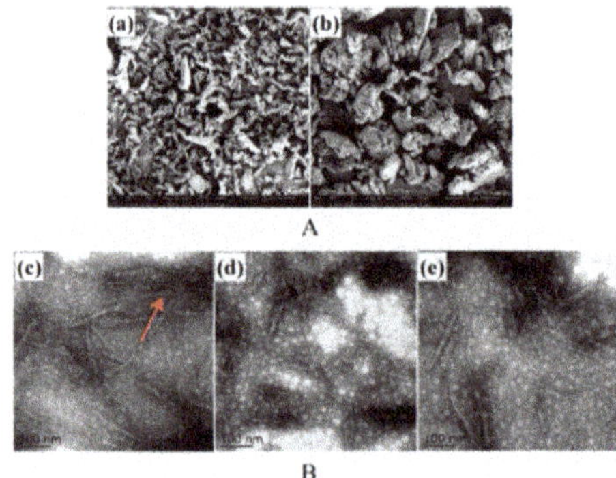

Figure 2. (**A**) Scanning electron microscopy (SEM) micrographs of (**a**) MCC and (**b**) MD-MCC; (**B**) transmission electron microscopy (TEM) micrographs of (**c**) AH-NC, (**d**) HH-NC, and (**e**) US-NC.

The TEM micrographs of AH-NC, HH-NC, and US-NC shown in Figure 2B illustrate their distinctive morphologies and dispersion states. Small bundles of needle-like cellulose fibers with nanoscale diameters were observed for AH-NC and branches of smaller bundles or partly individualized nanofibers were attached to the aggregates as well (arrows in Figure 2B(c)). The formation of such aggregates significantly reduced the surface area of the AH-NC particles and thus hindered their reinforcing ability. Conversely, the HH-NC and US-NC nanofibers were well-dispersed and more individualized.

The NC dispersion state was directly associated with the surface interactions between adjacent NC particles because different surface interactions exist in aqueous solution, including attraction forces (e.g., hydrogen bonding) and repulsion forces (e.g., electrostatic repulsion). Attraction and repulsion forces are expected to compete with each other and thus determine the distinctive dispersion state of aqueous NC from different preparation methods. Therefore, the uniform dispersion of HH-NC and US-NC in aqueous solution was ascribed to profound repulsion forces whereas the AH-NC aggregates were attributed to predominant attraction forces. These theories were confirmed by the zeta potential values shown in Table 1. As expected, the order of absolute zeta potential values of NC prepared from MD-MCC was AH-NC < HH-NC < US-NC, which corresponds to the observed dispersion states. Because of the sulfate anions present during acid hydrolysis, the zeta potential value of AH-NC was not substantially changed. The dispersion states also indicate that hydrogen bonding between nanofibers was lower for HH-NC and US-NC.

3.4. Length-Frequency and Diameter-Frequency Histograms

Length-frequency and diameter-frequency histograms (Figure 3) were prepared using the TEM data. Aggregation and overlapping nanofibers make it difficult to accurately measure dimensions; thus, only individual nanofibers with clearly identifiable ends were measured. The AH-NC exhibited a wide distribution of lengths whereas the diameter distribution had a narrower range of 2 to 8 nm with the maximum at 4.5 nm. The length distributions of HH-NC and US-NC were 25 to 275 nm and 75 to 350 nm, respectively. The diameter distributions of HH-NC and US-NC were 3 to 15 nm and 4 to 10 nm, respectively. The length and diameter distributions of HH-NC and US-NC were wider than those of AH-NC. Therefore, AH-NC had the smallest particles of the three preparation methods (Table 1).

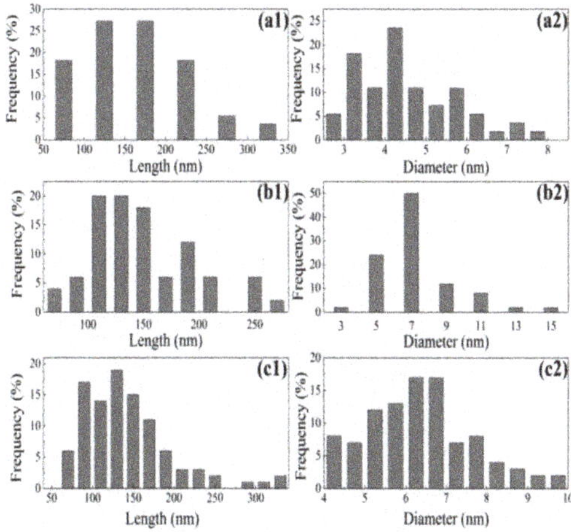

Figure 3. Length-frequency and diameter-frequency histograms of (**a**) AH-NC, (**b**) HH-NC, and (**c**) US-NC.

These results are comparable to those of other studies [2,34,35]. The average aspect ratios (L/D) of HH-NC and US-NC were 21.20 and 23.20, respectively, which were lower than that of AH-NC (36.73). According to a previous report [36], the reinforcing effect of NC is expected to improve as the aspect ratio increases, which may further improve the mechanical properties of biocomposites. The length and diameter of NC prepared from various sources has been reported: 100–300 nm length and 3–5 nm diameter from wood [37], 171.6 nm length and 14.6 nm diameter from cotton [38], 1160 nm length and 16 nm diameter from tunicate [39], and 100–1000 nm length and 10–50 nm diameter from bacterial cellulose [40].

The morphology and dimensions of NC depended on the preparation method and each method resulted in distinct features. Acid hydrolysis, a well-known method for preparing NC, results in a larger aspect ratio than the homogenization and ultrasonication methods. These mechanical process can break hydrogen bonds and disintegrate microfibers into nanofibrils, which form needle-like cellulose crystallites and consequently reduce the aspect ratio of NC.

3.5. XRD Analysis of MCC and NC

The effects of mechanical and chemical treatments on the crystalline structure of the cellulose samples were further characterized by XRD as shown in Figure 4. Both MCC and MD-MCC exhibited characteristic crystalline peaks near $2\theta = 15.1°$, $16.2°$, $21.0°$, $22.6°$, and $34.5°$; these peaks correspond to the ($1\bar{1}0$), (110), (012), (200), and (004) crystallographic planes, respectively, and are characteristic of the cellulose I structure [41,42]. The AH-NC had the same crystalline peaks as MCC and MD-MCC except that the diffraction peak at $2\theta = 34.5°$ was broader and flatter, and this peak disappeared after high-pressure homogenization and high-intensity ultrasonication. Generally, NC obtained via the three different preparation methods exhibited the characteristic cellulose I peaks, which indicates that acid hydrolysis, high-pressure homogenization, and high-intensity ultrasonication do not affect the main crystalline properties of cellulose.

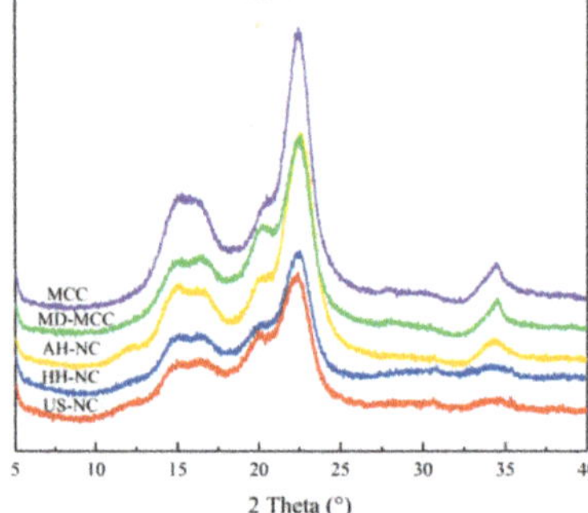

Figure 4. X-ray diffraction (XRD) patterns of MCC, MD-MCC, AH-NC, HH-NC, and US-NC.

The crystallinity of NC is an important parameter as it determines its reinforcing capability and mechanical strength in composite films [43]. Highly crystalline fibers are expected to be more effective at providing reinforcement for composite materials because of their increased stiffness and rigidity, which result in a higher Young's modulus. However, the crystallinity of MD-MCC and the prepared NC each decreased to a different extent relative to MCC (Table 2). The peak at 22.6° was less sharp for the alkali-treated MCC, which suggests that alkali swelling might destroy part of the crystalline structure [44]. Cationic modification of MCC disrupted both intermolecular and intramolecular hydrogen bonding and facilitated the formation of amorphous regions [6]. The results of this study indicate that the NC preparation methods were non-selective as they damaged both amorphous and crystalline cellulose; therefore, the crystallinity of NC decreased [2,7,45]. As shown in Table 2, the crystallinity of AH-NC (67.6%) was the highest of the three NC, which implies that the crystalline regions of MD-MCC were more resistant to the acid treatment than the mechanical treatments. Moreover, the crystallinity of US-NC (62.8%) was slightly higher than that of HH-NC (61.8%), which suggests that high-intensity ultrasonication was less aggressive than high-pressure homogenization.

Table 2. Crystallinity index (CI), degradation temperature, and mass loss of MCC, MD-MCC, AH-NC, HH-NC, and US-NC.

Samples	CI (%)	T_{on} (°C)	T_{max} (°C)	Weight loss at T_{max} (%)	Residue at 500 °C (%)
MCC	71.4	286	353	49.4	7.6
MD-MCC	68.2	278	374	56.6	9.4
AH-NC	67.6	249	298	50.8	20.6
HH-NC	61.8	228	349	52.9	14.2
US-NC	62.8	279	365	50.4	8.5

3.6. Fourier Transform Infrared (FTIR) Spectroscopy Analysis of MCC and NC

Fourier transform infrared spectroscopy was used to understand the changes in the chemical structures of the MCC and NC (Figure 5). Two main absorption regions, 2800–3600 cm^{-1} and 750–1750 cm^{-1}, were present in the spectra. The broad band centered at approximately 3332 cm^{-1} corresponds to O-H stretching of cellulose in the fiber [46]. The intensity of the O-H peak in the MD-MCC spectrum was considerably less than that of the MCC spectrum, which indicates that the number of hydroxyl groups in the MD-MCC was reduced by the etherification reaction [6]. The peaks near 2900 cm^{-1} are attributed to the C-H stretching vibration of cellulose and the peaks at 1644 cm^{-1} have been assigned to water absorption because of the strong cellulose-water interaction [47]. In addition to the characteristic peaks of the cellulose backbone, there was a small peak near 1479 cm^{-1} in the MD-MCC and NC spectra. This peak was not present in the MCC spectrum and thus was assigned to the trimethyl quaternary ammonium groups, which implies that trimethyl quaternary ammonium groups were successfully added on to cellulose chains even though the degree of substitution was low.

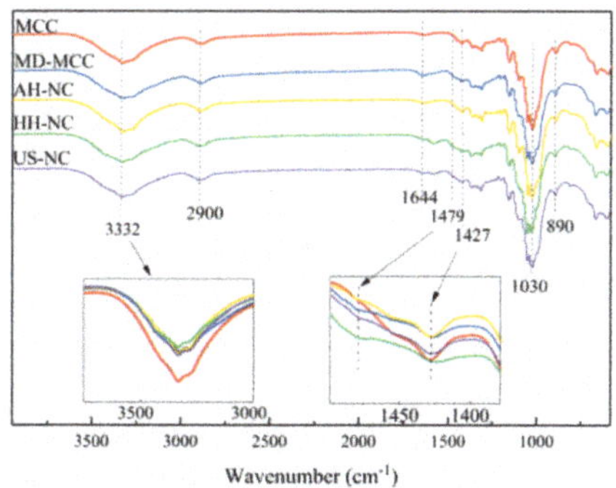

Figure 5. Fourier transform infrared (FTIR) spectra of MCC, MD-MCC, AH-NC, HH-NC, and US-NC.

The changes to the crystal structure of cellulose resulted in the intensity reduction of partial FTIR peaks attributed to the crystalline domains of cellulose [1]. The peak at 1427 cm^{-1} corresponds to the CH$_2$ bending vibration of crystalline cellulose and the NC spectra exhibit reduced intensity of this crystalline band compared to those of the MCC. The reduced intensity of this peak also supports the decreased crystallinity of NC after acid hydrolysis, high-pressure homogenization, and high-intensity ultrasonication as demonstrated by the XRD results. The reduced crystallinity of NC compared to MCC implies that there is less intermolecular and intramolecular hydrogen bonding in NC, which might increase the dispersion of NC in water [1]. The peak at 1030 cm^{-1} is related to C-O stretching of the pyranose ring skeleton [48]. The peak at 890 cm^{-1} is corresponds to β-glycosidic linkages between the glucose units of cellulose. Based on the above analysis, there was not a significant difference between the MCC and NC spectra, which indicates that neither the mechanical nor chemical treatments changed the main chemical structure of the fibers. This result agrees with the XRD analysis.

3.7. Thermal Stability Analysis

The thermal stability of each of MCC, MD-MCC, AH-NC, HH-NC, and US-NC was investigated by TGA, as shown in Figure 6. The onset of thermal decomposition temperature (T_{on}) corresponds to the beginning of degradation and the maximum decomposition temperature (T_{max}) corresponds to the temperature of the maximum rate of degradation. The T_{on}, T_{max}, mass loss at T_{max}, and the char residuals at 500 °C are given in Table 2. Generally, the thermal degradation of MCC and NC occurred in two steps, as shown in Figure 6. The initial mass loss, which was caused by the evaporation of absorbed water, was observed below 120 °C and was slightly different for the different cellulose samples.

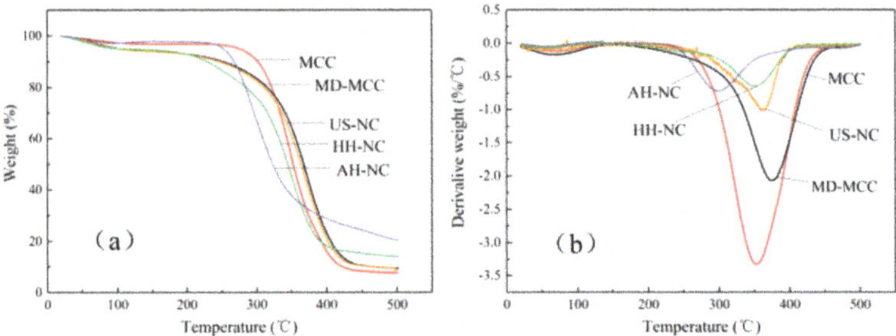

Figure 6. (**a**) Thermogravimetric analysis (TGA) and (**b**) derivative thermogravimetry (DTG) curves of MCC, MD-MCC, AH-NC, HH-NC, and US-NC.

As shown in Table 2, the T_{on} of each NC was less than that of MCC, which indicates that NC has decreased thermal stability. This decreased thermal stability might be attributed to the large number of cellulose chain segments that were damaged during the preparation of NC and formed low molecular chain segments and weak points in the cellulose chain on the surface of NC. At elevated temperatures, these low molecular chain segments and defects absorbed heat and thus began to degrade first, which resulted in the reduced thermal stability of NC. Furthermore, the nanoscale lateral dimensions of NC mean that NC has a higher surface to volume ratio than MCC and thus is heated more efficiently, which also decreases thermal stability [49]. Compared with AH-NC and US-NC, the lower thermal stability of HH-NC could be due to the increased damage to the crystalline region of cellulose during the high-pressure homogenization process as indicated by the XRD results [50].

The main thermal degradation stages of MCC, MD-MCC, AH-NC, HH-NC, and US-NC occurred in the range of 250–430 °C, 200–430 °C, 210–360 °C, 240–405 °C, and 250–405 °C, respectively; this degradation was mainly due to the thermal decomposition of the crystalline cellulose chains. The mass losses caused by the thermal decomposition of AH-NC, HH-NC, and US-NC at T_{max} were 50.8%, 52.9%, and 50.4%, respectively, and the corresponding T_{max} were 298 °C, 349 °C, and 365 °C. The mass losses of each type of NC were very similar but the corresponding T_{max} varied significantly, which indicates that acid hydrolysis, high-pressure homogenization, and high-intensity ultrasonication had different effects on the thermal stability of NC. The sulfate groups that were added to the cellulose chains during acid hydrolysis could facilitate the thermal degradation of cellulose [51], which would explain why the T_{max} of AH-NC was less than those of HH-NC and US-NC. Moreover, this result was consistent with the previous study by Wang et al. [36]. Above 450 °C, the thermal decomposition temperature of each cellulose sample leveled off and a slow thermal degradation profile was obtained. This continued degradation could be attributed to the carbonization of polysaccharide chains caused by the cleavage of C-C and C-H bonds.

The mass of char residue in AH-NC, HH-NC, and US-NC at 500 °C was 20.6%, 14.2%, and 8.5%, respectively, which are all greater than that of MCC (7.6%). The high char residue of AH-NC has been ascribed to the direct solid-to-gas phase transitions of decarboxylation catalyzed by the sulfate groups on the surface of AH-NC [52]. The high char yield of HH-NC might be due to its relatively high dehydration at low temperatures, which carbonized cellulose as confirmed by its relatively low degradation temperature. The low char residue yield of US-NC might be ascribed to its low crystalline content [2].

3.8. Mechanical Properties

The tensile strength (TS), elongation at break (EAB), and elastic modulus (EM) of starch films with and without the addition of 5% NC are shown in Table 3. TS is the measurement of maximum strength of a film against applied tensile stress, EAB represents the ability of a film to stretch, and EM indicates the rigidity of a film. The TS of the pure starch film was 6.32 MPa. The TS of the starch/NC films was considerably greater than that of the pure starch film. The TSs of the starch nanocomposite films prepared from AH-NC, HH-NC, and US-NC were 9.35 MPa, 11.74 MPa, and 10.75 MPa, respectively, which correspond to 1.47, 1.85, and 1.70 times the TS of pure starch film. Li et al. [22] reported that NC can improve the tensile strength of starch films when NC is uniformly distributed in the starch matrix. The strong adhesion at the starch/NC interface and the creation of a rigid NC percolating network within the starch matrix facilitate efficient stress transfer from the soft starch matrix to the rigid NC and thus improve the strength of the starch nanocomposite [53]. Moreover, cationic modification is a hydrophilic modification that introduces hydrophilic quaternary ammonium salt groups; therefore, the cationic NC has excellent compatibility and strong interactions with starch because hydrogen bonds can form between them, which greatly increases the TS of the nanocomposite films.

Table 3. Mechanical properties and water vapor permeability of starch/NC composite films.

Samples	TS (MPa) *	EAB (%) *	EM (MPa) *	WVP $(10^{-12}$ g·cm·cm^{-2}·s^{-1}·Pa^{-1}) *
starch	6.32 ± 0.86d	25.33 ± 0.85a	369.51 ± 13.67c	1.65 ± 0.12a
starch/AH-NC	9.35 ± 0.90c	20.1 ± 0.61b	492.69 ± 10.37b	1.42 ± 0.06b
starch/HH-NC	11.74 ± 0.52a	18.8 ± 0.83b	573.08 ± 13.24a	1.28 ± 0.04b
starch/US-NC	10.75 ± 0.79b	19.6 ± 0.76b	513.24 ± 24.25b	1.36 ± 0.88b

* Different lowercase letters in the same column indicate a statistically significant difference ($p < 0.05$).

The reinforcing effect of NC on the starch films is also different for the different preparation methods. This is potentially due to the different aspect ratios of the three types of NC, which could affect the dispersion of NC in the starch matrix [6]. Since the AH-NC has the highest aspect ratio, hydrogen bonding might prevent it from being uniformly dispersed throughout the starch matrix [28]. Conversely, HH-NC has a low aspect ratio and thus is expected to be well-dispersed in the starch film. This conjecture was verified using the SEM results of the films, which are discussed later in this section.

The effects of preparation method of NC on the EAB of starch/NC composite films were shown in Table 3. As expected, the incorporation of NC into the starch matrix substantially decreased the EAB and the results were in agreement with previous studies [54,55]. The decrease in EAB is attributed to the geometry and rigid nature of NC as well as the formation of a stiff NC network linked by hydrogen bonds and entanglements. Moreover, the addition of NC could hinder the plasticizing efficacy of glycerol and decrease the mobility of starch chains, which would result in brittle nanocomposite films [56,57]. Effects similar to those observed for NC have been reported for other biopolymers, such as whey protein isolate and agar [42,58]. Finally, EM, which indicates the rigidity of a film, increased significantly ($p < 0.05$) when 5% NC was added to the starch film.

3.9. Water Vapor Barrier Properties

The water vapor permeability (WVP) of a film is one of the most important properties for food packaging applications. Because of its hydrophilic nature, starch-based films usually have high WVP and thus poor water vapor barrier properties, which significantly restricts its application in food packaging. As shown in Table 3, the water vapor barrier properties of pure starch films could be improved by the incorporation of NC. The WVP of the pure starch film, 1.65×10^{-12} g·cm·cm^{-2}·s^{-1}·Pa^{-1}, decreased by approximately 14%, 22%, and 18% ($p < 0.05$) when 5% of AH-NC, HH-NC, and US-NC was incorporated into the starch matrix, respectively. This decrease in WVP as the result of the incorporation of NC agrees with previous reports [22,59].

The NC in the starch nanocomposite films functions as an impermeable barrier against WVP because strong hydrogen bonding interactions reduce the diffusion coefficient of the films by increasing the diffusion path for water vapor through the film [42]. Unexpectedly, the lower the aspect ratio of the NC in the starch matrix, the lower the WVP of the resultant starch nanocomposite films. This result might be ascribed to the agglomeration of high aspect ratio NC in the starch matrix, which creates diffusion pathways for water vapor transport within the film and thus facilitates WVP. Therefore, the uniform distribution of NC in the starch matrix played a more significant role than high aspect ratio in improving the water vapor barrier properties of starch nanocomposite films.

3.10. Surface Morphology

The surface morphologies of pure starch film and starch nanocomposite films with 5% NC are shown in Figure 7 at 10,000× magnification. The surface of the pure starch film was smooth with homogeneous morphology and compact structure because of the plasticization effect of glycerol [60]. Compared to the pure starch film, the incorporation of AH-NC increased the surface roughness, which is attributed to the agglomeration of AH-NC in the starch matrix [22]; however, it was difficult to observe the individual fibers in the film because of the small size and bundle structure of AH-NC in the starch matrix. Well-dispersed NC particles were observed in the starch/HH-NC and starch/US-NC films; this suggests that there was strong interfacial adhesion or good compatibility between the HH-NC or US-NC and the starch matrix, which could be attributed to the size and zeta potential of the NC, the chemical similarities of starch and NC, and hydrogen bonding between the two components [22]. Similar results have been reported for starch nanocomposite films containing pineapple leaf cellulose nanofibers [61]. The homogeneous distribution of NC in starch films could greatly improve their tensile strength and water vapor barrier properties.

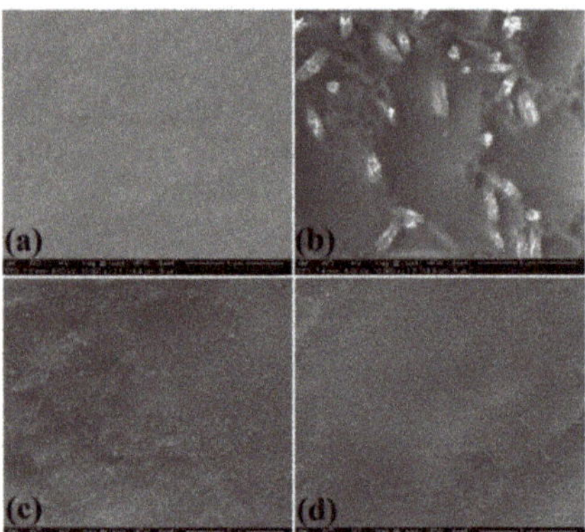

Figure 7. Surface morphology of (**a**) pure starch, (**b**) starch/AH-NC, (**c**) starch/HH-NC, and (**d**) starch/US-NC films.

4. Conclusions

High yields of NC with different aspect ratios were obtained efficiently from cationic MD-MCC by acid hydrolysis, high-pressure homogenization, and high-intensity ultrasonication and the effects of AH-NC, HH-NC, and US-NC on the fundamental properties of starch films were compared. The cationic modification of NC was confirmed by FTIR analysis. Both HH-NC and US-NC prepared from MD-MCC were stable suspensions because of their higher zeta potential compared to NC samples prepared from MCC. The AH-NC prepared from MD-MCC tended to aggregate because of the presence of sulfate groups and, thus, lack of surface charge. The XRD analysis revealed that the crystallinity of NC decreased for all preparation methods whereas the main chemical structure of fibers remained unchanged. The thermostability of the three types of NC decreased relative to that of MCC. The HH-NC exhibited the best dispersion in the starch matrix and demonstrated the best enhancement to the water vapor barrier properties of starch films. Cationic modification of MCC will be a promising strategy to improve the yields and dispersion of NC and ultimately enhance the properties of starch films. The starch/NC nanocomposite films, which are completely biodegradable and biocompatible, have immense potential for food-packaging applications.

Author Contributions: L.H. conceived, designed, and performed the experiments; L.H. and W.W. analyzed the data and wrote original draft; R.Z., H.D. and J.L. wrote review and edited the original draft; L.K. and H.H. contributed reagents/materials/analysis tools; H.H. managed the project and provided financial support.

Funding: This work was supported by Great Innovation Program of Agricultural Application Technology in Shandong province, the Key Research and Development Program of Shandong province (2018GNC113004, 2019GNC106035), Natural Science Foundation of Shandong province (ZR2017BC018), the Foundation (No. KF201808) of State Key Laboratory of Biobased Material and Green Papermaking in Qilu University of Technology (Shandong Academy of Sciences), and Funds of Shandong "Double Tops" Program.

Conflicts of Interest: The authors declare no conflict of interest.

References

1. Adsul, M.; Soni, S.K.; Bhargava, S.K.; Bansal, V. Facile approach for the dispersion of regenerated cellulose in aqueous system in the form of nanoparticles. *Biomacromolecules* **2012**, *13*, 2890–2895. [CrossRef]
2. Yang, X.; Han, F.; Xu, C.; Jiang, S.; Huang, L.; Liu, L.; Xia, Z. Effects of preparation methods on the morphology and properties of nanocellulose (NC) extracted from corn husk. *Ind. Crop. Prod* **2017**, *109*, 241–247. [CrossRef]
3. Abdul Khalil, H.P.; Davoudpour, Y.; Islam, M.N.; Mustapha, A.; Sudesh, K.; Dungani, R.; Jawaid, M. Production and modification of nanofibrillated cellulose using various mechanical processes: A review. *Carbohydr. Polym.* **2014**, *99*, 649–665. [CrossRef] [PubMed]
4. Wang, H.; Zhang, X.; Jiang, Z.; Li, W.; Yu, Y. A comparison study on the preparation of nanocellulose fibrils from fibers and parenchymal cells in bamboo (*Phyllostachys pubescens*). *Ind. Crop. Prod.* **2015**, *71*, 80–88. [CrossRef]
5. Missoum, K.; Martoïa, F.; Belgacem, M.N.; Bras, J. Effect of chemically modified nanofibrillated cellulose addition on the properties of fiber-based materials. *Ind. Crop. Prod.* **2013**, *48*, 98–105. [CrossRef]
6. Cheng, L.; Zhang, D.; Gu, Z.; Li, Z.; Hong, Y.; Li, C. Preparation of acetylated nanofibrillated cellulose from corn stalk microcrystalline cellulose and its reinforcing effect on starch films. *Int. J. Biol. Macromol.* **2018**, *111*, 959–966. [CrossRef]
7. Li, W.; Yue, J.; Liu, S. Preparation of nanocrystalline cellulose via ultrasound and its reinforcement capability for poly(vinyl alcohol) composites. *Ultrason. Sonochem.* **2012**, *19*, 479–485. [CrossRef]
8. Panyasiri, P.; Yingkamhaeng, N.; Lam, N.T.; Sukyai, P. Extraction of cellulose nanofibrils from amylase-treated cassava bagasse using high-pressure homogenization. *Cellulose* **2018**, *25*, 1757–1768. [CrossRef]
9. Deepa, B.; Abraham, E.; Cherian, B.M.; Bismarck, A.; Blaker, J.J.; Pothan, L.A.; Leao, A.L.; de Souza, S.F.; Kottaisamy, M. Structure, morphology and thermal characteristics of banana nano fibers obtained by steam explosion. *Bioresour. Technol.* **2011**, *102*, 1988–1997. [CrossRef]
10. Siró, I.; Plackett, D. Microfibrillated cellulose and new nanocomposite materials: A review. *Cellulose* **2010**, *17*, 459–494. [CrossRef]
11. Niu, F.; Li, M.; Huang, Q.; Zhang, X.; Pan, W.; Yang, J.; Li, J. The characteristic and dispersion stability of nanocellulose produced by mixed acid hydrolysis and ultrasonic assistance. *Carbohydr. Polym.* **2017**, *165*, 197–204. [CrossRef] [PubMed]
12. Salminen, R.; Reza, M.; Pääkkönen, T.; Peyre, J.; Kontturi, E. TEMPO-mediated oxidation of microcrystalline cellulose: Limiting factors for cellulose nanocrystal yield. *Cellulose* **2017**, *24*, 1657–1667. [CrossRef]
13. Zianor Azrina, Z.A.; Beg, M.D.H.; Rosli, M.Y.; Ramli, R.; Junadi, N.; Alam, A.K.M.M. Spherical nanocrystalline cellulose (NCC) from oil palm empty fruit bunch pulp via ultrasound assisted hydrolysis. *Carbohydr. Polym.* **2017**, *162*, 115–120. [CrossRef] [PubMed]

14. Miao, X.; Lin, J.; Tian, F.; Li, X.; Bian, F.; Wang, J. Cellulose nanofibrils extracted from the byproduct of cotton plant. *Carbohydr. Polym.* **2016**, *136*, 841–850.

15. Kian, L.K.; Jawaid, M.; Ariffin, H.; Karim, Z. Isolation and characterization of nanocrystalline cellulose from roselle-derived microcrystalline cellulose. *Int. J. Biol. Macromol.* **2018**, *114*, 54–63. [CrossRef] [PubMed]

16. Chen, L.; Wang, Q.; Hirth, K.; Baez, C.; Agarwal, U.P.; Zhu, J.Y. Tailoring the yield and characteristics of wood cellulose nanocrystals (CNC) using concentrated acid hydrolysis. *Cellulose* **2015**, *22*, 1753–1762. [CrossRef]

17. Yu, H.; Abdalkarim, S.Y.H.; Zhang, H.; Wang, C.; Tam, K.C. Simple process to produce high-yield cellulose nanocrystals using recyclable citric/hydrochloric acids. *ACS Sustain. Chem. Eng.* **2019**, *7*, 4912–4923. [CrossRef]

18. Guo, J.; Guo, X.; Wang, S.; Yin, Y. Effects of ultrasonic treatment during acid hydrolysis on the yield, particle size and structure of cellulose nanocrystals. *Carbohydr. Polym.* **2016**, *135*, 248–255. [CrossRef]

19. Wang, Q.Q.; Zhu, J.Y.; Reiner, R.S.; Verrill, S.P.; Baxa, U.; McNeil, S.E. Approaching zero cellulose loss in cellulose nanocrystal (CNC) production: Recovery and characterization of cellulosic solid residues (CSR) and CNC. *Cellulose* **2012**, *19*, 2033–2047. [CrossRef]

20. Wang, Q.; Zhao, X.; Zhu, J.Y. Kinetics of strong acid hydrolysis of a bleached kraft pulp for producing cellulose nanocrystals (CNCs). *Ind. Eng. Chem. Res.* **2014**, *53*, 11007–11014. [CrossRef]

21. Hietala, M.; Mathew, A.P.; Oksman, K. Bionanocomposites of thermoplastic starch and cellulose nanofibers manufactured using twin-screw extrusion. *Eur. Polym. J.* **2013**, *49*, 950–956. [CrossRef]

22. Li, M.; Tian, X.; Jin, R.; Li, D. Preparation and characterization of nanocomposite films containing starch and cellulose nanofibers. *Ind. Crop. Prod.* **2018**, *123*, 654–660. [CrossRef]

23. Hasani, M.; Cranston, E.D.; Westman, G.; Gray, D.G. Cationic surface functionalization of cellulose nanocrystals. *Soft Matter.* **2008**, *4*, 2238–2244. [CrossRef]

24. Lin, N.; Dufresne, A. Surface chemistry, morphological analysis and properties of cellulose nanocrystals with gradiented sulfation degrees. *Nanoscale* **2014**, *6*, 5384–5393. [CrossRef]

25. Ansari, F.; Salajkova, M.; Zhou, Q.; Berglund, L.A. Strong surface treatment effects on reinforcement efficiency in biocomposites based on cellulose nanocrystals in poly(vinyl acetate) Matrix. *Biomacromolecules* **2015**, *16*, 3916–3924. [CrossRef] [PubMed]

26. Raquez, J.M.; Murena, Y.; Goffin, A.L.; Habibi, Y.; Ruelle, B.; DeBuyl, F.; Dubois, P. Surface-modification of cellulose nanowhiskers and their use as nanoreinforcers into polylactide: A sustainably-integrated approach. *Compos. Sci. Tech.* **2012**, *72*, 544–549. [CrossRef]

27. Sirvio, J.A.; Honkaniemi, S.; Visanko, M.; Liimatainen, H. Composite films of poly(vinyl alcohol) and bifunctional cross-linking cellulose nanocrystals. *ACS Appl. Mater. Interfaces* **2015**, *7*, 19691–19699. [CrossRef]

28. Wang, D.; Yu, H.; Fan, X.; Gu, J.; Ye, S.; Yao, J.; Ni, Q. High aspect ratio carboxylated cellulose nanofibers cross-linked to robust aerogels for superabsorption-flocculants: Paving way from nanoscale to macroscale. *ACS Appl. Mater. Interfaces* **2018**, *10*, 20755–20766. [CrossRef]

29. Abe, K. Nanofibrillation of dried pulp in NaOH solutions using bead milling. *Cellulose* **2016**, *23*, 1257–1261. [CrossRef]

30. Chaker, A.; Boufi, S. Cationic nanofibrillar cellulose with high antibacterial properties. *Carbohydr. Polym.* **2015**, *131*, 224–232. [CrossRef]

31. Lu, Z.; Fan, L.; Zheng, H.; Lu, Q.; Liao, Y.; Huang, B. Preparation, characterization and optimization of nanocellulose whiskers by simultaneously ultrasonic wave and microwave assisted. *Bioresour. Technol.* **2013**, *146*, 82–88. [CrossRef] [PubMed]

32. Liu, C.; Sun, R.; Zhang, A.; Ren, J. Preparation of sugarcane bagasse cellulosic phthalate using an ionic liquid as reaction medium. *Carbohydr. Polym.* **2007**, *68*, 17–25. [CrossRef]

33. Sadeghifar, H.; Filpponen, I.; Clarke, S.P.; Brougham, D.F.; Argyropoulos, D.S. Production of cellulose nanocrystals using hydrobromic acid and click reactions on their surface. *J. Mater. Sci.* **2011**, *46*, 7344–7355. [CrossRef]

34. Xie, J.; Hse, C.Y.; De Hoop, C.F.; Hu, T.; Qi, J.; Shupe, T.F. Isolation and characterization of cellulose nanofibers from bamboo using microwave liquefaction combined with chemical treatment and ultrasonication. *Carbohydr. Polym.* **2016**, *151*, 725–734. [CrossRef] [PubMed]

35. Smyth, M.; García, A.; Rader, C.; Foster, E.J.; Bras, J. Extraction and process analysis of high aspect ratio cellulose nanocrystals from corn (*Zea mays*) agricultural residue. *Ind. Crop. Prod.* **2017**, *108*, 257–266. [CrossRef]

36. Wang, Z.; Yao, Z.; Zhou, J.; Zhang, Y. Reuse of waste cotton cloth for the extraction of cellulose nanocrystals. *Carbohydr. Polym.* **2017**, *157*, 945–952. [CrossRef]

37. Mariano, M.; El Kissi, N.; Dufresne, A. Cellulose nanocrystals and related nanocomposites: Review of some properties and challenges. *J. Polym. Sci. Polym. Phys.* **2014**, *52*, 791–806. [CrossRef]
38. Roohani, M.; Habibi, Y.; Belgacem, N.M.; Ebrahim, G.; Karimi, A.N.; Dufresne, A. Cellulose whiskers reinforced polyvinyl alcohol copolymers nanocomposites. *Eur. Polym. J.* **2008**, *44*, 2489–2498. [CrossRef]
39. Iwamoto, S.; Kai, W.; Isogai, A.; Iwata, T. Elastic modulus of single cellulose microfibrils from tunicate measured by atomic force microscopy. *Biomacromolecules* **2009**, *10*, 2571–2576. [CrossRef]
40. Liu, H.; Liu, D.; Yao, F.; Wu, Q. Fabrication and properties of transparent polymethylmethacrylate/cellulose nanocrystals composites. *Bioresour. Technol.* **2010**, *101*, 5685–5692. [CrossRef] [PubMed]
41. Sebe, G.; Ham-Pichavant, F.; Ibarboure, E.; Koffi, A.L.; Tingaut, P. Supramolecular structure characterization of cellulose II nanowhiskers produced by acid hydrolysis of cellulose I substrates. *Biomacromolecules* **2012**, *13*, 570–578. [CrossRef] [PubMed]
42. Shankar, S.; Rhim, J.W. Preparation of nanocellulose from micro-crystalline cellulose: The effect on the performance and properties of agar-based composite films. *Carbohydr. Polym.* **2016**, *135*, 18–26. [CrossRef] [PubMed]
43. Tang, Y.; Yang, S.; Zhang, N.; Zhang, J. Preparation and characterization of nanocrystalline cellulose via low-intensity ultrasonic-assisted sulfuric acid hydrolysis. *Cellulose* **2013**, *21*, 335–346. [CrossRef]
44. Oudiani, A.E.; Chaabouni, Y.; Msahli, S.; Sakli, F. Crystal transition from cellulose I to cellulose II in NaOH treated *Agave americana* L. fibre. *Carbohydr. Polym.* **2011**, *86*, 1221–1229. [CrossRef]
45. Cui, S.; Zhang, S.; Ge, S.; Xiong, L.; Sun, Q. Green preparation and characterization of size-controlled nanocrystalline cellulose via ultrasonic-assisted enzymatic hydrolysis. *Ind. Crop. Prod.* **2016**, *83*, 346–352. [CrossRef]
46. Benini, K.; Voorwald, H.J.C.; Cioffi, M.O.H.; Rezende, M.C.; Arantes, V. Preparation of nanocellulose from Imperata brasiliensis grass using Taguchi method. *Carbohydr. Polym.* **2018**, *192*, 337–346. [CrossRef]
47. Kian, L.K.; Jawaid, M.; Ariffin, H.; Alothman, O.Y. Isolation and characterization of microcrystalline cellulose from roselle fibers. *Int. J. Biol. Macromol.* **2017**, *103*, 931–940. [CrossRef]
48. Kargarzadeh, H.; Ahmad, I.; Abdullah, I.; Dufresne, A.; Zainudin, S.Y.; Sheltami, R.M. Effects of hydrolysis conditions on the morphology, crystallinity, and thermal stability of cellulose nanocrystals extracted from kenaf bast fibers. *Cellulose* **2012**, *19*, 855–866. [CrossRef]
49. Jiang, F.; Hsieh, Y.L. Chemically and mechanically isolated nanocellulose and their self-assembled structures. *Carbohydr. Polym.* **2013**, *95*, 32–40. [CrossRef]
50. Li, J.; Wei, X.; Wang, Q.; Chen, J.; Chang, G.; Kong, L.; Su, J.; Liu, Y. Homogeneous isolation of nanocellulose from sugarcane bagasse by high pressure homogenization. *Carbohydr. Polym.* **2012**, *90*, 1609–1613. [CrossRef]
51. Wang, L.-F.; Shankar, S.; Rhim, J.-W. Properties of alginate-based films reinforced with cellulose fibers and cellulose nanowhiskers isolated from mulberry pulp. *Food Hydrocoll.* **2017**, *63*, 201–208. [CrossRef]
52. Lu, P.; Hsieh, Y.-L. Preparation and properties of cellulose nanocrystals: Rods, spheres, and network. *Carbohydr. Polym.* **2010**, *82*, 329–336. [CrossRef]
53. Chi, K.; Catchmark, J.M. Improved eco-friendly barrier materials based on crystalline nanocellulose/chitosan/carboxymethyl cellulose polyelectrolyte complexes. *Food Hydrocoll.* **2018**, *80*, 195–205. [CrossRef]
54. Slavutsky, A.M.; Bertuzzi, M.A. Water barrier properties of starch films reinforced with cellulose nanocrystals obtained from sugarcane bagasse. *Carbohydr. Polym.* **2014**, *110*, 53–61. [CrossRef] [PubMed]
55. Pereda, M.; Amica, G.; Rácz, I.; Marcovich, N.E. Structure and properties of nanocomposite films based on sodium caseinate and nanocellulose fibers. *J. Food Eng.* **2011**, *103*, 76–83. [CrossRef]
56. Muscat, D.; Adhikari, R.; McKnight, S.; Guo, Q.; Adhikari, B. The physicochemical characteristics and hydrophobicity of high amylose starch–glycerol films in the presence of three natural waxes. *J. Food Eng.* **2013**, *119*, 205–219. [CrossRef]
57. Wang, W.; Zhang, H.; Jia, R.; Dai, Y.; Dong, H.; Hou, H.; Guo, Q. High performance extrusion blown starch/polyvinyl alcohol/clay nanocomposite films. *Food Hydrocoll.* **2018**, *79*, 534–543. [CrossRef]
58. Qazanfarzadeh, Z.; Kadivar, M. Properties of whey protein isolate nanocomposite films reinforced with nanocellulose isolated from oat husk. *Int. J. Biol. Macromol.* **2016**, *91*, 1134–1140. [CrossRef]
59. Fazeli, M.; Keley, M.; Biazar, E. Preparation and characterization of starch-based composite films reinforced by cellulose nanofibers. *Int. J. Biol. Macromol.* **2018**, *116*, 272–280. [CrossRef]
60. Shi, A.M.; Wang, L.J.; Li, D.; Adhikari, B. Characterization of starch films containing starch nanoparticles: Part 1: Physical and mechanical properties. *Carbohydr. Polym.* **2013**, *96*, 593–601.

61. Balakrishnan, P.; Sreekala, M.S.; Kunaver, M.; Huskic, M.; Thomas, S. Morphology, transport characteristics and viscoelastic polymer chain confinement in nanocomposites based on thermoplastic potato starch and cellulose nanofibers from pineapple leaf. *Carbohydr. Polym.* **2017**, *169*, 176–188. [CrossRef] [PubMed]

 *nanomaterials* 

Article

Reinforcement of Natural Rubber Latex Using Jute Carboxycellulose Nanofibers Extracted Using Nitro-Oxidation Method

Sunil K. Sharma [1], Priyanka R. Sharma [1,*], Simon Lin [1], Hui Chen [1], Ken Johnson [1], Ruifu Wang [1], William Borges [2], Chengbo Zhan [1] and Benjamin S. Hsiao [1,*]

[1] Department of Chemistry, Stony Brook University, Stony Brook, NY 11794-3400, USA; sunil.k.sharma@stonybrook.edu (S.K.S.); simon.lin@stonybrook.edu (S.L.); hui.chen.2@stonybrook.edu (H.C.); ken.johnson@stonybrook.edu (K.J.); Ruifu.Wang@stonybrook.edu (R.W.); chengbo.zhan@stonybrook.edu (C.Z.)

[2] Roslyn High School, Roslyn, NY 11576, USA; WBorges20@roslynschools.org

* Correspondence: priyanka.r.sharma@stonybrook.edu (P.R.S.); benjamin.hsiao@stonybrook.edu (B.S.H.); Tel.: +1-631-5423506 (P.R.S.); +1-631-6327793 (B.S.H.)

Received: 28 February 2020; Accepted: 29 March 2020; Published: 8 April 2020

Abstract: Synthetic rubber produced from nonrenewable fossil fuel requires high energy costs and is dependent on the presumed unstable petroleum price. Natural rubber latex (NRL) is one of the major alternative sustainable rubber sources since it is derived from the plant 'Hevea brasiliensis'. Our study focuses on integrating sustainably processed carboxycellulose nanofibers from untreated jute biomass into NRL to enhance the mechanical strength of the material for various applications. The carboxycellulose nanofibers (NOCNF) having carboxyl content of 0.94 mmol/g was prepared and integrated into its nonionic form (–COONa) for its higher dispersion in water to increase the interfacial interaction between NRL and NOCNF. Transmission electron microscopy (TEM) and atomic force microscopy (AFM) analyses of NOCNF showed the average dimensions of nanofibers were length (L) = 524 ± 203 nm, diameter (D) 7 ± 2 nm and thickness 2.9 nm. Furthermore, fourier transform infra-red spectrometry (FTIR) analysis of NOCNF depicted the presence of carboxyl group. However, the dynamic light scattering (DLS) measurement of NRL demonstrated an effective diameter in the range of 643 nm with polydispersity of 0.005. Tensile mechanical strengths were tested to observe the enhancement effects at various concentrations of NOCNF in the NRL. Mechanical properties of NRL/NOCNF films were determined by tensile testing, where the results showed an increasing trend of enhancement. With the increasing NOCNF concentration, the film modulus was found to increase quite substantially, but the elongation-to-break ratio decreased drastically. The presence of NOCNF changed the NRL film from elastic to brittle. However, at the NOCNF overlap concentration (0.2 wt. %), the film modulus seemed to be the highest.

Keywords: natural rubber latex; NOCNF; jute fibers; nitro-oxidation

---

1. Introduction

Synthetic and natural rubber are a staple commodity for numerous industrial applications [1–6]. The International Rubber Study Group (IRSG) [7] reported that the U.S. consumed 2.7 million metric tons of rubber ranging from automotive parts to sealants in 2013. These products consist of mostly synthetic rubber derived from petroleum sources and natural rubber derived from Hevea trees (*Hevea brasiliensis*). The use of petroleum-derived synthetic rubber causes several concerns [8,9]. The reliance on nonrenewable resources causes a dependent and unstable price for synthetic rubber costs. Synthetic rubber production also requires a higher energy consumption and is environmentally intensive when compared to using natural rubber [8]. Natural rubber in its unprocessed or raw form has low strength and that limit their applications. The strength of the natural rubber latex is improved mostly by the vulcanization process, where the long chains of rubber molecule are cross-linked through the chemical process which ultimately transform the natural rubber latex into a strong elastic product (natural rubber) with reversible deformability, good mechanical strength, excellent dynamic properties and fatigue resistance [10]. The mechanical properties of natural rubber latex can be improved by addition of varying types of reinforcing fillers.

Nanocellulose is a most abundant, inexpensive and renewable nanomaterial that has potential in many different applications including pharmaceuticals, food, energy storage, water purification, biomedical, 3D printing, anti-bacterial, carbon nanotubes stabilizer, electronics and tissue engineering. Owing to its exceptional mechanical properties, nontoxicity, biodegradability and tunable chemistry of surface hydroxyl groups, nanocellulose has garnered tremendous levels of attention over the past decades [11–13]. There are many methods reported for preparation of nanocellulose from various biomass sources, for example, carboxymethylation, acid hydrolysis to produce cellulose nanocrystals, 2,2,6,6-Tetramethylpiperidin-1-yl)oxyl (TEMPO) oxidation, nitro-oxidation etc.

A potential reinforcing agent for the latex rubbers is the derivatives of natural cellulose polymers [14–19]. These polymers are called carboxycellulose and have been widely used for biomedical applications such as surgical sutures [20–22]. Recent developments of carboxycellulose in the nanoscale have further expanded their uses in making strengthened nanocomposite materials [23–25]. Since carboxycellulose nanofibers is derived from cellulose microfibril building blocks, it is readily able to be extracted from a variety of biomass materials [26,27]. Some of these biomass materials include jute, spinifex, agave and agricultural wastes [28]. Our study primarily focuses on jute-derived carboxycellulose nanofibers extracted using the recently developed nitro-oxidation method [28–32]. The nitro-oxidation method is found to be a simple, cost-effective process to extract the carboxycellulose nanofibers from any type of raw biomass that does not require any pretreatment steps. However, the other methods—TEMPO oxidation and carboxymethylation processes—are fully efficient in extracting the carboxycellulose nanofibers from delignified pulp, which requires prior treatment of raw biomass. Nitro-oxidation produces carboxycellulose nanofibers with residual lignin and hemicellulose impurities; however, it requires less chemicals, processing time and steps for their extraction. Additionally, the unused effluent of the reaction has potential to be converted into nitrogen-rich plant fertilizer. The nitro-oxidation method involves the reaction of nitric acid with sodium nitrite to create nitroxonium ions (NO^+) which attacks the hydroxyl group on cellulose to produce a nitrite ester ($R–CH_2–O–NO$). The nitrite ester then decomposes and generates nitroxyl (HNO) and aldehyde groups which is further oxidized into carboxyl (COOH) groups. This oxidation cycle continues at the presence of excess HNO and HNO_2 to create the saturated carboxyl groups which provide the function sites for further chemical reaction. Since this is a recently developed method [28,31,33–35], there are also interests in applying these carboxycellulose nanofibers materials for further testing.

The primary focus of this study is to integrate this low-cost nitro-oxidized carboxycellulose nanofibers (NOCNF) into natural rubber latex sources to observe the enhancement effect of the samples at various concentrations. Since latex are primarily composed of cis 1,4 polyisoprene emulsions in water [36], we chose the carboxylate functional group ($–COO^-$) to induce a high dispersity during the integration stage. The carboxylate functional group is hydrophilic which along with dispersity, could allow for better interfacial interactions between the fibers and the isoprene molecules. Other latex enhancement studies also indicate the use of a hydrophilic carboxycellulose nanofibers state to be effective for making enhanced nanocomposites with enhanced tensile modulus [23].

2. Methodologies

2.1. Materials

Untreated jute fibers were provided by Toptrans Bangladesh Ltd. (Dhaka, Bangladesh). Fibers were cut to 3–5 cm in length and further grinded by an IKA MF 10 basic grinder at 1000 rpm (IKA Works Inc., Wilmington, NC, USA). Analytical-grade nitric acid (ACS reagent 60%) and sodium nitrite (ACS reagent ≥ 97%) were purchased from Sigma-Aldrich (Allentown, PA, USA); sodium bicarbonate was purchased from Fischer Scientific (Fairlawn, NJ, United States). Processed polygen liquid latex (NRL) nonvulcanized with 60% concentration was obtained from UK suppliers (ReAgent, Runcorn, UK).

2.2. Experimental Method

2.2.1. Preparation of Carboxycellulose Nanofibers (NOCNF)

Fifteen grams of untreated grinded jute fibers were placed in a 3 L three-neck, round-bottom flask with 210 mL of 60 % nitric acid. Fibers in the flask were allowed to completely soak before adding 14.4 g of sodium nitrite. The addition of sodium nitrite forms red gas inside the flask due to generation of NO_2 gas. Hence, the mouths of the round bottom flask were sealed with stoppers sealed with parafilm. The reaction was performed at 40 °C for 16 h and was then quenched by adding 1 L distilled (DI) water. The supernatant liquid was discarded to remove excess acid and decantation with 70% ethanol and DI water in the ratio of 80:20 was performed 4–5 times until the suspended fibers stopped settling down. The fibers suspension was then transferred to a dialysis bag (Spectral/Por, molecular weight cut-off (MWCO): 6–8 kDA) and equilibrated until the conductivity of the water reached below 5 μS. The fibers were then treated with 8% sodium bicarbonate up to pH 7.5, to transform the initially generated carboxyl group (COOH) to carboxylate groups (COO$^-$). To remove the excess bicarbonate from the fibers, the suspension was again dialyzed using the dialysis bag until the conductivity of the water reached below 5 μS. The 0.2 wt. % of fibers suspension then passed through homogenizer (GEA Niro Soavi Panda Plus Bench top homogenizer, Columbia, MD, USA), at 250 bar for one cycle.

2.2.2. Nanocomposite Preparation

A 50 mL closed glass vial was used to integrate NRL and NOCNF. Ten milliliters of 60% solid NRL was measured for each 0–0.4 wt. % sample. A 0.26 wt. % NOCNF suspension with different volume was added into NRL solution to prepare the solution containing different (0.1, 0.2 and 0.4 wt. %) of NOCNF. The solutions were set to stir for 16 h followed by 1 h of sonication. Afterwards the prepared solution was casted evenly on a petri dish and degassed to prevent bubbles forming.

2.2.3. Characterization of Carboxycellulose Nanofibers (NOCNF)

Fourier Transform Infra-Red Spectrometry (FTIR)

The FTIR curve was measured with a PerkinElmer Spectrum One instrument (product model, city, country) with the transmission mode set between 450 and 4000 cm^{-1}. Three scans were taken per sample with a resolution of 4 cm^{-1}.

Conductometric Titration Method

The carboxylate (–COO$^-$) group in NOCNF was determined by measuring the conductivity throughout a base titration experiment. A calculated volume containing 0.3 g of nanofibers was dispersed in 55 mL of DI water. The pH was set between 2.5 and 3.0 by adding 0.1 M HCl. The solution was then titrated with 0.4 M NaOH at a rate of 0.1 mL/min until pH reached 11. Throughout the titration, the pH was measured along with the conductivity. The carboxylate content was calculated using the conductivity and pH curves.

Lignin and Hemicellulose Analysis in Raw Jute Fibers and NOCNF

Lignin and hemicellulose (total sugar) analysis of the samples was performed by Celignis (Limerick, Ireland). The following analytical procedures were used: (1) acid hydrolysis of samples, (2) determination of acid-soluble lignin (ASL) using Ultraviolet–Visible (UV-Vis) spectroscopy, (3) gravimetric determination of klason lignin (KL) and (4) chromatographic analysis of hydrolysate. A detailed explanation of the analytical procedure is provided in Supplementary Materials.

Transmission Electron Microscopy (TEM)

The TEM image was obtained with a FEI Tecnai G2 Spirit Bio TWIN instrument (Columbia, MD, United States). The instrument is equipped with a digital camera which allowed it to take photographic film. The instrument is also equipped with tilt stage and electron diffraction capabilities. The samples were prepared using a 10 μL aliquot sample of 1 mg of NOCNF in 10 mL DI water deposited on carbon coated Copper grids (300 mesh, Ted Pella Inc., Redding, CA, United States). The prepared grid was then stained with 2 wt. % aqueous uranyl acetate solution.

Atomic Force Microscopy (AFM)

AFM of NOCNF was performed using a Bruker Dimension ICON scanning probe microscope (Bruker Corporation, Billerica, MA, USA) equipped with a Bruker OTESPA tip (tip radius (max.) = 10 nm). In this measurement, a 10 μL of 0.005 wt. % NOCNF suspension was deposited on the surface of a silica plate, where the air-dried sample was measured in the tapping mode.

Zeta Potential Measurements

Zeta potential of the NOCNF sample was measured by Zeta probe Analyzer (Colloid Dynamics). This instrument consisted of a built-in titration setup equipped with pH electrode and Electrokinetic Sonic Amplitude (ESA) sensor probe. Before analyzing the sample, the pH electrode was calibrated using three different pH buffer standards (pH = 4.01, 7.01 and 10.01), followed by a standard titration solution. The ESA sensor was calibrated using the standard zeta probe polar solution (KSiW solution). Upon the completion of calibration test, the NOCNF suspension (0.26 wt. %, 250 mL) was filled in the sample holder, where the ESA sensor was then introduced into the sample under magnetic stirring to analyze the zeta potential.

Dynamic Light Scattering (DLS)

The DLS of NRL sample was measured using Nano Brook 90 Plus particle size analyzer (Brookhaven, Holtsville, NY, USA) The DLS graph was set to lognormal plot and the data is an average of four total runs to obtain the polydispersity index (PDI).

Contact Angle Measurement

Static contact angle of NRL and composite films prepared by NOCNF and NRL were measured using the FDS-contact angle measurement instrument (Model no. OCA 15 EC). Films of NRL alone and composite films containing NRL and NOCNF were prepared by solvent casting method. A flat portion of film was cut and fixed onto sample holder. A 10 μL drop of deionized water was dropped onto film through a syringe needle operated automatically by syringe pump. The contact angle was measured 20 s after the drop casting to ensure that the water droplet reached its equilibrium position.

Tensile Test

The tensile data was obtained using the INSTRON model 4442 device (Instron, Norwood, MA, USA). Clamps were calibrated with a 1.75 cm gap and rectangular samples of 2 cm wide by 5 cm long were clamped evenly in the device. Elongation rate was set to 60 mm/min with data recording every second. Each data plot obtained is an average of three sets of experiments.

Scanning Electron Microscopy (SEM)

A Zeiss LEO 1550 SFEG-SEM instrument (White Plains NY, USA) was used to record SEM images of the samples. The instrument was comprised of an in-lens secondary electron detector in addition to the standard E-T detector, and a Rutherford backscatter electron detector. It was also equipped with an EDS (energy dispersive X-ray spectroscopy) system, provides elemental compositions and X-ray maps of the various phases of the materials examined. Images of surface morphology of NRL and NRL composite films were taken to observe the NRL film surface change on addition of NOCNF.

3. Results and Discussion

3.1. Characterization of NOCNF

The characterization on the surface functionalization of NOCNF extracted from untreated jute fibers was first carried out by FTIR and conductometric titration. Figure 1i demonstrates the FTIR spectra of jute fibers and prepared NOCNF. The characteristic peaks of cellulose are ascribed at (i) 3327 cm^{-1} to O–H stretching vibrations, and at (ii) 2904 cm^{-1} to CH and CH$_2$ stretching. The prominent peak in NOCNF at 1591 cm^{-1} presented the carboxylate groups (–COONa) appeared in NOCNF, which represents the oxidation of anhydroglucose unit at C6 position. Additional peaks at 1372, 1150, 1100, and 1030 cm^{-1} were due to stretching and bending vibrations in glycosidic bonds in cellulose. Other peaks in the FTIR of jute fibers such as 1512; 1732, 1456, 1235 and 808 cm^{-1} are because of aromatic symmetrical streching of C=C bonds in the lignin and in hemicellulose units respectively. Interestingly, the peaks belonging to lignin and hemicellulose completely disappear or significantly reduce in NOCNF, indicating that the nitro-oxidation was resonably effective in removing the lignin and hemicellulsoe impurities from raw jute fibers. The quantitative determination of lignin and hemicelulose in NOCNF was also perfomed to find the exact amount of lignin and hemicellulsoe, which is explained in the next section.

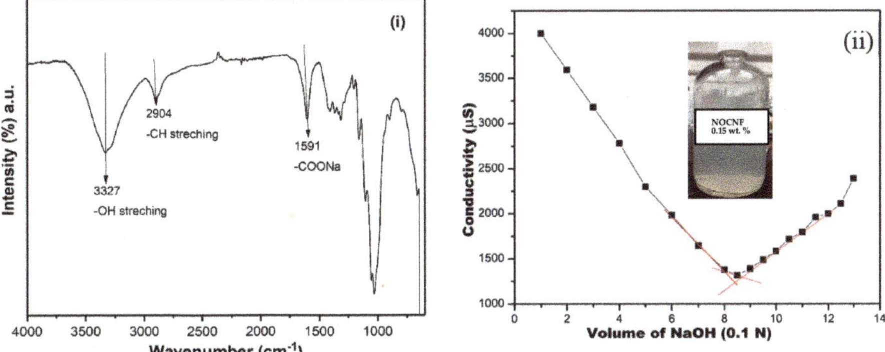

Figure 1. (**i**) Fourier transform infrared spectrometry (FTIR) of carboxycellulsoe nanofibers (NOCNF) and jute fibers, and (**ii**) conductometric titration graph to determine the carboxylate group on NOCNF (volume of NaOH consumed = 0.705 mL), inset the photograph of NOCNF suspension.

The quantitative determination of carboxylate groups in NOCNF was performed by conductometric titration method. The following equation (Equation (1)) was used to determine the degree of oxidation (*DO*):

$$DO = \frac{MX\,(V_2 - V_1)}{w} \tag{1}$$

where *M* is the molarity of NaOH in mol/L, V_2 and V_1 is the final and initial volume of NaOH in mL, *w* is the weight of the NOCNF dried fibers added in grams.

The conductometric titration plot of NOCNF shown in Figure 1ii indicates its calculated DO value which is 0.94 mmol/g. This DO value shows that NOCNF contains a moderate degree of oxidation, which is further confirmed by zeta potential measurement. The zeta potential measurement demonstrates the presence of −115 ± 4 mV charge on the NOCNF surface. NOCNF showed good dispersion in water (inset photograph in Figure 1ii) because of the repulsion caused in between the fibers due to similar charges.

The lignin and hemicellulose analysis of raw jute fibers is presented in Table 1. It shows that the total hemicellulose (sugar content) and total lignin content (klason lignin (KL) + acid soluble lignin (ASL)) in the raw jute fibers was 68.9% and 17.55%, respectively. However, the total hemicellulose and lignin content in NOCNF was found as 65% and 1.94% respectively. The results indicate that the reasonable hemicellulose and residual lignin are still present in the NOCNF after the nitro-oxidation.

Table 1. Characteristic of carboxycellulsoe nanofibers (NOCNF) obtained from jute.

Sample	Carboxylate Content (mmol/g)	Zeta Potential (mV)	Residual Lignin (%) KL/ASL [a]	Residual hemicellulose (%)	Length/Width (nm)	Thickness (nm)
NOCNF	0.94	−115 ± 4	0.58/1.36	65	524 ± 203/ 7 ± 2	2.9

[a] KL = klason lignin, ASL = acid soluble lignin.

The TEM image of NOCNF is shown in Figure 2i. The average fiber length observed for NOCNF was 524 ± 203 nm and width were in the range of 7 ± 2 nm. However, the AFM of the NOCNF indicated the average fibers thickness was 2.9 nm. In this study, the NOCNF obtained has greater width and thickness as compared to the cross section of most of the cellulose fibers where width is in the range of 4–5 nm and thickness ~1.5 nm [37]. This is probably due to the chosen nitro-oxidation conditions are relatively mild, where the presence of residual hemicellulose and lignin contents were still high.

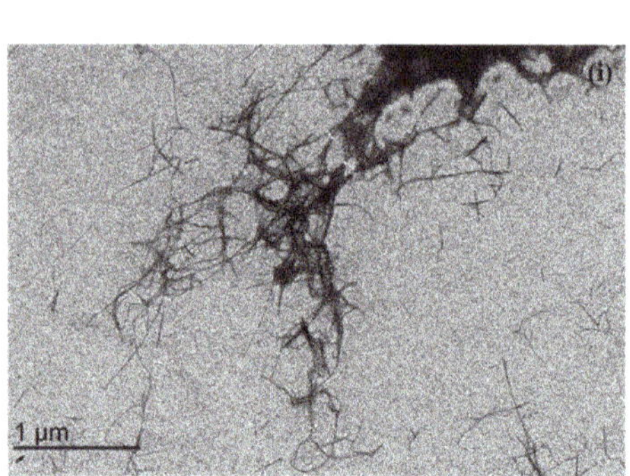

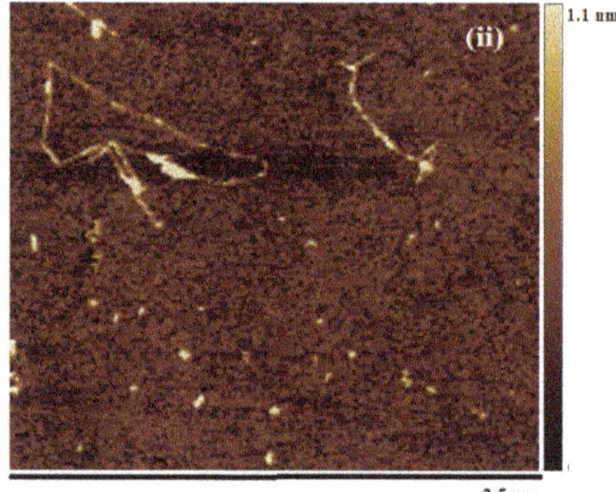

Figure 2. (i) Transmission electron microscopy (TEM) and (ii) atomic force microscopy (AFM) of NOCNF extracted from raw jute fibers.

The dynamic light scattering (DLS) data measurement of NRL is presented in Figure 3. The data shows that the effective diameter of NRL molecule is 637 nm with polydispersity value of 0.005. This indicates that the NRL is composed of polyisoprene molecules with almost the same size. The TEM measurement of NOCNF indicated its fibers length in the range of 524 ± 203 nm, which is almost like the NRL particles' size. We have assumed that similar sizes of two interacted molecules NOCNF and NRL will provide the better chances of their physical interaction.

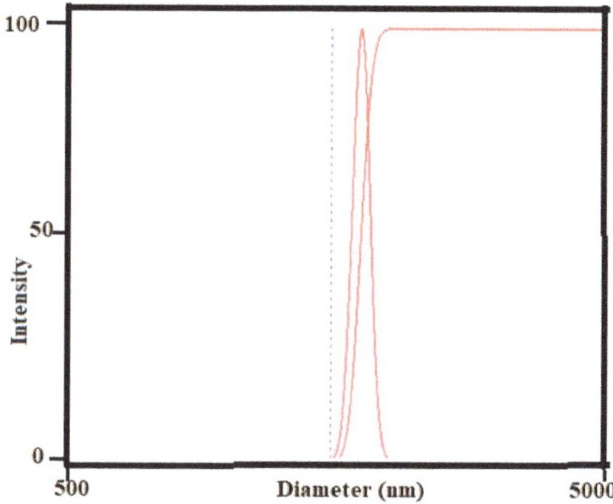

Figure 3. Dynamic light scattering (DLS) data of natural rubber latex (NRL) (average diameter = 637 nm with size polydispersity (PDI) = 0.005.

3.2. Characterization of Natural Rubber Latex (NRL) and Composite Films

The contact angle measurement was performed on the films prepared by NRL and the composite films made of NRL and varying content of NOCNF (0.1, 0.2 and 0.4 wt. %), which is shown in Figure 4. The contact angle measurement of pure NRL has shown the average angles (left = 63.8° and right = 65°) clearly indicate that it has hydrophobic surface. An earlier study of the chemically crosslinked NRL film (using potassium persulphate (KPS) as an initiator) reported that the contact angle should be in the range of 94°, which was clearly more hydrophobic than the noncrosslinked film in the present study [38]. However, the small addition of NOCNF (0.1 wt. %) into NRL has changed the average contact angle of composite membrane to be around 49.5 ° (left = 51° and right = 48°). The decrease in contact angle indicates the appearance of hydrophilic behavior in composite film because of the presence of NOCNF which has more hydrophilic groups such as hydroxyl and carboxylate. The further addition of NOCNF (0.2 and 0.4 wt. %) into NRL has further reduced the contact angle of composite films (e.g., film with 0.2 wt. % NOCNF: left = 48.4° and right = 44°; film with 0.4 wt. % NOCNF: left = 48° and right = 43°). Interestingly, not much change in the contact angle of composite films containing 0.2 and 0.4 wt. % of NOCNF was observed. This is probably because 0.2 wt.% is the overlapping concentration of NOCNF [31], where some aggregations might have occurred in the NRL matrix at and above overlapping concentration (i.e., 0.2 and 0.4 wt. %) of NOCNF, which resulted in decrease in hydrophobicity of the composite films.

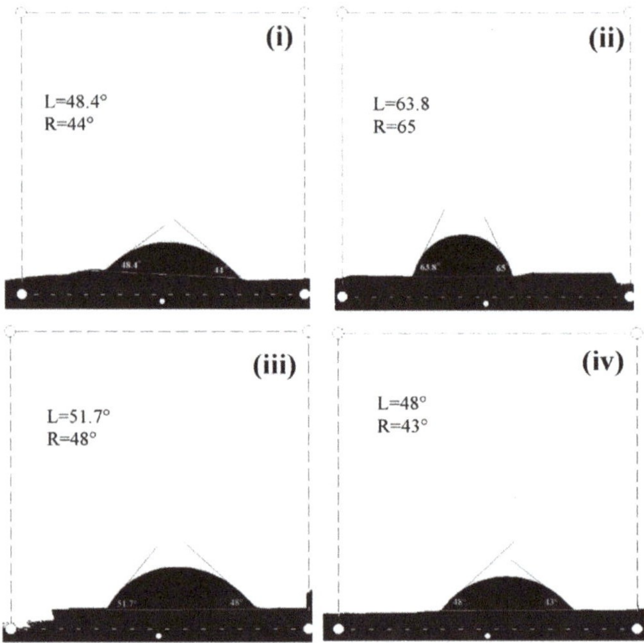

Figure 4. Contact angle measurements of (**i**) NRL film (control), and composite films made of NRL and NOCNF with varying concentration of NOCNF (**ii**) 0.1 wt. %, (**iii**) 0.2 wt. %, (**iv**) 0.4 wt.%.

3.3. SEM Images

SEM images of film made of NRL and composite films consisting of NRL and varying concentrations NOCNF are presented in Figure 5. The image of NRL in Figure 5i indicates that the film possesses uniform surface roughness. This was due to the agglomeration of similar granular shape NRL particles. This was consistent with the size characterization of NRL using the DLS technique (Figure 3, NRL possessed an average particle size of 640 nm with the polydispersity of 0.005). The morphology of NRL film on addition of 0.1 wt. % of NOCNF (nanofibers) has not shown any significant changes on the surface appearance. However, the addition of 0.2 wt. % and 0.4 wt. % nanofibers into NRL (Figure 5iii,iv) resulted in significant changes in surface appearance in terms of roughness.

The roughness of the latex film decreases on increasing the amount of nanofibers concentration above 0.1 wt. % that could be because of the well distribution of NOCNF with the latex particles. These data correlate well with the contact angle measurement, as the drastic decrease in contact angle from 63° to 43° was observed for the NRL film on addition of 0.1 and 0.2 wt. % of NOCNF. However, the cracks start beginning in composite films containing the (highest) 0.4 wt. % of NOCNF. The probable reason for the crack could be the phase separation in between NRL and NOCNF because of their corresponding hydrophobic and hydrophilic behavior.

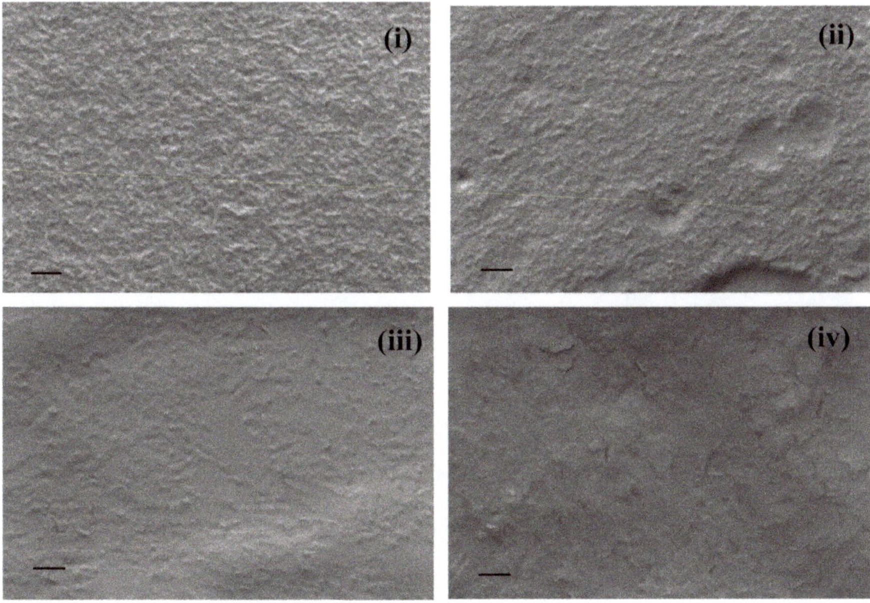

Figure 5. Scanning electron microscopy (SEM) images taken at scale bar = 200 nm on (**i**) NRL film (control), and composite films made of NRL and NOCNF with varying concentration of NOCNF (**ii**) 0.1 wt.%, (**iii**) 0.2 wt.%, (**iv**) 0.4 wt.%. Red circles indicate the cracks in the film.

3.4. Mechanical Properties of Latex and Composite Films

The mechanical properties of NRL and composite films containing NRL and varying concentrations of NOCNF is shown in Table 2. The stress–strain curve was plotted for all the composite films and is presented in Figure 6. The stress on the y-axis was calculated by dividing the load by the cross-sectional area of the film. The thickness and length of the film is measured using a caliper and all nanocomposites showed the same thickness of 0.08 cm^2. There could be a slight variation of thickness due to the increased amount of NOCNF added, but the difference would be negligible since the NRL solid amount contributes to most of the sample's volume. The stress–strain curves for the NRL and composite films also indicates their ultimate tensile strength (UTS).

Table 2. Young's modulus (Ym), ultimate tensile strength (UTS), and maximum elongation (λ_{max}). Each value is the average of three replicates samples. NRL—pure natural rubber latex, NRL 0.1—NRL containing 0.1 wt. % of NOCNF; NRL 0.2—NRL containing 0.2 wt. % of NOCNF; NRL 0.4—NRL containing 0.4 wt. % of NOCNF.

Sample	Ym (kPa)	UTS (MPa)	λ_{max} (%)
NRL	3.3	0.77	234
NRL 0.1	79.6	2.5	31.4
NRL 0.2	2080	5.2	2.5
NRL 0.4	1770	6.2	3.5

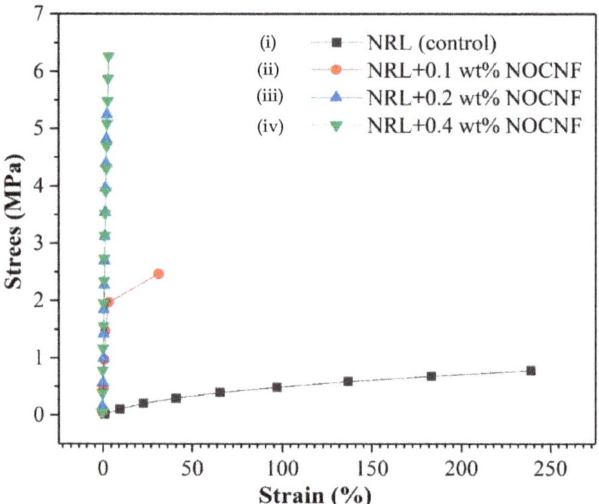

Figure 6. Stress–strain curves on (i) NRL film (control), and composite films made of NRL and NOCNF with varying concentration of NOCNF (ii) 0.1 wt.%, (iii) 0.2 wt.%, (iv) 0.4 wt.%.

It was found that the NRL film exhibited the UTS value of 0.77 Mpa. However, the addition of NOCNF increased the UTS value of the film quite notably (e.g., the films with 0.1, 0.2 and 0.4 wt. % of NOCNF showed the UTS value of 2.5, 5.2 and 6.2 MPa, respectively). We note that the ultimate tensile strength reported for the pure NOCNF extracted from jute fibers using the nitro-oxidation method was 108 MPa [31] and the chemically crosslinked NRL film typically exhibited the UTS value of 27 Mpa [39]. In the above study [39], vulcanized (chemically crosslinked) NRL was reinforced by addition of cellulose nanocrystals (CNC) where the 3 wt. % of CNC was found to be most effective to increase the ultimate tensile strength of NRL (by about 29%). In this study, the authors have used the term cellulose nanofibers to describe nanocellulose isolated from coconut spathe using the acid hydrolysis method. We believe that the length of such nanocellulose particles extracted by the acid hydrolysis approach should be shorter, the cross-sectional dimensions larger and the crystallinity higher than those in NOCNF, and they should be termed CNC. It is interesting to note that the overlap concentration of CNC usually varies between 1.5 wt. % to 3 wt. %, depending on the source of the biomass [40]. We hypothesize the ultimate tensile strength of vulcanized CNC-NRL also takes place near the overlap concentration of CNC.

On the other side, the maximum elongation (λ_{max}, %) observed in the tested films showed an opposite trend. For example, the pure NRL film exhibited λ_{max} at about 234 %, where 0.1 wt.% of NOCNF in the composite film decreased the λ_{max} value to 31.4 %. The increase in the NOCNF content further decreased the λ_{max} value (e.g., 0.2 and 0.4 wt. % of NOCNF films showed the λ_{max} value of around 2.5% and 3.5%), rendering the films to be quite brittle. These results were consistent with the SEM images (Figure 5), which showed that the addition of NOCNF decreased the roughness of the NRL film and increased the content of cracks owing to phase separation between NOCNF and NRL.

The ductile–brittle transition was also noticeable from the Young's modulus (Ym) evaluation shown in Figure 7. The pure noncrosslinked NRL was very ductile, showing an Ym value of merely 3.5 KPa, where 0.1 wt. % of NOCNF addition increased the Ym value to 79.6 KPa and 0.2 wt. % of NOCNF addition increased the Ym value maximum to 2080 kPa. The further increase of NOCNF content decreased the Ym value to 1770 kPa. As a result, the higher NOCNF content would not lead to any property enhancement, where the best content of NOCNF addition appeared to occur near its overlap concentration (0.2 wt. %).

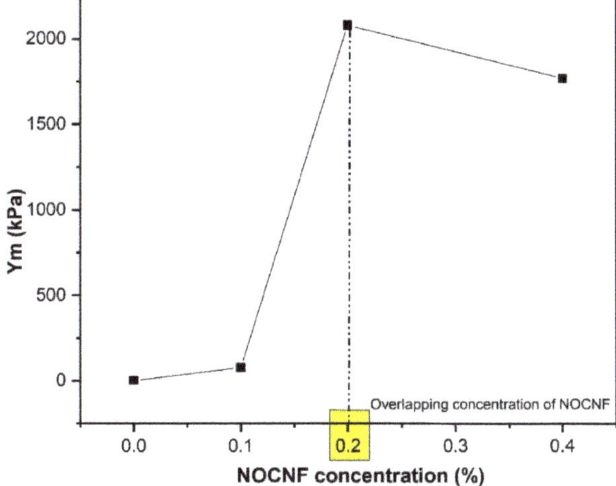

Figure 7. Graph represents the relationship between the Young's modulus (kPa) and the NOCNF concentration in the composite films.

4. Conclusions

This study showed that the nitro-oxidized carboxycellulose nanofibers (NOCNF) extracted from raw jute fibers could be incorporated into the NRL matrix to increase the mechanical properties even in a nonvulcanized state. The optimal amount of the NOCNF for the overall property improvement seems to take place around the overlap concentration of NOCNF (around 0.2 wt. %). This is not surprising as the overlap concentration of nanocellulose represents the transition point from a viscous state to a gel state. Addition of NOCNF into non-vulcanized rubber has changed the NRL film from elastic to brittle. The more schematic study on latex composite preparation can explore the use of these inexpensive and sustainable nanofibers material into the preparation of other rubber-based (e.g., Guayule) composite materials.

Supplementary Materials: The following are available online at http://www.mdpi.com/2079-4991/10/4/706/s1.

Author Contributions: Methodology, H.C. and K.J.; software, S.L.; validation, P.R.S., S.K.S. and B.S.H.; formal analysis, R.W. and W.B.; data curation, C.Z.; writing—original draft preparation, S.K.S. and S.L.; writing—review and editing, P.R.S. and B.S.H.; supervision, P.R.S., S.K.S. and B.S.H.; funding acquisition, B.S.H. All authors have read and agreed to the published version of the manuscript.

Funding: The financial support for this work was provided by a grant from the Polymer Program of the Division of Materials Science of the National Science Foundation, United States (DMR-1808690).

Acknowledgments: Authors would like to acknowledge ThINC facility at AERTC, Stony Brook University for the AFM, TEM, DMA, SEM characterizations of the samples.

Conflicts of Interest: The authors declare no conflict of interest.

References

1. Gangadhar, V.; Babu, S.; Cadambi, R.M.; Rao, A.R.; Venkataram, N. Design and Analysis of Rubber Pallet for Industrial Application. *Mater. Today-Proc.* **2017**, *4*, 10886–10893. [CrossRef]
2. Sasikala, A.; Kala, A. Thermal Stability And Mechanical Strength Analysis of EVA and Blend of EVA With Natural Rubber. *Mater. Today-Proc.* **2018**, *5*, 8862–8867. [CrossRef]
3. Fumagalli, M.; Berriot, J.; de Gaudemaris, B.; Veyland, A.; Putaux, J.L.; Molina-Boisseau, S.; Heux, L. Rubber materials from elastomers and nanocellulose powders: Filler dispersion and mechanical reinforcement. *Soft Matter* **2018**, *14*, 2638–2648. [CrossRef]

4. Pascual-Villalobos, M.J.; López, M.D. New application of guayule resin in controlled release formulations. *Ind. Crop. Prod.* **2013**, *43*, 44–49. [CrossRef]

5. Roy, K.; Jatejarungwong, C.; Potiyaraj, P. Development of highly reinforced maleated natural rubber nanocomposites based on sol-gel-derived nano alumina. *J. Appl. Polym. Sci.* **2018**, *135*, 46248. [CrossRef]

6. Fedorko, G.; Molnar, V.; Dovica, M.; Toth, T.; Fabianova, J.; Strohmandl, J.; Neradilova, H.; Hegedus, M.; Belusko, M. Analysis of defects in carcass of rubber-textile conveyor belts using metrotomography. *J. Ind. Text.* **2018**, *47*, 1812–1829. [CrossRef]

7. IRSG. *Rubber Industry Report*; International Rubber Study Group: Singapore, 2013.

8. Rasutis, D.; Soratana, K.; McMahan, C.; Landis, A.E. A sustainability review of domestic rubber from the guayule plant. *Ind. Crop. Prod.* **2015**, *70*, 383–394. [CrossRef]

9. Riyajan, S.A.; Patisat, S. A Novel Packaging Film from Cassava Starch and Natural Rubber. *J. Polym. Environ.* **2018**, *26*, 2845–2854. [CrossRef]

10. Kohjiya, S.; Ikeda, Y. Introduction. In *Chemistry, Manufacture and Applications of Natural Rubber*; Kohjiya, S., Ikeda, Y., Eds.; Woodhead Publishing: Cambridge, UK, 2014; pp. xvii–xxvi. [CrossRef]

11. Ummartyotin, S.; Manuspiya, H. A critical review on cellulose: From fundamental to an approach on sensor technology. *Renew. Sustain. Energy Rev.* **2015**, *41*, 402–412. [CrossRef]

12. Wang, L.; Zuo, X.; Raut, A.; Isseroff, R.; Xue, Y.; Zhou, Y.; Sandhu, B.; Schein, T.; Zeliznyak, T.; Sharma, P.; et al. Operation of proton exchange membrane (PEM) fuel cells using natural cellulose fiber membranes. *Sustain. Energy Fuels* **2019**, *3*, 2725–2732. [CrossRef]

13. Hoeng, F.; Denneulin, A.; Bras, J. Use of nanocellulose in printed electronics: A review. *Nanoscale* **2016**, *8*, 13131–13154. [CrossRef] [PubMed]

14. Hosseinmardi, A.; Annamalai, P.K.; Wang, L.Z.; Martin, D.; Amiralian, N. Reinforcement of natural rubber latex using lignocellulosic nanofibers isolated from spinifex grass. *Nanoscale* **2017**, *9*, 9510–9519. [CrossRef] [PubMed]

15. Silva, M.J.; Sanches, A.O.; Medeiros, E.S.; Mattoso, L.H.C.; McMahan, C.M.; Malmonge, J.A. Nanocomposites of natural rubber and polyaniline-modified cellulose nanofibrils. *J. Therm. Anal. Calorim.* **2014**, *117*, 387–392. [CrossRef]

16. Kato, H.; Nakatsubo, F.; Abe, K.; Yano, H. Crosslinking via sulfur vulcanization of natural rubber and cellulose nanofibers incorporating unsaturated fatty acids. *RSC Adv.* **2015**, *5*, 29814–29819. [CrossRef]

17. Pingot, M.; Szadkowski, B.; Zaborski, M. Effect of carbon nanofibers on mechanical and electrical behaviors of acrylonitrile-butadiene rubber composites. *Polym. Adv. Technol.* **2018**, *29*, 1661–1669. [CrossRef]

18. Rashid, E.S.A.; Julkapli, N.B.M.; Yehya, W.A.H. Reinforcement effect of nanocellulose on thermal stability of nitrile butadiene rubber (NBR) composites. *J. Appl. Polym. Sci.* **2018**, *135*, 46594. [CrossRef]

19. Li, X.X.; Cho, U.R. Mechanical Performance and Oil Resistance Behavior of Modified Starch/Cellulose with Silica by Adsorption Method Filled into SBR Rubber Latex. *Polym. Korea* **2018**, *42*, 492–497. [CrossRef]

20. Jeschke, M.G.; Sandmann, G.; Schubert, T.; Klein, D. Effect of oxidized regenerated cellulose/collagen matrix on dermal and epidermal healing and growth factors in an acute wound. *Wound Repair Regen.* **2005**, *13*, 324–331. [CrossRef]

21. Dineen, P. Antibacterial Activity of Oxidized Regenerated Cellulose. *Surg. Gynecol. Obstet.* **1976**, *142*, 481–486.

22. Wu, H.L.; Williams, G.R.; Wu, J.Z.; Wu, J.R.; Niu, S.W.; Li, H.Y.; Wang, H.J.; Zhu, L.M. Regenerated chitin fibers reinforced with bacterial cellulose nanocrystals as suture biomaterials. *Carbohydr. Polym.* **2018**, *180*, 304–313. [CrossRef]

23. Abraham, E.; Deepa, B.; Pothan, L.A.; John, M.; Narine, S.S.; Thomas, S.; Anandjiwala, R. Physicomechanical properties of nanocomposites based on cellulose nanofibre and natural rubber latex. *Cellulose* **2013**, *20*, 417–427. [CrossRef]

24. Guo, W.W.; Wang, X.; Zhang, P.; Liu, J.J.; Song, L.; Hu, Y. Nano-fibrillated cellulose-hydroxyapatite based composite foams with excellent fire resistance. *Carbohydr. Polym.* **2018**, *195*, 71–78. [CrossRef]

25. Tominaga, Y.; Sato, K.; Hotta, Y.; Shibuya, H.; Sugie, M.; Saruyama, T. Improvement of thermal conductivity of composite film composed of cellulose nanofiber and nanodiamond by optimizing process parameters. *Cellulose* **2018**, *25*, 3973–3983. [CrossRef]

26. Isogai, A. Wood nanocelluloses: Fundamentals and applications as new bio-based nanomaterials. *J. Wood Sci.* **2013**, *59*, 449–459. [CrossRef]

27. Gopakumar, D.A.; Pasquini, D.; Henrique, M.A.; de Morais, L.C.; Grohens, Y.; Thomas, S. Meldrum's Acid Modified Cellulose Nanofiber-Based Polyvinylidene Fluoride Microfiltration Membrane for Dye Water Treatment and Nanoparticle Removal. *ACS Sustain. Chem. Eng.* **2017**, *5*, 2026–2033. [CrossRef]

28. Sharma, P.R.; Joshi, R.; Sharma, S.K.; Hsiao, B.S. A Simple Approach to Prepare Carboxycellulose Nanofibers from Untreated Biomass. *Biomacromolecules* **2017**, *18*, 2333–2342. [CrossRef]

29. Sharma, P.R.; Chattopadhyay, A.; Sharma, S.K.; Geng, L.; Amiralian, N.; Martin, D.; Hsiao, B.S. Nanocellulose from Spinifex as an Effective Adsorbent to Remove Cadmium(II) from Water. *ACS Sustain. Chem. Eng.* **2018**, *6*, 3279–3290. [CrossRef]

30. Sharma, P.R.; Chattopadhyay, A.; Zhan, C.; Sharma, S.K.; Geng, L.; Hsiao, B.S. Lead removal from water using carboxycellulose nanofibers prepared by nitro-oxidation method. *Cellulose* **2018**, *25*, 1961–1973. [CrossRef]

31. Sharma, P.R.; Zheng, B.; Sharma, S.K.; Zhan, C.; Wang, R.; Bhatia, S.R.; Hsiao, B.S. High Aspect Ratio Carboxycellulose Nanofibers Prepared by Nitro-Oxidation Method and Their Nanopaper Properties. *ACS Appl. Nano Mater.* **2018**, *1*, 3969–3980. [CrossRef]

32. Sharma, P.R.; Chattopadhyay, A.; Sharma, S.K.; Hsiao, B.S. Efficient Removal of UO22+ from Water Using Carboxycellulose Nanofibers Prepared by the Nitro-Oxidation Method. *Ind. Eng. Chem. Res.* **2017**, *56*, 13885–13893. [CrossRef]

33. Zhan, C.; Sharma, P.R.; Geng, L.; Sharma, S.K.; Wang, R.; Joshi, R.; Hsiao, B.S. Structural characterization of carboxyl cellulose nanofibers extracted from underutilized sources. *Sci. China Technol. Sci.* **2019**, *62*, 971–981. [CrossRef]

34. Kumar, R.; Kumari, S.; Surah, S.S.; Rai, B.; Kumar, R.; Sirohi, S.; Kumar, G. A simple approach for the isolation of cellulose nanofibers from banana fibers. *Mater. Res. Express* **2019**, *6*, 105601. [CrossRef]

35. Kumar, R.; Rai, B.; Kumar, G. A Simple Approach for the Synthesis of Cellulose Nanofiber Reinforced Chitosan/PVP Bio Nanocomposite Film for Packaging. *J. Polym. Environ.* **2019**, *27*, 2963–2973. [CrossRef]

36. Lin, S.-S. Degradation Behaviors of Natural, Guayule, and Synthetic Isoprene Rubbers. *Rubber Chem. Technol.* **1989**, *62*, 315–331. [CrossRef]

37. Geng, L.; Naderi, A.; Mao, Y.; Zhan, C.; Sharma, P.; Peng, X.; Hsiao, B.S. Rheological Properties of Jute-Based Cellulose Nanofibers under Different Ionic Conditions. In *Nanocelluloses: Their Preparation, Properties, and Applications*; American Chemical Society: Washington, DC, USA, 2017; Volume 1251, pp. 113–132.

38. Sukhlaaied, W.; Riyajan, S.-A. A Novel Environmentally Compatible Bio-Based Product from Gelatin and Natural Rubber: Physical Properties. *J. Polym. Environ.* **2018**, *26*, 2708–2719. [CrossRef]

39. Venugopal, B.; Gopalakrishnan, J. Reinforcement of natural rubber using cellulose nanofibres isolated from Coconut spathe. *Mater. Today-Proc.* **2018**, *5*, 16724–16731. [CrossRef]

40. Wu, Q.; Meng, Y.; Wang, S.; Li, Y.; Fu, S.; Ma, L.; Harper, D. Rheological behavior of cellulose nanocrystal suspension: Influence of concentration and aspect ratio. *J. Appl. Polym. Sci.* **2014**, *131*. [CrossRef]

 *nanomaterials*

Article

Chitosan Nanofiber and Cellulose Nanofiber Blended Composite Applicable for Active Food Packaging

Le Van Hai [1,2], Lindong Zhai [1], Hyun Chan Kim [1], Pooja S. Panicker [1], Duc Hoa Pham [1] and Jaehwan Kim [1,*

[1] CRC for Nanocellulose Future Composites, Inha University, Incheon 22212, Korea; levanhai121978@gmail.com (L.V.H.); duicaofei@naver.com (L.Z.); kim_hyunchan@naver.com (H.C.K.); pooja.panicker7@gmail.com (P.S.P.); phamduchoa.tdt@gmail.com (D.H.P.)

[2] Pulp and Paper Technology Department, Phutho College of Industry and Trade, Phutho 290000, Vietnam

*   Correspondence: jaehwan@inha.ac.kr; Tel.: +82-32-874-7325

Received: 19 August 2020; Accepted: 2 September 2020; Published: 4 September 2020

Abstract: This paper reports that, by simply blending two heterogeneous polysaccharide nanofibers, namely chitosan nanofiber (ChNF) and cellulose nanofiber (CNF), a ChNF–CNF composite was prepared, which exhibited improved mechanical properties and antioxidant activity. ChNF was isolated using the aqueous counter collision (ACC) method, while CNF was isolated using the combination of TEMPO oxidation and the ACC method, which resulted in smaller size of CNF than that of ChNF. The prepared composite was characterized in terms of morphologies, FT-IR, UV visible, thermal stability, mechanical properties, hygroscopic behaviors, and antioxidant activity. The composite was flexible enough to be bent without cracking. Better UV-light protection was shown at higher content of ChNF in the composite. The high ChNF content showed the highest antioxidant activity in the composite. It is the first time that a simple combination of ChNF–CNF composites fabrication showed good mechanical properties and antioxidant activities. In this study, the reinforcement effect of the composite was addressed. The ChNF–CNF composite is promising for active food packaging application.

Keywords: cellulose nanofiber; chitosan nanofiber; composite; mechanical properties; antioxidant activity

---

1. Introduction

Nowadays, the use of renewable materials instead of plastics is essential since the proliferation of plastics in the environment has been known to create various health and ecological problems. An example of extensive use of plastics is packaging. Recently, packaging was identified as an essential element to address the key challenge of sustainable food consumption [1]. When a food product is thrown away, the packaging is also discarded, leading to an additional environmental burden. Thus, petroleum-based packaging materials need to be replaced with renewable materials.

Chitosan, the second most abundant material after cellulose on Earth, is renewable, biodegradable, biocompatible, non-toxic and capable of transporting antioxidants [2–4]. Chitosan has been investigated for various applications including drug delivery, artificial skin, wound-dressing and biomedical and pharmaceutical applications [5]; contact lens and, water filtration [6]; food packaging [7–9]; and fruit preservations such as tomatoes, carrots and raw shrimps [10–13]. Chitosan does not cause any intrinsic food contamination such as phthalate leaching, thus has emerged as suitable alternatives for commercial plastics. Chitosan can be formed as a nanofiber. Chitosan nanofiber (ChNF) is beneficial to various applications since it has high aspect ratio, good chemical/physical interaction and flexibility. Most ChNFs can be fabricated by electro-spinning process including dissolving and purifying steps [14]. Instead of dissolving, ChNF can be isolated from its raw materials by physical methods, for example, by using supermasscolloider and high water jet pressure, above 200 MPa [15,16]. ChNF is considered for many applications such as removal of Arsenate [17], biomedical applications [18,19] and filtration membranes [14,20].

Cellulose, the most abundant polymer on Earth, has been used for long time. However, it has become a very interesting subject in recent years due to its possibility for substituting petroleum-based materials, and it is readily available around the world. Cellulose has been used for a wide variety of applications such as paper, packaging, composites, textiles, biomedical and pharmaceutical applications [21,22]. Cellulose nanofiber (CNF) is a nano-sized fiber in the range of ten to a couple hundred nanometers. Since CNF has merits over cellulose nanocrystals, its market is remarkably increasing for various applications [23]. It can be prepared mainly by mechanical and chemical methods. 2,2,6,6-tetramethylpiperidine-1-oxylradical-oxidation (TEMPO-oxidation) is a chemical method to extract CNF from various cellulose resources [24,25].

Cellulose and chitosan have been studied for food packaging materials [3,4,21,22]. It is well-known that chitosan has antibacterial, antioxidant and good food preservation properties. Cellulose has also been used as a food packaging material for long time. Early studies explored cellulose–chitosan composites for food packaging materials [26,27]. However, the blending of CNFs and ChNF has not been employed in any advance research, which could be applicable for food packaging. Thus, in this research, two types of nanofibers, namely CNF and ChNF, were blended for a potential active food packaging material. IThe ChNF was isolated using a physical treatment, so-called, aqueous counter collision (ACC) method [28]. CNF was also prepared from softwood pulp using a combination of chemical method, TEMPO-oxidation, and the ACC method to further decrease its size. We intended to distinguish the ChNF and CNF size to blend them with different morphologies. The prepared ChNF and CNF were directly blended to prepare ChNF–CNF composites. The advantages of blending CNF and ChNF are simple and benign preparation. Furthermore, by distributing various sizes of ChNF and CNF, physical and functional properties of the composite can be controlled. Active food packaging or smart packaging for food products refers to packaging that has functionalities in protecting the products. Those functionalities include preserving freshness and antimicrobial activity. Previous studies have reported that chitosan exhibits the functionalities suitable for active food packaging, for example antioxidant behavior and antimicrobial activity [3,7,10]. Thus, owing to functionalities of chitosan, the ChNF–CNF composite can be an active food packaging material.

To evaluate the morphology of CNF and ChNF, several techniques are available, for example scanning electron microscope (SEM), atomic force microscope (AFM), transmission electron microscope (TEM) and particle size analyzer. The size distributions of CNF and ChNF are very broad depending on the isolation methods. ChNF prepared by electrospinning exhibited a diameter ranging from 70 to 330 nm [18]; 260 nm with beads [29]; from 128 to 153 nm without beads [17]; and between 3 nm and few microns, at which the large diameter of nanofiber was due to self-assembly of ChNF [30]. The ChNF produced by grinder and high-pressure homogenizer yielded around 88 nm in diameter [31]. When chitosan nanoparticles were prepared by dissolving then slowly precipitating them in sodium tripolyphosphate solution, the chitosan nanoparticles exhibited a diameter of around 164 nm [32]. In the case of CNF, the TEMPO-oxidized CNF exhibited its width between 3 nm and few microns in length [25]. Its morphology was also investigated by AFM and, after centrifugal fractionation, the average width of the CNF was reduced to 2.0 ± 0.6 nm [33]. Depending on the treatment conditions, not only the size distribution, but also the physical properties including the thermal properties and crystallinity index can be varied. To the best of our knowledge, there has not been any research focused on the conversion of chitosan to ChNF by using ACC method. Furthermore, no research has been attempted to explore ChNF and CNF composites applicable for active food packaging. ChNF–CNF composites can be easily prepared just by blending ChNF and CNF.

Therefore, in this paper, we investigated the effect of ACC treatment conditions on the properties of ChNF, and prepared ChNF–CNF composites by simply blending two nanofibers, which can be applicable for an active food packaging material. The prepared ChNF–CNF composites as well as

ChNF were characterized in terms of morphology, hygroscopic behavior and chemical interaction, as well as thermal, optical, mechanical and antioxidant properties.

2. Materials and Methods

2.1. Materials

Low molecular weight Chitosan was purchased from Sigma Aldrich. The chitosan samples were dipped in deionized (DI) water for at least 30 min before subjected to ACC treatment. Softwood bleached kraft pulp was received from Chungnam National University, Daejeon, South Korea. 2,2,6,6-tetramethylpiperidine-1-oxylradical (TEMPO 98%), sodium bromide (NaBr 99%) and hydrochloric acid (HCl 37%) were purchase from Sigma-Aldrich, St. Louis, MO, USA. Sodium hypochlorite solution (NaClO 12%) was purchase from Yakuri Pure Chemicals Co. Ltd. Uji, Japan. Sodium hydroxide anhydrous (NaOH 98%) was purchase from Daejung, South Korea.

2.2. Chitosan Nanofiber Preparation

ACC is a water jet system that uses two high-pressure (200 MPa) water jets colliding with each other to produce a high shear force so as to isolate nanofibers from the original suspension [28,34]. ACC is a benign and environmentally friendly isolation method. Thus, ACC was selected for ChNF isolation by using an ACC machine (ACCNAC–100, CNNT, Korea). The nozzle size of two water jets was 160 μm in diameter. Chitosan suspension of 1% concentration was fed to the ACC machine to extract the ChNF. Different passes of chitosan suspension through the ACC machine was done at 10, 15 and 30 passes. The number of pass indicates how many times the suspension goes through the ACC chamber.

2.3. Cellulose Nanofiber Preparation

To isolate CNF, a combination of chemical method, TEMPO-oxidation and the ACC method was adopted to further decrease the size of CNF. Dried bleached softwood kraft pulp was dipped in DI water for at least 30 min followed by disintegration under high speed food mixer for 10 min. The pulp was then subjected to TEMPO-oxidation treatment by using the chemicals: TEMPO 0.013 g/g, NaBr 0.13 g/g and NaClO 12% 4 mL/g-cellulose. The pH 10 was adjusted by addition of 0.1 M NaOH, and the reaction time was set to 90 min. After the reaction, the TEMPO-oxidized cellulose was neutralized with HCl 0.1 M, followed by the addition of methanol to stop the reaction. The TEMPO-oxidized cellulose was washed several times with DI water.

The TEMPO-oxidized softwood cellulose was first homogenized by using a homogenizer (IKA T25, IKA, Staufen, Germany) for 10 min at 10,000 rpm to strip off any bundled fibers before going to ACC for smooth treatment. The cellulose suspension was passed through the ACC machine for 10 passes. The transparency and morphology of the prepared CNF were determined by using UV spectroscopy, FE-SEM and AFM.

2.4. ChNF–CNF Composites Preparation

To prepare the composites, the CNF was used as a matrix and ChNF was blended to reinforce the composites. The blended suspensions were mixed by using the homogenizer (IKA 25) for 10 min at 10,000 rpm. The mixture was then cast on a polycarbonate substrate by using a doctor blade in a clean room and left to dry on air. The weight percent of ChNF was changed to 3%, 5%, 7%, 10%, 15% and 20%, and the composites were named as CTS3, CTS5, CTS7, CTS10, CTS15 and CTS20, respectively. The thickness of the prepared ChNF–CNF composites was between 40 and 45 μm.

2.5. Characterizations

2.5.1. Morphology

Morphologies of the prepared ChNF as well as ChNF–CNF composites were investigated by using a field emission scanning electron microscopy (FE-SEM, S-4,000, Hitachi, Japan) and an atomic force microscopy (AFM, Veeco 3100, USA). Since the morphologies of CNF are well reported [21–25], they are not repeated in this paper.

2.5.2. FTIR Spectra

FTIR spectra of the ChNF–CNF composites and ChNF were determined using a FTIR spectroscopy (Cary 630, Agilent Technol. Santa Clara, CA, USA) with a diamond crystal that has the wavelength range from 650 to 4000 cm^{-1}. The specimens were tested for the absorbance between 650 and 4000 cm^{-1} with accumulation of 32 scans and the data were collected at a resolution of 4 cm^{-1}.

2.5.3. X-ray Diffraction (XRD)

The crystallinity index (CrI) of ChNF and ChNF–CNF composites were measured using an X-ray diffractometer (XRD, X'Pert PRO MRD, Malvern). It is hard to make chitosan particles into a thin film of 45–70 g/m^2. Thus, the pure chitosan was used in powder form. A thin ChNF film of 40–45 μm was cast on a polycarbonate substrate by using the doctor blade, and 2×2 cm^2 specimens were prepared to evaluate the CrI. The CrI of chitosan was calculated using the following equation [7].

$$\text{CrI } (\%) = (I_{002} - I_{am})/I_{002} * 100 \ (\%) \tag{1}$$

where I_{002} is XRD peak at $2\theta = 19.7°$ and I_{am} is diffraction pattern of amorphous area at $2\theta = 15°$. The prepared ChNF–CNF composites were directly used for XRD.

2.5.4. Mechanical Properties

Tensile test was performed to evaluate the mechanical properties of the prepared ChNF–CNF composites by using a tensile test machine [35]. Specimens were cut to the size of 0.5×5 cm^2. The length between grips was 3 cm. The specimen thickness was varied from 40 to 45 μm. Dried samples were kept in a condition chamber (30% RH and 25 °C) for at least 8 h before the tensile test, and five specimens were tested for each case. The ChNF specimen was prepared by casting it and tested also for comparison.

2.5.5. Thermogravimetric Analysis

Thermal stability of the prepared ChNF and ChNF–CNF composites were analyzed by using Thermogravimetric analyzer (TGA, STA 409PC, NETZSCH, Selb, Germany). Seven milligrams of the sample were prepared and thermally induced starting from 30 °C until 500 °C.

2.5.6. Viscosity

The viscosity of the prepared ChNF suspension was investigated by using a viscometer (LV DV2T, Brookfield viscometer, USA). The spindle LV–04 (64), speed 0.1 rpm, T = 23.5 °C and time recorded from 1 to 5 min were chosen for the test condition, and 0.8 wt% of ChNF suspensions were provided. The ChNF yield was investigated by using centrifugation at 7000 rpm for 1 h.

2.5.7. UV–Transmittance

The UV transmittance of the prepared ChNF and ChNF–CNF composites were investigated by using a UV spectrometer (HP 845x, Hewlett-Packard, Hayward, CA, USA). Suspensions of the specimens were used to measure the UV transmittance at the wavelength range of 200–800 nm.

2.5.8. Hygroscopic Behaviors

Water contact angle (WCA) measurement of the pure CNF and ChNF–CNF composites was carried out. A drop of 5 µL was deposited on a thin film and the images were taken by AMcap software and then analyzed using ImageJ tool. Water vapor transmittance rate (WVTR) was tested according to ASTM standard E 96-95 [36]. The sample was kept in a humidity chamber at 25 °C and 50% RH. The WVTR was taken hourly for up to 8 h continuously

2.5.9. Antioxidant Property

The antioxidant activity of the prepared ChNF–CNF composites was tested by using ABTS free radical [37]. The antioxidant activity analysis was based on the discolored radicals of ATBS after 40 min under UV light measurement at 734 nm. The UV absorption was adjusted to absorption at 0.8; 20 mg of CNF, ChNF and ChNF–CNF composites in 2 mL of ABTS were used for the measurement after 40 min. In other words, various chitosan contents (at 0.03, 0.05, 0.07, 0.1, 0.15 and 0.2 mg/mL of ABTS) were used. All antiradical tests were carried out twice for each sample. Then, 7 mM ABTS was dissolved in DI water and mixed with 2.45 mM potassium persulfate and kept in a dark drawer for 16 h. After that, the ABTS suspension was diluted with methanol to adjust the absorption of 0.8 at wavelength 734 nm. The antioxidant activity of CNF and ChNF–CNF composites was calculated by the following equation:

$$AO(\%) = 100 * (1 - A_a/A_o) \tag{2}$$

where AO is the antiradical activity, A_o is the absorption of the control ABTS solution and A_a is the absorption of the ABTS solution with sample in steady state.

3. Results

3.1. Chitosan Nanofibers

Figure 1 shows the photograph of ChNF suspensions treated with 10, 15 and 30 passes of the ACC treatment. After ACC treatment, the suspensions were changed to ivory color, and as the ACC pass increased the suspension color turned to a milky. The viscosity of ChNF suspensions was measured and Figure 2 shows the result. The viscosity of ChNF suspensions increased with the number of ACC pass. The higher is the number of passes at the ACC chamber, the higher is the viscosity of ChNF. Higher viscosity means a higher fibrillation of chitosan, thus reducing its size to nanofibers. Note that the low viscosity ChNF samples (10 and 15 ACC passes) easily flowed when the samples were placed upside down, while the 30 passes case did not flow when it was placed upside down. Interestingly, the viscosity values decreased with the time especially, for the 15 and 30 passes cases. This might be associated with the broken inner bonds in ChNF and the layer separation in the suspension. Figure 2b shows the viscosity change with the number of ACC pass when the time is 5 min. The viscosity linearly increased with the number of ACC pass.

The yield of ChNF was evaluated in each sample by centrifugation. The result shows that, after 10, 15 and 30 passes of the ACC treatment, the yields of ChNF were 13.2%, 15.5% and 27.7%, respectively. From the viscosity and yield data, it is clear that the higher is the number of ACC passes, the better is the isolation of ChNF.

The morphology changes of chitosan after the ACC treatment were observed by FE-SEM and AFM. Figure 3a–d shows the morphologies of the original chitosan and ChNFs with 10, 15 and 30 passes of the ACC treatment, respectively. The chitosan particles were around 50–100 µm. After the ACC treatment, it was changed to nanofibers, as shown in Figure 3b–d. After 30 passes, the ChNF size was reduced to 38 ± 16.5 nm in width and length of several microns. The width of ChNF was calculated in 100 measurements from the AFM images. This size of ChNF is larger and longer than that of CNF. Thus, in the preparation of ChNF–CNF composites, 30 passed ChNF was chosen.

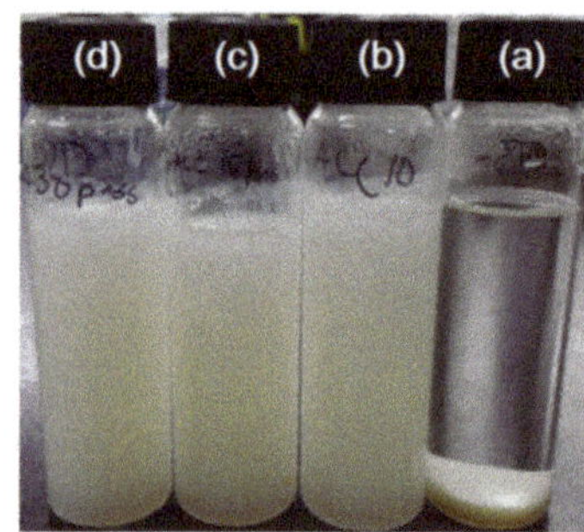

Figure 1. Photograph of ChNF suspensions: (**a**) the original chitosan; (**b**) after 10 ACC passes; (**c**) after 15 ACC passes; and (**d**) after 30 ACC passes.

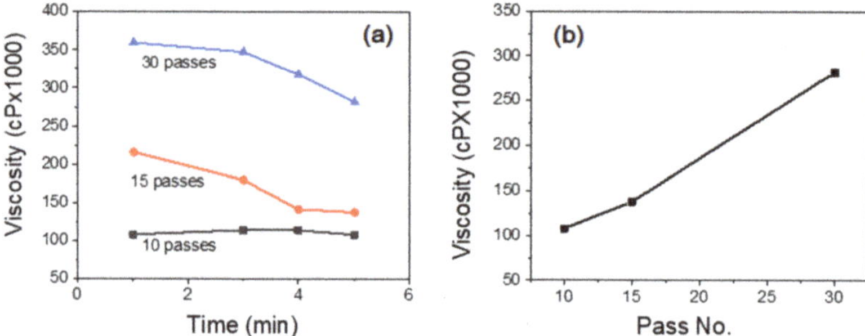

Figure 2. Effect of ACC passes on the viscosity of ChNF suspension: (**a**) viscosity change with time, and (**b**) viscosity change with ACC pass number.

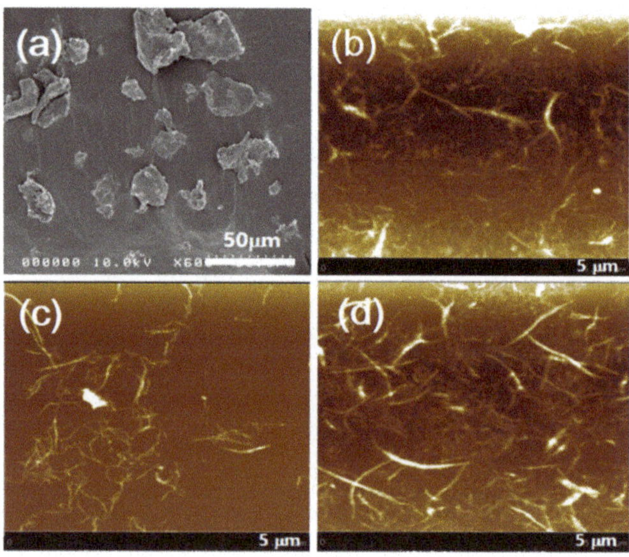

Figure 3. Morphologies of chitosan and its nanofibers: (**a**) SEM image of the original chitosan particles; and AFM images of ChNFs with different ACC passes: (**b**) 10 passes; (**c**) 15 passes; and (**d**) 30 passes (AFM images are in 5 μm × 5 μm).

3.2. ChNF–CNF Composites

3.2.1. Morphologies

The morphologies of the ChNF–CNF composite structure were investigated. Figure 4 shows the SEM images of the ChNF–CNF composite and CNF film. Figure 4a shows a smooth surface of the CNF film and Figure 4b is the cross-section SEM image of the CNF film, which exhibits the layer-by-layer structure of CNF. After blending with ChNF, the surface of ChNF–CNF composite turned out to be rough. The cross-section SEM image of the composite also shows a layer-by-layer structure similar to the CNF film. It was shown that the ChNF was well blended with CNF to maintain the layer-by-layer structure.

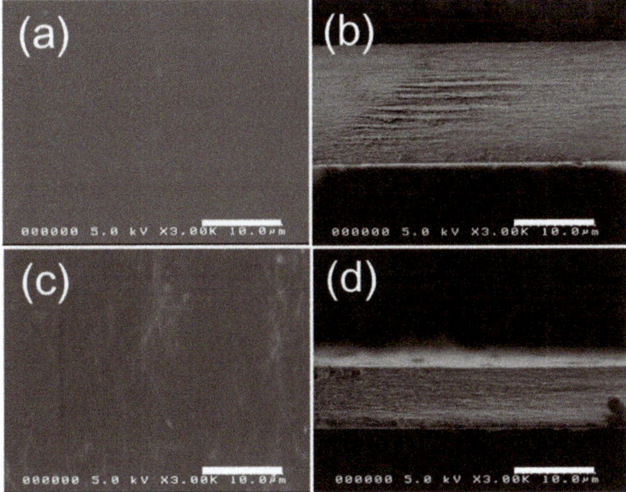

Figure 4. FE-SEM images of: (**a**) surface of CNF film; (**b**) cross section of CNF film; (**c**) surface of ChNF–CNF composite; and (**d**) cross section of ChNF–CNF composite.

In addition, to consider ChNF–CNF composites for packaging, the composites should be bendable. Thus, a quick bending test was performed with the ChNF–CNF composite. The composites were so flexible that they could be bent in any direction without cracking. The thickness of the composites was in the range of 40–50 μm. Figure 5 shows flexibility of the composite in bending and rolling deformation.

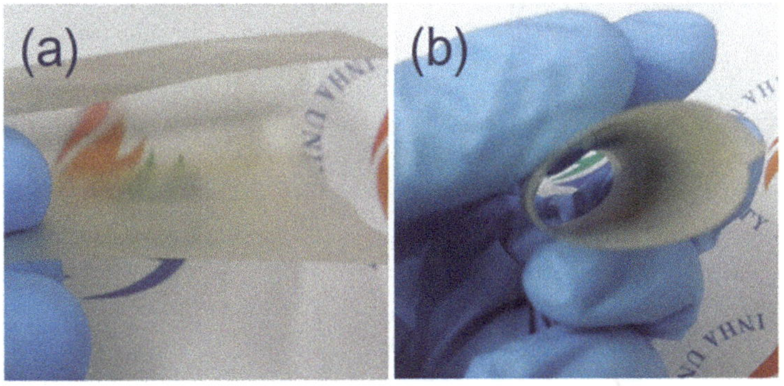

Figure 5. Flexibility of ChNF–CNF composite by: (**a**) bending; and (**b**) rolling.

3.2.2. FTIR

Figure 6 shows the FTIR spectra of the CNF, ChNF and ChNF–CNF composites with different ChNF concentration. The peak in the range of 3200–3500 cm^{-1} of the ChNF–CNF composites was assigned to the hydrogen-bonded O–H stretching in both CNF and ChNF. Chitosan showed broader O–H bonded as compared to cellulose with higher intensity. Due to the combination of these two materials, with 20% chitosan, the intensity of the peak at 3200–3500 cm^{-1} appeared similar to pure chitosan. The CNF showed stronger and higher peak of hydrogen-bonding than the ChNF. When ChNF and CNF were blended, peaks at 3000 and 3500 cm^{-1} were reduced, which might be due to the intra-bonding between two materials. The peak at 1596 cm^{-1} assigned for –C=O stretching is clear for CNF. The peak at 1736 cm^{-1} was assigned for the C=O stretch of the –COOH group. This was only observed in the CNF at first, and the intensity of this peak decreased due to the addition of ChNF. At last, this peak became weak and disappeared when the ChNF concentration increased. This phenomenon indicates the interaction between ChNF and CNF. The peak at 1596 cm^{-1} only appeared for cellulose and composites but not clearly shown for chitosan. This peak might be associated with the aromatic ring stretching of lignin or hemicellulose remained small amount in the bleached kraft pulp. From the FT-IR analysis, the structure of CNF and ChNFs were confirmed in the composites.

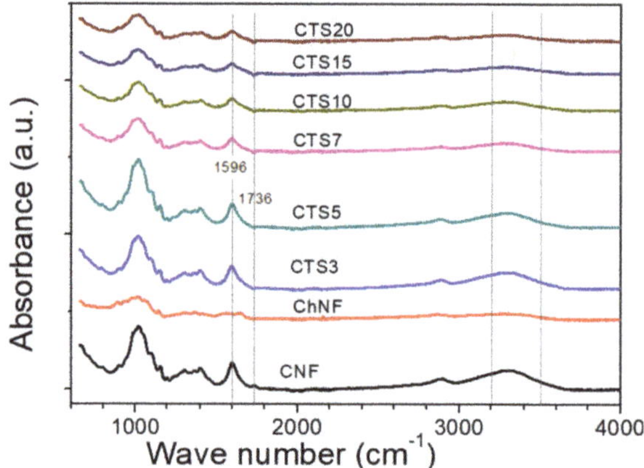

Figure 6. FTIR of CNF, ChNF and ChNF–CNF composites.

3.2.3. Crystallinity Index

The crystallinity index (CrI) of the original chitosan and ChNF were obtained by XRD and calculated according to Equation 1 as shown in Figure 7a. The crystalline peaks of chitosan appeared at 9.7° and 19.7°. The peak at 19.7° decreased by the ACC treatment, while the strong peak at 9.7° increased, which might be associated with the incorporation of bound water molecules of α-chitin [38]. It means that a higher ACC treatment leads to more water molecules bound on hydrophilic surface of ChNF. The CrI values of the original chitosan and ChNFs with 10, 15 and 30 ACC passes were shown to be 65.4%, 44.0%, 44.0% and 36.2%, respectively. The CrI of chitosan highly degraded after ACC treatment. Note that 30 ACC passed ChNF was used for the ChNF–CNF composite preparation.

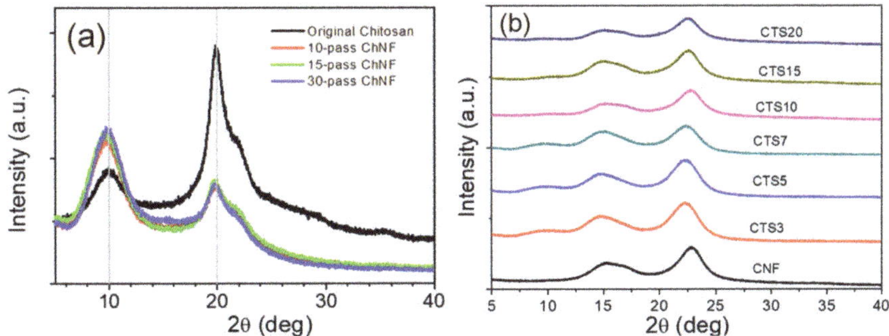

Figure 7. Crystallinity index of: (**a**) ChNF with number of ACC passes; and (**b**) CNF and ChNF–CNF composites.

The XRD patterns of the ChNF–CNF composites are shown in Figure 7b with different ACC passes. As one can see, cellulose peaks at around 15.8° and 22.7° were clearly shown and as the ChNF content increased, chitosan peaks were slightly appeared by interfering with cellulose peaks. Thus, CrI of the composites was calculated mainly by the dominant cellulose peaks at 22.7°. The CrI values for low ChNF content composites slight improved at low chitosan content (ChNF3 and ChNF5) from 63% to 65%, and then it slightly decreased to 58% with addition of ChNF content. This might be due to good miscibility between ChNF and CNF.

3.2.4. Mechanical Properties

Tensile test was performed for ChNF, CNF and ChNF–CNF composites and the results are shown in Table 1. The tensile strength of the composite at first was 174.5 MPa (no ChNF), reached 224.0 MPa when the ChNF content was 10% and then decreased thereafter. Note that mechanical properties of the pristine ChNF is much lower than the CNF, i.e. more flexible than the CNF. Similarly, the yield strength of the composite was initially 111.7 MPa and reached its maximum of 149.0 MPa when the ChNF content was 10%. It was shown that the elongation at break of the composite increased from 2.02% to 4.17% and then decreased. It was shown that, when reinforced by ChNF, the tensile strength, yield strength and elongation at break of the composite improved. However, Young's modulus of the composite was not improved and slightly decreased due to the low modulus of ChNF. The modulus of ChNF is 7.3 GPa, while that of CNF is 16.9 GPa.

Table 1. Tensile test results of ChNF, CNF and their composites.

Samples	Young's Modulus (GPa)	Yield Strength (MPa)	Tensile Strength (MPa)	Elongation at Break (%)
CNF	16.9 + 2.6	111.7 + 57	174.5 + 56.0	2.02
CTS3	13.9 + 2.1	100.2 + 23	183.1 + 60.8	2.06
CTS5	15.2 + 4.0	124.6 + 17	197.5 + 32.0	2.46
CTS7	13.2 + 0.5	136.2 + 80	216.5 + 95.7	2.85
CTS10	12.9 + 0.9	149.0 + 20	224.0 + 27.0	4.17
CTS15	13.5 + 2.3	131.8 + 23	198.0 + 31.4	3.40
CTS20	15.8 + 2.7	104.4 + 22	157.1 + 30.5	1.79
ChNF *	7.3 + 0.70	87.5 + 26.8	133.4 + 17.9	3.69

* when number of ACC passes is 30.

Nevertheless, the overall mechanical properties of the composite were improved, which might be due to the reinforcement of ChNF to CNF in the composite. Most previous studies produced cellulose–chitosan composites by dissolving chitosan or cellulose and blending them [2,32]. However, in this research, we adopted the non-dissolving method to prepare ChNF–CNF composites, and a remarkable improvement of the mechanical properties of the ChNF–CNF composite was observed. There could be several mechanisms that can explain the mechanical properties improvement of the composite. The first one is improvement of the bonding sites between ChNF and CNF due to different size of ChNF and CNF. Note that the ChNF size was larger than that of CNF. The long length of ChNF in the CNF matrix and high surface areas of CNF created many reaction sites and bonding areas on their surfaces. Furthermore, the CNF is relatively stiff and ChNF is flexible. The mismatched mechanical properties of ChNF and CNF could give a room for managing the composite mechanical properties. Figure 8 shows the possible concept of reinforcement in the composite. By blending these two heterogeneous polysaccharide nanofibers, we can manage the mechanical properties of the composite. It is hoped that this idea could be a way to produce cellulose–chitosan composites with better mechanical properties without dissolving process of chitosan or cellulose.

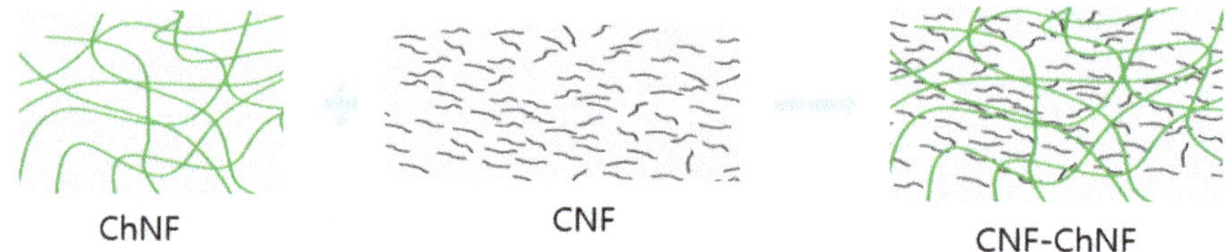

Figure 8. A possible reinforcement mechanism of ChNF and CNF composite.

3.2.5. Thermal Stability

The thermal stability of ChNF, CNF and their composites were examined by TGA, and the results are shown in Figure 9. At 30 to 100 °C of the evaporation stage, there was no significant difference in the first stage of water evaporation in the composites except ChNF. The CNF and their composites started to degrade around 200 °C and reached around 320 °C for the first phase of degradation, while ChNF started to degrade near 250 °C. The composites showed almost the same thermal degradation range as the CNF. It can be seen that they are stable up to 250 °C. Note that the CNF and ChNF exhibited larger residuals than the composites. The residuals after 320 °C was around 40–50% for CNF and ChNF. The reason for larger residuals of the CNF and ChNF than the composites is under investigation.

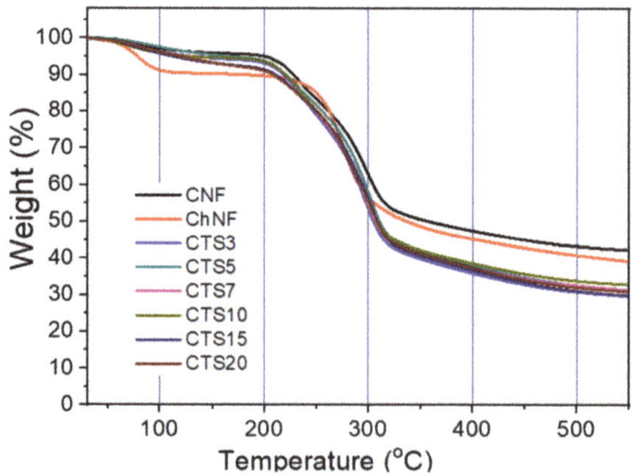

Figure 9. TGA of ChNF–CNF composites.

3.2.6. UV Transmittance

UV protection of ChNF–CNF composites is an important factor for food packaging applications. UV transmission of the composites was evaluated, and the results are shown in Figure 10. As compared to the neat CNF, the ChNF–CNF composites showed better UV-protection. Higher CTS content showed better the UV protection. In the range of UV-A, UV-B, and UV-C, CTS10, CTS15 and CTS20 exhibited higher UV protection than CTS3, CTS5 and CTS7. However, transparency of the composites is sacrificed to improve the UV protection. Thus, good UV protection with maintaining good transparency is future work.

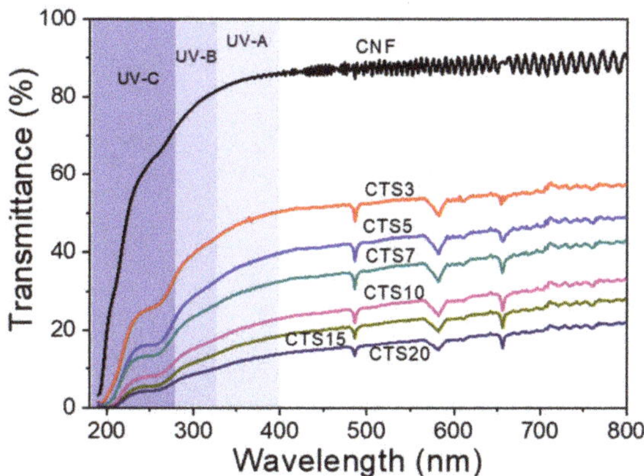

Figure 10. UV-transmittance of ChNF–CNF composites.

3.2.7. Hygroscopic Behaviors

WCA and WVTR were tested to investigate the hygroscopic behaviors of the composites. Table 2 shows the results. The WVTR was a little increased at first by the addition of ChNF up to 10% and then saturated thereafter. This result shows a good agreement with the previous research for CNF and acetylated CNF [22]. Regarding the WCA of the composites, it was rather a bit decreased by adding ChNF, which indicates hydrophilic behavior of the composites. This hydrophilic behavior might be associated with the hydroxyl groups appeared on the surface of the composites. Enhancing this behavior is future work in this research.

Table 2. Water vapor transmission rate and water contact angle of CNF and ChNF–CNF composites.

Sample	Water Vapor Transmission Rate (g /m^2.day)	Water Contact Angle (°)
CNF	164.98 + 3.09	47
CTS3	164.69 + 1.85	47
CTS5	167.33 + 5.32	45
CTS7	171.18 + 0.89	45
CTS10	175.29 + 1.08	41
CTS15	173.45 + 0.67	41
CTS20	173.04 + 6.06	38

3.2.8. Antioxidant Property

The antioxidant abilities of CNF, ChNF and ChNF–CNF composites were investigated and expressed in AOA%/100 mg, as shown in Figure 11. The antioxidant activities of CNF, ChNF, CTS3, CTS5, CTS7, CTS10, CTS15 and CTS20 were 16.9%, 27.9%, 19.4%, 23.4%, 23.4%, 33.4%, 47.4% and 52.0%, respectively. The CTS20 showed the highest antioxidant activity among all cases. Interestingly, although the CNF and ChNF have low antioxidant activities, when they were blended together, the antioxidant activity of the composites increased. Cellulose and chitosan are both polysaccharides, and, when they are combined, the reducing ends of the materials may open, thus increasing the antioxidant activity. Nanofibrillation can also increase the hydroxyl groups, which are good for antioxidant activities. The antioxidant activity is the interaction of free radicals with the hydroxyl groups, and free amino groups of the chitosan and cellulose [39,40]. The improvement of antioxidant activities of the composites is agreeable with the previous research [39], which showed a steady increase of the antioxidant activity of the composites by increasing the chitosan content.

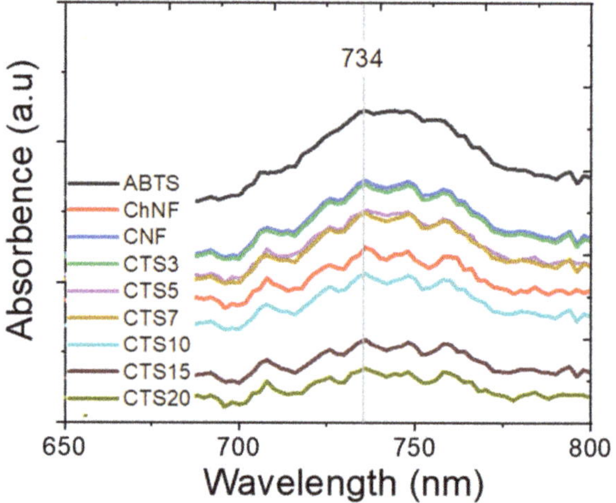

Figure 11. Antioxidant of ChNF–CNF composites.

4. Conclusions

In this research, ChNF was isolated by using the ACC method and increasing the number of ACC passes reduced the size and increased the yield of ChNF. Isolating CNF by using the combination of TEMPO oxidation and the ACC method resulted in smaller size than that of ChNF. By blending two heterogeneous polysaccharide nanofibers, we were able to prepare ChNF–CNF composites. ChNF was well blended with CNF so as to maintain the layer-by-layer structure of the composites. The composites were so flexible that they could be bent in any direction without cracking. From the FT-IR analysis, the structure of CNF and ChNFs were confirmed in the composites. The composites showed remarkable improvement of the mechanical properties, which might be due to the reinforcement of ChNF to CNF in the composites. The composites showed almost the same thermal degradation range of the CNF. Better UV-light protection was shown at higher content of ChNF in the composites. The WVTR was slightly increased at first by the addition of ChNF up to 10% and then saturated thereafter. The WCA of the composites was rather slightly decreased by adding ChNF, and enhancing this behavior is future work. The high ChNF content showed the highest antioxidant activity in the composites. Although the CNF and ChNF had low antioxidant activities, their blended composites had increased the antioxidant activity. It is the first time a simple combination of ChNF–CNF composites fabrication showed good mechanical properties and antioxidant activities. Reinforcement of ChNF into CNF was a good selection in terms of antioxidant activity, UV protection and mechanical properties. The ChNF–CNF composite could be useful for active food packaging application.

Author Contributions: Conceptualization, J.K. and L.V.H.; methodology, H.C.K.; validation, L.Z. and L.V.H.; investigation, P.S.P.; resources, L.V.H.; data curation, D.H.P.; writing—original draft preparation, L.V.H.; writing—review and editing, J.K.; visualization, D.H.P.; supervision, J.K.; and funding acquisition, J.K. All authors have read and agreed to the published version of the manuscript.

Nanomaterials **2020**, *10*, 1752

Funding: This research was supported by National Research Foundation of Korea (NRF-2015R1A3A2066301).

Conflicts of Interest: The authors declare no conflict of interest.

References

1. Guillard, V.; Gaucel, S.; Fornaciari, C.; Angellier-Coussy, H.; Buche, P.; Gontard, N. The Next Generation of Sustainable Food Packaging to Preserve Our Environment in a Circular Economy Context. *Front. Nutr.* **2018**, *5*, 1–13. [CrossRef] [PubMed]
2. Hirano, S.; Zhang, M.; Chung, B.G.; Kim, S.K. The N-acylation of chitosan fibre and the N-deacetylation of chitin fibre and chitin-cellulose blended fibre at a solid state. *Carbohyd. Polym.* **2000**, *41*, 175–179. [CrossRef]
3. Yadav, S.; Mehrotra, G.K.; Dutta, P.K. Chitosan based ZnO nanoparticles loaded gallic-acid films for active food packaging. *Food Chem.* **2020**, *334*, 127605. [CrossRef] [PubMed]
4. Mujtaba, M.; Morsi, R.E.; Kerch, G.; Elsabee, M.Z.; Kaya, M.; Labidi, J.; Khawar, K.M. Current advancements in chitosan-based film production for food technology: A review. *Int. J. Biol. Macromol.* **2019**, *121*, 889–904. [CrossRef] [PubMed]
5. Rinaudo, M. Chitin and chitosan: Properties and applications. *Prog. Polym. Sci.* **2006**, *31*, 603–632. [CrossRef]
6. Ravi Kuma, M.N.V. A review of chitin and chitosan applications. *React. Funct. Polym.* **2000**, *46*, 1–27. [CrossRef]
7. Kittur, F.S.; Kumar, K.R.; Tharanathan, R.N. Functional packaging properties of chitosan films. *Z Lebensm Unters Forsch A* **1997**, *206*, 44–47. [CrossRef]
8. Shahidi, F.; Arachchi, J.K.V.; Jeon, Y.J. Food applications of chitin and chitosan. *Trends Food Sci. Technol.* **1999**, *10*, 37–51. [CrossRef]
9. Tripathi, S.; Mehrotra, G.K.; Dutta, P.K. Chitosan based antimicrobial films for food packaging applications. *Euro. Polym.* **2008**, *93*, 1–7. [CrossRef]
10. Elghaouth, A.; Ponnampalam, R.; Castaigne, F.; Arul, J. Chitosan coating to extend the storage life of Tomatoes. *Hortscience* **1992**, *27*, 1016–1018.
11. Cheah, L.H.; Page, B.B.C.; Shepherd, R. Chitosan coating for inhibition of sclerotinia rot of carrots. *New Zeal. J. Crop. Hot.* **1997**, *25*, 89–92. [CrossRef]
12. Simpson, B.K.; Gagne, N.; Ashie, I.N.A.; Noroozi, E. Utilization of chitosan for preservation of raw shrimo (Pandalus borealis). *Food Biotechnol.* **1997**, *11*, 25–44. [CrossRef]
13. Jianglian, D.; Shaoying, Z. Application of chitosan based coating in fruit and vegetable preservation: A Review. *Food Process. Technol.* **2013**, *4*, 1–4. [CrossRef]
14. Arkoun, M.; Daigle, F.; Heuzey, M.C.; Aji, A. Antibacterial electrospun chitosan-based nanofibers: A bacterial membrane perforator. *Food Sci. Nutr.* **2017**, *5*, 865–874. [CrossRef] [PubMed]
15. Dutta, A.K.; Kawamoto, N.; Sugino, G.; Izawa, H.; Morimoto, M.; Saimoto, H.; Ifuku, S. Simple preparation of chitosan nanofibers from dry chitosan powder by the Star Burst System. *Carbohyd. Polym.* **2013**, *97*, 363–367. [CrossRef] [PubMed]
16. Ifuku, S. Chitin and chitosan nanofibers: Preparation and chemical modifications. *Molecules* **2014**, *19*, 18367–18380. [CrossRef] [PubMed]
17. Min, L.L.; Zhong, L.B.; Zheng, Y.M.; Liu, Q.; Yuan, Z.H.; Yang, L.M. Functionalized chitosan electrospun nanofiber for effective removal of trace arsenate from water. *Sci. Rep.* **2016**, *6*, 32480. [CrossRef]
18. Jayakumar, R.; Prabaharan, M.; Nair, S.V.; Tamura, H. Novel chitin and chitosan nanofibers in biomedical applications. *Biotechnol. Adv.* **2010**, *28*, 142–150. [CrossRef]
19. Aruma, K.; Ifuku, S.; Osaka, T.; Okamoto, Y.; Minamil, S. Preparation and biomedical applications of chitin and chitosan nanofibers. *J. Biomed. Nanotech.* **2014**, *10*, 2891–2920. [CrossRef]
20. Ali, F.A.A.; Haider, S.; Al-Måy, W.A.; Al-zeghaye, Y. Fabrication of chitosan nanofibers membrane and its treatment. 2013. Available online: www.jeaconf.org/UploadedFiles/.../a2a3a6d5-6810-4b3e-93a0-09140c4740c4.pdf (accessed on 4 September 2020).
21. Bacakova, L.; Pajorova, J.; Bacakova, M.; Skogberg, A.; Kallio, P.; Kolarova, K.; Svorcik, V. Versatile Application of Nanocellulose: From Industry to Skin Tissue Engineering and Wound Healing. *Nanomaterials* **2019**, *9*, 164. [CrossRef]
22. Ferrer, A.; Pal, L.; Hubbe, M. Nanocellulose in packaging: Advances in barrier layer technologies. *Ind. Crop. Prod.* **2017**, *95*, 574–582. [CrossRef]

23. Future Markets. The Global Market for Cellulose Nanofibers to 2027. Future Markets Inc.: Edinburgh, UK, 2018.

24. Isogai, A.; Kato, Y. Preparation of Polyuronic Acid from Cellulose by TEMPO-mediated Oxidation. *Cellulose* **1998**, *5*, 153. [CrossRef]

25. Saito, T.; Kimura, S.; Nishiyama, Y.; Isogai, A. Cellulose nanofiber prepared by TEMPO-mediated oxidation of native cellulose. *Biomacromolecules* **2007**, *8*, 2485–2491. [CrossRef] [PubMed]

26. Noorbakhsh-Soltani, S.M.; Zerafat, M.M.; Sabbaghi, S. A comparative study of gelatin and starch-based nano-composite films modified by nano-cellulose and chitosan for food packaging applications. *Carbohyd. Polym.* **2018**, *189*, 48–55. [CrossRef]

27. Fernandes, S.C.M.; Freire, C.S.R.; Silvestre, A.J.D.; Neto, C.P.; Gandini, A.; Berglund, L.A.; Salmen, L. Transparent chitosan films reinforced with a high content of nanofibrillated cellulose. *Carbohyd. Polym.* **2010**, *81*, 394–401. [CrossRef]

28. Hai, L.V.; Zhai, L.; Kim, H.C.; Kim, J.W.; Choi, E.S.; Kim, J. Cellulose nanofibers isolated by TEMPO-oxidation and aqueous counter collision methods. *Carbohyd. Polym.* **2018**, *191*, 65–70.

29. Sohofi, N.; Tavanai, H.; Morshed, M.; Abdolmaeki, A. Electrospinning of 100% Carboxymethyl Chitosan Nanofibers. *J. Eng. Fiber. Fabr.* **2014**, *9*, 87–92. [CrossRef]

30. Pillai, C.K.S.; Sharma, C.P. Electrospinning of chitin and chitosan nannofibers. *Trends Biomater. Artif. Organs* **2009**, *22*, 179–201.

31. Fadaie, M.; Mirzaei, E. Nanofibrillated chitosan/Polycaprolactonebionanocompsoite scaffold with improved tensile strength and cellular behavior. *Nanomed. J.* **2018**, *5*, 77–89.

32. Hassan, A.E.; Hassan, M.L.; Abo-zeid, R.E.; El-wkil, N.A. Novel nanofibrillated cellulose/chitosan nanoparticles nanocomposites films and their use for paper coating. *Ind. Crop. Prod.* **2016**, *93*, 219–226. [CrossRef]

33. Zhai, L.; Kim, H.C.; Kim, J.W.; Kim, J. Simple centrifugal fractionation to reduce the size distribution of cellulose nanofibers. *Sci. Rep.* **2020**, *10*, 11744. [CrossRef] [PubMed]

34. Kondo, T.; Kose, R.; Naito, H.; Kasai, W. Aqueous counter collision using paired water jets as a novel means of preparing bio-nanofibers. *Carbohyd. Polym.* **2014**, *112*, 284–290. [CrossRef] [PubMed]

35. Kafy, A.; Kim, H.C.; Zhai, L.; Kim, J.W.; Hai, L.V.; Kang, T.J.; Kim, J. Cellulose long fibers fabricated from cellulose nanofibers and its strong and tough characteristics. *Sci. Rep.* **2017**, *7*, 17683. [CrossRef] [PubMed]

36. ASTM. Standard Test Methods for Water Vapor Transmission of Materials. E96-95. Available online: https://www.astm.org/DATABASE.CART/HISTORICAL/E96-95.htm (accessed on 4 September 2020).

37. Cao, T.L.; Yang, S.Y.; Song, K.B. Development of burdock root Inulin/Chitosan Blend films Containing Oregano and Thyme Essential oils. *Int. J. Mol. Sci.* **2017**, *19*, 31.

38. Al Sagheer, F.A.; Al-Sughayer, M.A.; Muslim, S.; Elsabee, M.Z. Extraction and characterization of chitin and chitosan from marine sources in Arabian Gulf. *Carbohyd. Polym.* **2009**, *77*, 410–419. [CrossRef]

39. Tamer, T.M.; Valachova, K.; Mohyeldin, M.S.; Soltes, L. Free radial scavenger activity of chitosan and its aminated derivative. *J. Appl. Pharm. Sci.* **2016**, *6*, 195–201. [CrossRef]

40. Yen, M.T.; Yang, J.H.; Mau, J.L. Antioxidant properties of chitosan from crab shells. *Carbohyd. Polym.* **2008**, *74*, 840–844. [CrossRef]

Article

Nano-Cellulose/MOF Derived Carbon Doped CuO/Fe₃O₄ Nanocomposite as High Efficient Catalyst for Organic Pollutant Remedy

Hailong Lu [1,2], Lili Zhang [1,2], Jinxia Ma [1], Nur Alam [2], Xiaofan Zhou [1,*] and Yonghao Ni [2,*]

[1]   National-Provincial Joint Engineering Research Center of Electromechanical Product Packaging, Jiangsu Co-Innovation Center of Efficient Processing and Utilization of Forest Resources, Nanjing Forestry University, Nanjing 210037, China; 15298393737@163.com (H.L.); luo913302@163.com (L.Z.); jxma@njfu.edu.cn (J.M.)

[2]   Department of Chemical Engineering, University of New Brunswick, Fredericton, NB E3B 5A3, Canada; malam6@unb.ca

[*]   Correspondence: zxiaofan@njfu.com.cn (X.Z.); yonghao@unb.ca (Y.N.); Tel.: +86-25-85428932 (X.Z.); +1-506-451-6857 (Y.N.); Fax: +86-25-85428932 (X.Z.); +1-506-453-4767 (Y.N.)

Received: 20 January 2019; Accepted: 11 February 2019; Published: 16 February 2019

Abstract: Metal–organic framework (MOF)-based derivatives are attracting increased interest in various research fields. In this study, nano-cellulose MOF-derived carbon-doped CuO/Fe_3O_4 nanocomposites were successfully synthesized via direct calcination of magnetic Cu-BTC MOF (HKUST-1)/Fe_3O_4/cellulose microfibril (CMF) composites in air. The morphology, structure, and porous properties of carbon-doped CuO/Fe_3O_4 nanocomposites were characterized using SEM, TEM, powder X-ray diffraction (PXRD), X-ray photoelectron spectroscopy (XPS), and vibrating sample magnetometry (VSM). The results show that the as-prepared nanocomposite catalyst is composed of Fe_3O_4, CuO, and carbon. Compared to the CuO/Fe_3O_4 catalyst from HKUST-1/Fe_3O_4 composite and CuO from HKUST-1, this carbon-doped CuO/Fe_3O_4 nanocomposite catalyst shows better catalytic efficiency in reduction reactions of 4-nitrophenol (4-NP), methylene blue (MB), and methyl orange (MO) in the presence of $NaBH_4$. The enhanced catalytic performance of carbon-doped CuO/Fe_3O_4 is attributed to effects of carbon preventing the aggregation of CuO/Fe_3O_4 and providing high surface-to-volume ratio and chemical stability. Moreover, this nanocomposite catalyst is readily recoverable using an external magnet due to its superparamagnetic behavior. The recyclability/reuse of carbon-doped CuO/Fe_3O_4 was also investigated.

Keywords: nano-cellulose; MOF; carbon-doped CuO/Fe_3O_4 nanocatalyst; catalytic reduction; pollutant remedy

---

1. Introduction

Recently, metal nanoparticles (NPs) were widely used in the fields of biomedicine and chemical reactions due to their high selectivity and catalytic efficiency [1–3]. Noble-metal nanoparticles (gold, silver, etc.) [4–6] and non-noble-metal nanoparticles (copper, zinc, and their oxides, sulfides, etc.) [7–10] are particularly noticeable. For example, Jiang et al. [11] reported that CuO and Au domains could greatly improve the photocatalytic activity and stability of Cu_2O cubes in the photocatalytic degradation of methyl orange (MO). Rodríguez et al. [12] reported that potassium poly(heptazine imide) (PHIK)/Ti-based metal–organic framework (MIL-125-NH₂) composites had superior photocatalytic activity in rhodamine B (RhB) degradation under blue-light irradiation. Among the applications, metal nanoparticles can also be used for treating wastewater and drinking water due to their large surface areas and high activity [13,14]. With growing focus on the development of cost-effective, efficient catalysts, more attention is being paid to non-noble-metal catalysts, such as metal-oxide catalysts.

Metal NPs with nanometer-scale dimensions are unstable and tend to aggregate due to their high surface energy and surface area, which can lead to the loss of catalytic efficiency [15–17]. The concept of immobilization/stabilization of metal NPs onto support/substrates is one of the effective methods to overcome aggregation of NPs [18,19]. There are many types of substrates/matrices that can be used to support metal NPs, such as carbon [20,21], silica [22], metal oxide [23], polymers, etc. [24–26]. Carbon, as a support for metal nanoparticles, provides multiple accessible channels for diffusion/transport to take advantage of the excellent catalytic functionalities of metal nanoparticles [27].

In past decades, metal–organic frameworks (MOFs) attracted much attention due to their porous structures and potential applications in gas storage, molecule separation, chemical sensing, drug delivery, and catalysis [28–30]. Recently, MOF-based derivative catalysts received more attention because MOF-derived materials have advantages in terms of tailorable morphologies, hierarchical porosity, and easy functionalization with other heteroatoms and metal oxides [31,32]. For instance, Ji et al. used ZIF-67 as a precursor to synthesize a $Pt@Co_3O_4$ composite to improve the catalytic activity of CO oxidation [33]. Yang et al. reported that ZnO nanoparticles prepared via calcination of MOF-5 in air at 600 °C showed excellent photocatalytic degradation of rhodamine B [34]. Niu et al. synthesized a hybrid catalyst consisting of Cu/Cu_2O NPs supported on porous carbon for the catalytic reduction of 4-nitrophenol (4-NP) using HKUST-1 as a precursor [27].

The direct pyrolysis/thermolysis treatment of MOFs is a simple and controllable method to prepare various metal oxides in a one-step process. By following it, we can successfully synthesize carbon-doped CuO/Fe_3O_4 composite catalysts for organic pollutant reduction. Herein, we prepared carbon-doped CuO/Fe_3O_4 composite catalysts via direct calcination of HKUST-1/Fe_3O_4/CMF composites under air. We then applied the as-prepared carbon-doped CuO/Fe_3O_4 composite catalysts for the catalytic reduction of 4-NP. Its catalytic performance, in comparison with a CuO/Fe_3O_4 composite from HKUST-1/Fe_3O_4 composite and CuO, is remarkably better, which is attributed to the fact that carbon doping can (1) minimize the aggregation of CuO/Fe_3O_4, (2) provide high surface-to-volume ratio and chemical stability for the catalyst in contact with the target pollutants, and (3) enhance the catalytic activity of the CuO/Fe_3O_4 catalyst. Furthermore, the carbon-doped CuO/Fe_3O_4 composite catalyst has excellent efficiency in the reduction of methylene blue (MB) and methyl orange (MO). The features of the carbon-doped CuO/Fe_3O_4 composite catalyst are as follows: CuO acts as the effective catalyst, with the doped carbon having the three functions discussed, and the magnetic Fe_3O_4 supports easy reuse/recycling of the catalyst using a magnet.

2. Experimental

2.1. Materials

Cellulose microfibrils (CMF) (~4.0 wt.%) from Cellulose Lab; trimesic acid (H_3BTC), copper (II) acetate monohydrate, ethanol, and hydrochloric acid (HCl, 37%, w/w) from Sigma-Aldrich; methylene blue, ferric nitrate nonahydrate ($Fe(NO_3)_3\cdot9H_2O$), ferrous sulfate ($FeSO_4\cdot7H_2O$), citric acid anhydrous, and sodium hydroxide (NaOH, 50%, w/w) from Fisher Scientific were used in this study. All other chemicals were of analytical grade and used without further purification.

2.2. Synthesis of Magnetic Carbon-Doped CuO/Fe₃O₄ Nanocomposite Catalyst

Fe_3O_4/cellulose microfibril (CMF) composites were prepared using chemical co-precipitation of aqueous ferrous and ferric ions, and the procedures were similar to those described in the literature [35]. CMF (7.5 g, ~4%, w/w) was diluted to 0.3 wt.% with distilled water and stirred for 5 min at 1000 rpm. The diluted CMF suspension was further treated in a sonicator (QSON-ICA) for 5 min to disperse the individual nanocellulose, followed by heating at 65 °C and bubbling with nitrogen for

10 min under magnetic stirring (500 rpm) to remove the dissolved oxygen from the CMF suspension. Then, citric acid (1 mg) and diluted HCl solution (1 mL, 2 mol/L) were added to the above solution, followed by the successive additions of 1.3 mmol $Fe(NO_3)_3 \cdot 9H_2O$ and 0.65 mmol $FeSO_4 \cdot 7H_2O$ solid samples with nitrogen bubbling and magnetic stirring. The function of citric acid was to protect the as-prepared Fe_3O_4 NPs from being oxidized by the dissolving oxygen in water. A dilute NaOH solution (10 mL, 2 mol/L) was added drop-wise into the above system, followed by bubbling with nitrogen and magnetic stirring (500 rpm) for 120 min at room temperature. The obtained Fe_3O_4/CMF composites (using a magnet) were washed three times with distilled water in a centrifugation step (4000 rpm, 10 min).

Copper (II) acetate monohydrate was dissolved in distilled water. Fe_3O_4/CMF nanocomposites and H_3BTC were dispersed in ethanol and then treated in a sonicator (QSON-ICA) for 5 min. The ethanol solution containing Fe_3O_4/CMF composite and H_3BTC was added to the aqueous system. The color of the solution changed from light blue to blue-black immediately, and the system was stirred continuously for 4 h. The final HKUST-1/Fe_3O_4/CMF composite was washed in a centrifugation step (4000 rpm, 10 min) with distilled water and absolute ethanol three times. Then, the drying was done at 50 °C under vacuum.

The synthesized HKUST-1/Fe_3O_4/CMF composites were placed in a ceramic boat, then transferred to a muffle roaster, which was operated under the conditions of 500 °C, air atmosphere, and a heating rate of 20 °C/min, before being held at 500 °C for 5 h; the procedures were similar to those described in the literature [36]. After cooling to room temperature, the sample was washed several times with distilled water and absolute ethanol in a centrifugation step (4000 rpm, 10 min), and finally dried in a vacuum at 60 °C for 5 h.

2.3. Characterization

Transmission electron microscope (TEM) analyses were conducted with a JEOL JEM 2011 transmission electron microscope at 200 kV. Scanning electron microscope (SEM) analyses were conducted with a JEOL JSM 6400 scanning electron microscope. The powder X-ray diffraction (PXRD) patterns of the prepared samples were collected using an X-ray diffractometer with Cu-Kα radiation (40 kV, 30 mA). The patterns were recorded in the region of 2θ from 10° to 80° with a scan step of 0.02°. The chemical binding energies of the respective ions in the samples were measured using X-ray photoelectron spectroscopy (XPS, ESCALa-b220i-XL electron spectrometer, Thermo Fisher Scientific K-Alpha, UK) under 300-W Al-Kα radiation. The magnetic properties were measured with a vibrating sample magnetometer (VSM), a physical property measurement system, at 300 K, as a function of the applied magnetic field between −80 and 80 kOe. The ultraviolet–visible (UV–Vis) diffuse reflectance data were collected with a UV–Vis spectrophotometer (Evolution 201, Thermo Scientific) equipped with an integrated sphere.

2.4. Evaluation of Catalytic Performance

The catalytic reduction of 4-nitrophenol (4-NP) using $NaBH_4$ was chosen as the model reaction for investigating the catalytic performance of magnetic carbon-doped CuO/Fe_3O_4 nanocomposite catalysts. Typically, catalytic reduction of a 30-mL 4-NP solution (1 mmol/L) was carried out in a beaker (100 mL) by continuously stirring at room temperature. Upon the addition of $NaBH_4$ (10 mM) into the 4-NP solution, its color changed immediately from light yellow to dark yellow due to the formation of 4-nitrophenolate ions (formed from the high alkalinity of $NaBH_4$) [37]. Then, the dark-yellow color faded with time (due to the conversion of 4-NP to 4-aminophenol (4-AP)) after the addition of carbon-doped CuO/Fe_3O_4 nanocomposite catalyst (5 mg) (Scheme 1). UV–Vis adsorption spectra were recorded using a UV–Vis spectrophotometer in the range of 250–550 nm. When the reaction completed, the catalyst was easily separated from the solution using an external magnet due to its good magnetic performance.

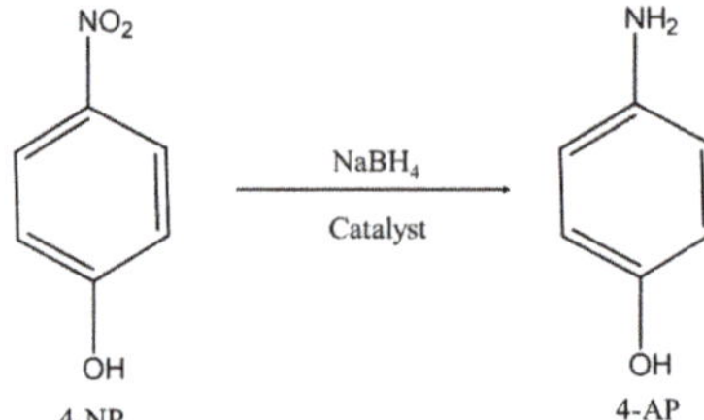

Scheme 1. Catalytic reduction of 4-nitrophenol to 4-aminophenol.

During the degradation process of organic dyes, 30 mL of dye solution (10 mg/L) and $NaBH_4$ (10 mM) were mixed in a beaker (100 mL) by continuously stirring at room temperature. Then, 5 mg of carbon-doped CuO/Fe_3O_4 nanocomposite catalyst was added to the reaction system, and the catalytic process was monitored using the UV–Vis spectrophotometer. When the reaction ended, the catalyst was readily separated from the solution using an external magnet.

3. Results and Discussion

3.1. Structure and Morphological Characterization

MOFs can be readily converted to metal-oxide composites, which take advantage of their original morphology and porosity. The as-prepared HKUST-1, HKUST-1/Fe_3O_4, and HKUST-1/Fe_3O_4/CMF composite samples were then calcinated to obtain the nanoporous metal-oxide particles, generating nanoporous CuO, CuO/Fe_3O_4, and carbon-doped CuO/Fe_3O_4 composite catalysts, respectively. Pyrolysis/thermolysis of HKUST-1/Fe_3O_4/CMF composites led to the formation of porous carbon-doped CuO/Fe_3O_4 composites.

The size and the morphology of the synthesized materials were investigated using a transmission electron microscope (TEM) and scanning electron microscope (SEM). Figure 1 shows the SEM and TEM images of HKUST-1/Fe_3O_4/CMF composites and carbon-doped CuO/Fe_3O_4 composites after calcination in air. HKUST-1 crystals and Fe_3O_4 nanoparticles were uniformly loaded onto the surface of CMF (Figure 1). The nanocellulose MOF-derived porous carbon-doped CuO/Fe_3O_4 composite was then obtained through a calcination process under air at 500 °C for 5 h. The Brunauer–Emmett–Teller surface area (S_{BET}) of the carbon-doped CuO/Fe_3O_4 sample was 38.7 m^2/g, which was much higher than that of the CuO/Fe_3O_4 sample (0.042 m^2/g) obtained from the calcination of HKUST-1/Fe_3O_4. The resultant carbon-doped CuO/Fe_3O_4 composite inherited largely the original porous morphology of HKUST-1. Based on the mapping images, the conclusion was that both Cu and Fe elements dispersed well in the carbon-doped CuO/Fe_3O_4 composite catalyst.

XPS analysis was employed to investigate the elemental composition on the surface of different composites. For the XPS spectrum of carbon-doped CuO/Fe_3O_4 composite catalyst (Figure 2a), the main peaks were C*1s*, O*1s*, Fe*2p*, and Cu*2p*, centered at around 285 eV, 530 eV, 720 eV, and 930 eV, respectively. The C*1s* XPS pattern of the sample is shown in Figure 2b. The spectrum contains two peaks at 285 eV and 288.5 eV, which may be attributed to carbon (sp^2 hybridization) in the sample, and the Cu–O–C bonds or Fe–O–C bonds, respectively. Figure 2c shows the Cu*2p* core-level XPS spectrum of the composite catalyst. The Cu*2p1* and Cu*2p3* binding energies for the composite catalyst were 952.8 and 932.7 eV, respectively, indicating the presence of CuO in the composite catalyst. Similar results were reported in the literature [38,39]. The Fe*2p3* and Fe*2p1* binding energies (Figure 2d) for the composite catalyst were 710.7 and 725.4 eV, respectively, which agrees with published results [35,40], confirming the presence of Fe_3O_4 in the composite catalyst.

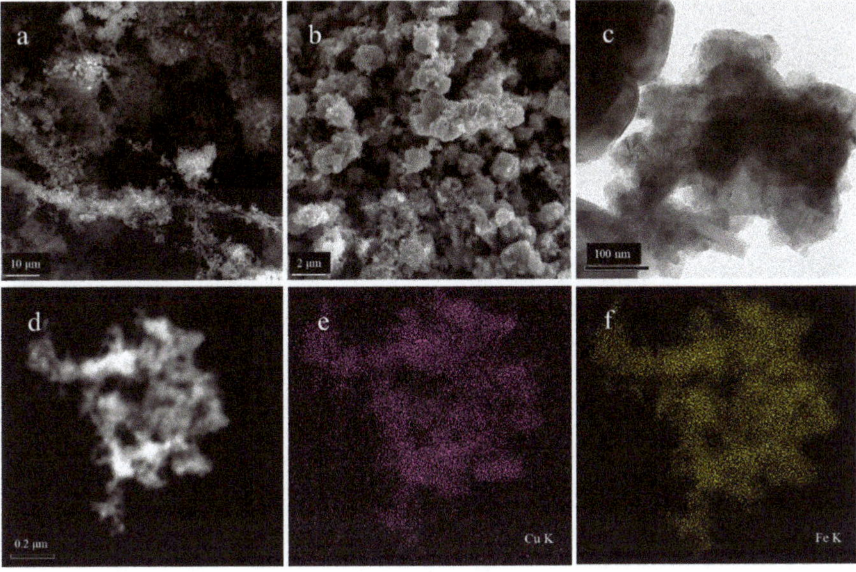

Figure 1. SEM images of (**a**) HKUST-1/Fe$_3$O$_4$/cellulose microfibril (CMF) composites and (**b**) carbon-doped CuO/Fe$_3$O$_4$ composite catalysts after calcination; TEM images of (**c**) carbon-doped CuO/Fe$_3$O$_4$ composite catalysts; (**d–f**) energy-dispersive X-ray spectroscopy (EDX) mapping of carbon-doped CuO/Fe$_3$O$_4$ composite catalysts.

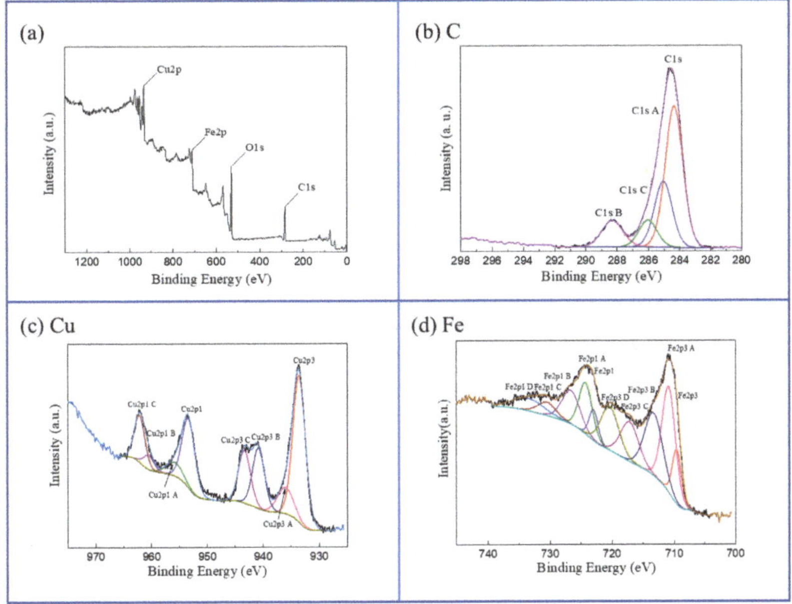

Figure 2. (**a**) High-resolution X-ray photoelectron spectroscopy (XPS) survey spectra of the carbon-doped CuO/Fe$_3$O$_4$ composite catalysts. (**b**) High-resolution XPS scans over C*1s* peaks of the carbon-doped CuO/Fe$_3$O$_4$ composite catalysts. (**c**) High-resolution XPS scans over Cu*2p* peaks of carbon-doped CuO/Fe$_3$O$_4$ composite catalysts. (**d**) High-resolution XPS scans over Fe*2p* peaks of carbon-doped CuO/Fe$_3$O$_4$ composite catalysts.

The crystalline nature and the composition of the as-synthesized products were characterized using PXRD. The crystalline phases of CuO, such as (110), (11-1), (111), (20-2), (020), (202), (11-3), (31-1), and (220) are shown in Figure 3a, which are consistent with those reported in the literature [41,42]. These results support the conclusion that HKUST-1 was transformed into CuO via calcination under the present conditions. In the XRD pattern of carbon-doped CuO/Fe_3O_4 composite, the diffraction peaks and relative intensities were indexed to Fe_3O_4 (JCPD NO. 19-0629) and CuO (JCPD NO. 05-0661). This indicates that CuO and Fe_3O_4 were both obtained via calcining HKUST-1/Fe_3O_4/CMF composites under air. No other C-related impurity peaks were detected; a similar result was reported in the previous research [43].

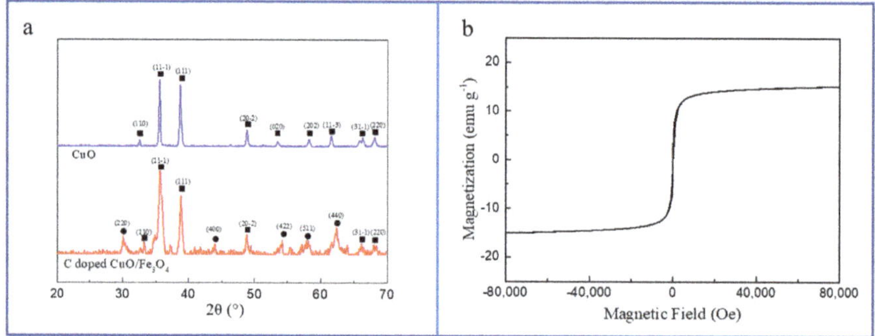

Figure 3. (**a**) Powder X-ray diffraction (PXRD) patterns of carbon-doped Fe_3O_4/CuO composite catalyst. (**b**) Magnetization curve at 300 K of carbon-doped Fe_3O_4/CuO composite catalyst.

The magnetic behavior of the carbon-doped Fe_3O_4/CuO composite catalyst sample was evaluated at 300 K. Its magnetization saturation value (Ms) was 15.1 emu·g^{-1}, suggesting a good magnetic property (Figure 3b). Thus, the magnetic carbon-doped Fe_3O_4/CuO composite catalyst could be readily separated from the reaction system using a magnet for the recycling process.

3.2. Catalytic Reduction

In this study, we investigated the application of carbon-doped CuO/Fe_3O_4 composite catalysts for the catalytic reduction of 4-nitrophenol. In Figure 4a, the results are shown for the catalytic reduction of 4-NP to 4-AP in the presence of $NaBH_4$ and carbon-doped CuO/Fe_3O_4 composite catalysts. The reduction process was followed based on the UV–Vis spectrophotometry. It shows that the absorbance at 400 nm (4-NP) decreased gradually as a function of time, while the absorbance at 290 nm (due to 4-AP) increased, confirming the catalytic reduction of 4-NP to 4-AP [44]. The catalytic reduction was almost complete within 10 min at room temperature, and the color of the solution changed from yellow to colorless. Similar results were also reported in the literature; Bordbar et al. [45] found that, using CuO/ZnO nanocomposites, catalytic reduction of 4-NP to 4-AP (using $NaBH_4$ as the reducing agent) was completed in several minutes.

The kinetics of calcined CMF, CuO, CuO/Fe_3O_4, and carbon-doped CuO/Fe_3O_4 composites are shown in Figure 4b. In the absence of catalyst, the reduction of 4-NP by $NaBH_4$ was negligible. In each case, the pseudo-first-order kinetic prevailed. The calcined CMF also had a negligible effect on the reduction of 4-NP. The rate constants (k) for CuO, CuO/Fe_3O_4, and carbon-doped CuO/Fe_3O_4 composite samples were 1.3×10^{-3} s^{-1}, 3.6×10^{-3} s^{-1} and 6.5×10^{-3} s^{-1}, respectively. In the case of CuO, the catalytic activity was the lowest due to the aggregation of CuO nanoparticles, while the catalytic efficiency of carbon-doped CuO/Fe_3O_4 composites was much better than that of CuO/Fe_3O_4, demonstrating that carbon doping is effective for enhancing the catalytic activity of the catalysts.

The catalytic reduction of 4-NP by $NaBH_4$ using metal-oxide nanoparticles (CuO) has two steps [46]: (1) borohydride ions are adsorbed onto the nanoparticle surface, forming active surface-hydrogen, while 4-NP is also adsorbed onto the nanoparticle surface; (2) active hydrogen attacks the positively charged nitrogen in the nitro group of 4-NP, followed by the addition of two hydrogen atoms, producing 4-AP.

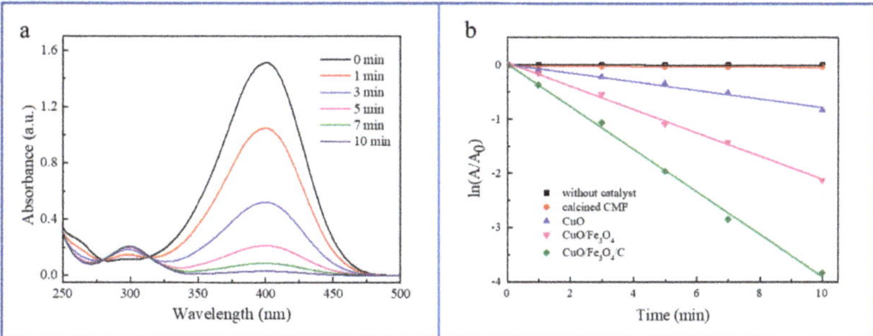

Figure 4. (**a**) Time-dependent ultraviolet–visible (UV–Vis) absorption spectra of 4-nitrophenol (4-NP) reduced by $NaBH_4$ in the presence of carbon-doped CuO/Fe_3O_4 composite catalysts; (**b**) the first-order kinetic plot (absorbance at 400 nm, $\ln(A_t/A_0)$) versus reaction time for the reduction of 4-nitrophenol; A_t and A_0 represent the absorbance values of 4-NP at 400 nm at designated time t and t = 0, respectively.

Cationic and anionic organic dyes were chosen to further investigate the catalytic properties of carbon-doped CuO/Fe_3O_4 composite catalyst. As shown in Figure 5a, for cationic dye (methylene blue), in the presence of $NaBH_4$ and carbon-doped CuO/Fe_3O_4 composite catalyst, the absorbance at 660 nm (MB) gradually decreased as a function of time; furthermore, the catalytic reduction was completed within 6 min at room temperature (the color of the solution was colorless).

The results from methyl orange (an anionic dye) are shown in Figure 5c. Under otherwise the same conditions, the color change (from orange to colorless) was slower than that for MB (Figure 5a, from blue to colorless). The pseudo-first-order rate law was also valid here (Figure 5d). For the carbon-doped CuO/Fe_3O_4 composite catalyst, the rate constant (k) was 2.4×10^{-3} s^{-1} for MO, while it was 12.9×10^{-3} s^{-1} for MB.

We compared the catalytic performance of the carbon-doped CuO/Fe_3O_4 composite catalyst for the reduction of 4-NP, MB, and MO, with other related ones from the literature (Table 1). As shown, the carbon-doped CuO/Fe_3O_4 composite catalyst showed much improved results, and the pseudo-first-order rate constant (k) for the nanocomposite catalyst from the present study was indeed consistently higher than that reported in the literature. The improved results may be attributed to the unique original morphologies associated with HKUST-1.

Table 1. Comparison of the catalytic performance for the reduction of 4-nitrophenol, methylene blue (MB), and methyl orange (MO) using the carbon-doped CuO/Fe_3O_4 composite catalyst and other reported catalysts in the literature. NP—nanoparticle.

Pollutant	Samples	Concentration of Pollutant	Time (min)	k (s^{-1})	Reference
4-Nitrophenol	Agnp-PSAC	0.02 mM	25	3.9×10^{-3}	[47]
	CF-AuNPs	1 mM	11	0.17×10^{-3}	[48]
	CuO	0.1 mM	32	5.8×10^{-3}	[49]
	$CuO/Fe_3O_4/C$	1 mM	10	6.5×10^{-3}	This work
Methylene Blue	$Au/Fe_3O_4@C$	0.01 mM	10	5.5×10^{-3}	[50]
	AuNPs	1 mM	9	4.0×10^{-3}	[51]
	$Fe_3O_4@$polydopamine-Ag	0.1 mM	-	7.2×10^{-3}	[52]
	$CuO/Fe_3O_4/C$	0.03 mM	6	12.9×10^{-3}	This work
Methyl Orange	$Ag-\gamma-Fe_2O_3$	0.3 mM	30	1.53×10^{-3}	[53]
	Ag-CuO	0.006 mM	120	0.3×10^{-3}	[54]
	Ag@Fe	0.1 mM	14	1.6×10^{-3}	[55]
	$CuO/Fe_3O_4/C$	0.03 mM	25	2.4×10^{-3}	This work

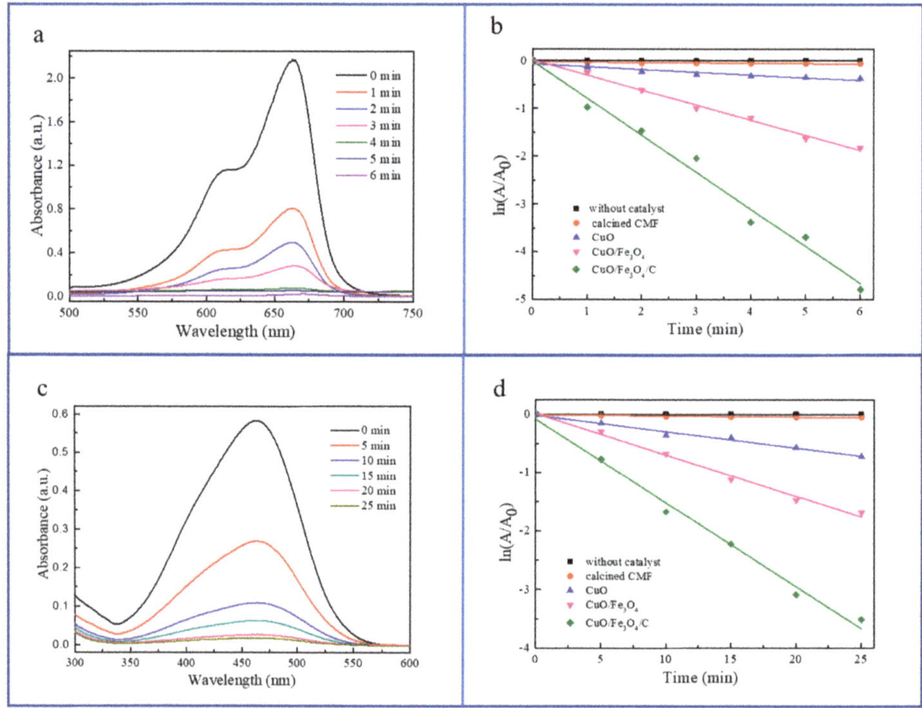

Figure 5. (**a**) Time-dependent UV–Vis absorption spectra of methylene blue (MB) with the carbon-doped CuO/Fe_3O_4 composite catalyst in the presence of $NaBH_4$; (**b**) the corresponding first-order kinetic plot (absorbance at 660 nm, $\ln(A_t/A_0)$) versus reaction time for the reduction of MB; A_t and A_0 represent the absorbance of MB (660 nm) at designated time t and t = 0, respectively; (**c**) time-dependent UV–Vis absorption spectra of methyl orange (MO) with the carbon-doped CuO/Fe_3O_4 composite catalyst in the presence of $NaBH_4$; (**d**) the corresponding first-order kinetic plot (absorbance at 460 nm, $\ln(A_t/A_0)$) versus reaction time for the reduction of MO; A_t and A_0 represent the absorbance of MO (460 nm) at designated time t and t = 0, respectively.

From the viewpoint of practical application, the recycling/reuse of the catalyst is of critical importance. In the present study, after the catalytic degradation experiments, the magnetic carbon-doped CuO/Fe_3O_4 composite catalyst was readily separated from the reaction system using an external magnet. The used catalysts were collected, and rinsed with distilled water several times. After a thorough washing process, the recovered magnetic catalyst was reused in the subsequent run of catalytic reduction of 4-NP under identical conditions, and the same process was repeated five times. The results are shown in Figure 6. The catalytic performance of the magnetic carbon-doped CuO/Fe_3O_4 composite catalyst decreased only slightly (the 4-NP reduction ratio decreased from 100% to 96%) after five cycles. Similar results were obtained in the reuse/recycling experiments of the as-prepared magnetic carbon-doped CuO/Fe_3O_4 composite catalyst during the catalytic reduction of MB and MO. Therefore, the as-prepared magnetic carbon-doped CuO/Fe_3O_4 composite catalyst is a promising system for practical applications.

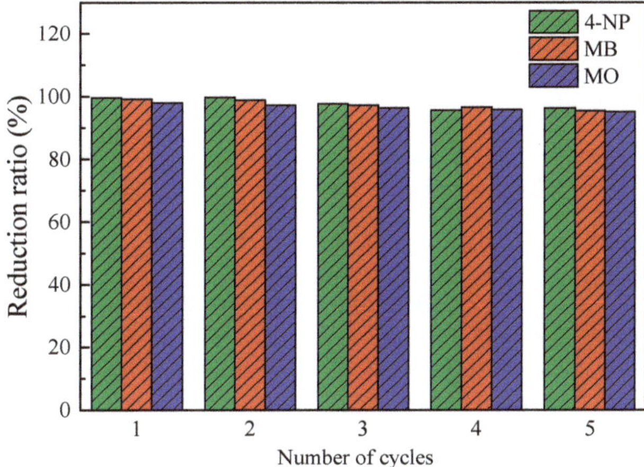

Figure 6. Reduction conversion ratio of 4-NP, MB, and MO after five successive cycles using the carbon-doped CuO/Fe_3O_4 catalyst.

4. Conclusions

In this study, a nano-cellulose/MOF-derived carbon-doped CuO/Fe_3O_4 composite catalyst was successfully fabricated through pyrolysis/thermolysis of the HKUST-1/Fe_3O_4/CMF composite. The resultant carbon-doped CuO/Fe_3O_4 composite catalyst took advantage of the original porous morphology of HKUST-1; consequently, the carbon-doped CuO/Fe_3O_4 composite catalyst exhibited high catalytic activity for the reduction of 4-NP and organic dyes (MB and MO). In addition, the carbon-doped CuO/Fe_3O_4 composite catalyst showed good reusability/recyclability after five cycles. Notably, this strategy can be extended to the preparation of other functional MOF-based derivatives.

Author Contributions: Conceptualization, H.L. and L.Z.; Methodology, H.L.; Software, H.L. and N.A.; Validation, N.A. and J.M.; Formal Analysis, H.L. and L.Z.; Investigation, H.L.; Resources, X.Z. and Y.N.; Data Curation, J.M.; Writing-Original Draft Preparation, H.L.; Writing-Review & Editing, X.Z. and Y.N.; Visualization, L.Z. and J.M.; Supervision, Y.N.; Project Administration, X.Z. and Y.N.; Funding Acquisition, X.Z. and Y.N.

Funding: This research was funded by the National Natural Science Foundation of China (31570576), the financial support from the Natural Science Foundation of Jiangsu Provincial University (16KJA220005), the Canada Research Chairs program of the Government of Canada, the Postgraduate Research and Practice Innovation Program of Jiangsu Province (KYCX17_0835), the Doctorate Fellowship Foundation of Nanjing Forestry University, and the Priority Academic Program Development (PAPD) of Jiangsu Higher Education Institutions. The APC was funded by the Natural Science Foundation of Jiangsu Provincial University (16KJA220005).

Acknowledgments: The authors acknowledge Mike Johnson (Department of Chemistry and Institute for Research in Materials, Dalhousie University, Halifax, NS, Canada) for the acquisition and interpretation of the PPMS data.

Conflicts of Interest: The authors declare no conflicts of interest.

References

1. Almasi, H.; Jafarzadeh, P.; Mehryar, L. Fabrication of novel nanohybrids by impregnation of CuO nanoparticles into bacterial cellulose and chitosan nanofibers: Characterization, antimicrobial and release properties. *Carbohydr. Polym.* **2018**, *186*, 273–281. [CrossRef] [PubMed]
2. Palomo, J.M.; Filice, M. Biosynthesis of Metal Nanoparticles: Novel Efficient Heterogeneous Nanocatalysts. *Nanomaterials* **2016**, *6*, 84. [CrossRef] [PubMed]
3. Tran, C.D.; Makuvaza, J.; Munson, E.; Bennett, B. Biocompatible Copper Oxide Nanoparticle Composites from Cellulose and Chitosan: Facile Synthesis, Unique Structure, and Antimicrobial Activity. *ACS Appl. Mater. Interfaces* **2017**, *9*, 42503–42515. [CrossRef] [PubMed]

4. Müller, A.; Peglow, S.; Karnahl, M.; Kruth, A.; Junge, H.; Brüser, V.; Scheu, C. Morphology, optical properties and photocatalytic activity of photo-and plasma-deposited au and Au/Ag core/shell nanoparticles on titania layers. *Nanomaterials* **2018**, *8*, 502. [CrossRef] [PubMed]

5. Corma, A.; Garcia, H. Supported gold nanoparticles as catalysts for organic reactions. *Chem. Soc. Rev.* **2008**, *37*, 2096–2126. [CrossRef] [PubMed]

6. Gangula, A.; Podila, R.; Karanam, L.; Janardhana, C.; Rao, A.M. Catalytic Reduction of 4-Nitrophenol using Biogenic Gold and Silver Nanoparticles Derived from *Breynia rhamnoides*. *Langmuir* **2011**, *27*, 15268–15274. [CrossRef] [PubMed]

7. Huang, Z.; Cui, F.; Kang, H.; Chen, J.; Zhang, X.; Xia, C. Highly Dispersed Silica-Supported Copper Nanoparticles Prepared by Precipitation−Gel Method: A Simple but Efficient and Stable Catalyst for Glycerol Hydrogenolysis. *Chem. Mater.* **2008**, *20*, 5090–5099. [CrossRef]

8. Ma, J.; Sun, Z.; Wang, Z.; Zhou, X. Preparation of ZnO−cellulose nanocomposites by different cellulose solution systems with a colloid mill. *Cellulose* **2016**, *23*, 3703–3715. [CrossRef]

9. Aguirre, M.E.; Zhou, R.; Eugene, A.J.; Guzman, M.I.; Grela, M.A. Cu_2O/TiO_2 heterostructures for CO_2 reduction through a direct Z-scheme: Protecting Cu_2O from photocorrosion. *Appl. Catal. B Environ.* **2017**, *217*, 485–493. [CrossRef]

10. Zhou, R.; Guzman, M.I. Photocatalytic Reduction of Fumarate to Succinate on ZnS Mineral Surfaces. *J. Phys. Chem. C* **2016**, *120*, 7349–7357. [CrossRef]

11. Jiang, D.; Zhang, Y.; Li, X. Synergistic effects of CuO and Au nanodomains on Cu_2O cubes for improving photocatalytic activity and stability. *Chin. J. Catal.* **2019**, *40*, 105–113. [CrossRef]

12. Rodríguez, N.A.; Savateev, A.; Grela, M.A.; Dontsova, D. Facile Synthesis of Potassium Poly(heptazine imide) (PHIK)/Ti-Based Metal–Organic Framework (MIL-125-NH2) Composites for Photocatalytic Applications. *ACS Appl. Mater. Interfaces* **2017**, *9*, 22941–22949. [CrossRef] [PubMed]

13. Hu, J.; Chen, G.; Lo, I.M.C. Removal and recovery of Cr(VI) from wastewater by maghemite nanoparticles. *Water Res.* **2005**, *39*, 4528–4536. [CrossRef] [PubMed]

14. Simeonidis, K.; Mourdikoudis, S.; Kaprara, E.; Mitrakas, M.; Polavarapu, L. Inorganic engineered nanoparticles in drinking water treatment: A critical review. *Environ. Sci. Water Res. Technol.* **2016**, *2*, 43–70. [CrossRef]

15. Keller, A.A.; Wang, H.; Zhou, D.; Lenihan, H.S.; Cherr, G.; Cardinale, B.J.; Miller, R.; Ji, Z. Stability and Aggregation of Metal Oxide Nanoparticles in Natural Aqueous Matrices. *Environ. Sci. Technol.* **2010**, *44*, 1962–1967. [CrossRef] [PubMed]

16. Marambio-Jones, C.; Hoek, E.M.V. A review of the antibacterial effects of silver nanomaterials and potential implications for human health and the environment. *J. Nanopart. Res.* **2010**, *12*, 1531–1551. [CrossRef]

17. Su, X.; Kanjanawarut, R. Control of Metal Nanoparticles Aggregation and Dispersion by PNA and PNA−DNA Complexes, and Its Application for Colorimetric DNA Detection. *ACS Nano* **2009**, *3*, 2751–2759. [CrossRef]

18. Zhang, Q.; Lee, I.; Ge, J.; Zaera, F.; Yin, Y. Surface-Protected Etching of Mesoporous Oxide Shells for the Stabilization of Metal Nanocatalysts. *Adv. Funct. Mater.* **2010**, *20*, 2201–2214. [CrossRef]

19. Zhu, Q.-L.; Li, J.; Xu, Q. Immobilizing Metal Nanoparticles to Metal–Organic Frameworks with Size and Location Control for Optimizing Catalytic Performance. *J. Am. Chem. Soc.* **2013**, *135*, 10210–10213. [CrossRef]

20. Bak, W.; Soo Kim, H.; Chun, H.; Cheol Yoo, W. Facile synthesis of metal/metal oxide nanoparticles inside a nanoporous carbon matrix (M/MO@C) through the morphology-preserved transformation of metal–organic framework. *Chem. Commun.* **2015**, *51*, 7238–7241. [CrossRef]

21. Xia, W.; Zou, R.; An, L.; Xia, D.; Guo, S. A metal–organic framework route to in situ encapsulation of $Co@Co_3O_4@C$ core@bishell nanoparticles into a highly ordered porous carbon matrix for oxygen reduction. *Energy Environ. Sci.* **2015**, *8*, 568–576. [CrossRef]

22. Ennas, G.; Falqui, A.; Marras, S.; Sangregorio, C.; Marongiu, G. Influence of Metal Content on Size, Dispersion, and Magnetic Properties of Iron−Cobalt Alloy Nanoparticles Embedded in Silica Matrix. *Chem. Mater.* **2004**, *16*, 5659–5663. [CrossRef]

23. Obena, R.P.; Lin, P.C.; Lu, Y.W.; Li, I.C.; del Mundo, F.; Arco, S.D.; Nuesca, G.M.; Lin, C.-C.; Chen, Y.-J. Iron Oxide Nanomatrix Facilitating Metal Ionization in Matrix-Assisted Laser Desorption/Ionization Mass Spectrometry. *Anal. Chem.* **2011**, *83*, 9337–9343. [CrossRef] [PubMed]

24. Kaushik, M.; Moores, A. Review: Nanocelluloses as versatile supports for metal nanoparticles and their applications in catalysis. *Green Chem.* **2016**, *18*, 622–637. [CrossRef]

25. Budarin, V.L.; Clark, J.H.; Luque, R.; Macquarrie, D.J.; White, R.J. Palladium nanoparticles on polysaccharide-derived mesoporous materials and their catalytic performance in C–C coupling reactions. *Green Chem.* **2008**, *10*, 382–387. [CrossRef]

26. Zhang, Z.; Wang, Z. Diatomite-Supported Pd Nanoparticles: An Efficient Catalyst for Heck and Suzuki Reactions. *J. Org. Chem.* **2006**, *71*, 7485–7487. [CrossRef]

27. Niu, H.; Liu, S.; Cai, Y.; Wu, F.; Zhao, X. MOF derived porous carbon supported Cu/Cu$_2$O composite as high performance non-noble catalyst. *Microporous Mesoporous Mater.* **2016**, *219*, 48–53. [CrossRef]

28. Burtch, N.C.; Jasuja, H.; Walton, K.S. Water Stability and Adsorption in Metal–Organic Frameworks. *Chem. Rev.* **2014**, *114*, 10575–10612. [CrossRef]

29. Duan, C.; Meng, J.; Wang, X.; Meng, X.; Sun, X.; Xu, Y.; Zhao, W.; Ni, Y. Synthesis of novel cellulose-based antibacterial composites of Ag nanoparticles@ metal-organic frameworks@ carboxymethylated fibers. *Carbohydr. Polym.* **2018**, *193*, 82–88. [CrossRef]

30. Leo, P.; Orcajo, G.; Briones, D.; Calleja, G.; Sánchez-Sánchez, M.; Martínez, F. A recyclable Cu-MOF-74 catalyst for the ligand-free O-arylation reaction of 4-nitrobenzaldehyde and phenol. *Nanomaterials* **2017**, *7*, 149. [CrossRef]

31. Shen, K.; Chen, X.; Chen, J.; Li, Y. Development of MOF-Derived Carbon-Based Nanomaterials for Efficient Catalysis. *ACS Catal.* **2016**, *6*, 5887–5903. [CrossRef]

32. Li, H.; Ke, F.; Zhu, J. MOF-derived ultrathin cobalt phosphide nanosheets as efficient bifunctional hydrogen evolution reaction and oxygen evolution reaction electrocatalysts. *Nanomaterials* **2018**, *8*, 89. [CrossRef] [PubMed]

33. Ji, W.; Xu, Z.; Liu, P.; Zhang, S.; Zhou, W.; Li, H.; Zhang, T.; Li, L.; Lu, X.; Wu, J.; et al. Metal–Organic Framework Derivatives for Improving the Catalytic Activity of the CO Oxidation Reaction. *ACS Appl. Mater. Interfaces* **2017**, *9*, 15394–15398. [CrossRef] [PubMed]

34. Yang, S.J.; Im, J.H.; Kim, T.; Lee, K.; Park, C.R. MOF-derived ZnO and ZnO@C composites with high photocatalytic activity and adsorption capacity. *J. Hazard. Mater.* **2011**, *186*, 376–382. [CrossRef] [PubMed]

35. An, X.; Cheng, D.; Dai, L.; Wang, B.; Ocampo, H.J.; Nasrallah, J.; Jia, X.; Zou, J.; Long, Y.; Ni, Y. Synthesis of nano-fibrillated cellulose/magnetite/titanium dioxide (NFC@Fe$_3$O$_4$@TNP) nanocomposites and their application in the photocatalytic hydrogen generation. *Appl. Catal. B Environ.* **2017**, *206*, 53–64. [CrossRef]

36. Zhang, Y.-F.; Qiu, L.-G.; Yuan, Y.-P.; Zhu, Y.-J.; Jiang, X.; Xiao, J.-D. Magnetic Fe$_3$O$_4$@C/Cu and Fe$_3$O$_4$@CuO core–shell composites constructed from MOF-based materials and their photocatalytic properties under visible light. *Appl. Catal. B Environ.* **2014**, *144*, 863–869. [CrossRef]

37. An, X.; Long, Y.; Ni, Y. Cellulose nanocrystal/hexadecyltrimethylammonium bromide/silver nanoparticle composite as a catalyst for reduction of 4-nitrophenol. *Carbohydr. Polym.* **2017**, *156*, 253–258. [CrossRef] [PubMed]

38. Boruban, C.; Esenturk, E.N. Activated carbon-supported CuO nanoparticles: A hybrid material for carbon dioxide adsorption. *J. Nanopart. Res.* **2018**, *20*, 59. [CrossRef]

39. Gu, C.; Qi, R.; Wei, Y.; Zhang, X. Preparation and performances of nanorod-like inverse CeO$_2$–CuO catalysts derived from Ce-1,3,5-Benzene tricarboxylic acid for CO preferential oxidation. *React. Kinet. Mech. Catal.* **2018**, *124*, 651–667. [CrossRef]

40. Dong, X.-L.; Mou, X.-Y.; Ma, H.-C.; Zhang, X.-X.; Zhang, X.-F.; Sun, W.-J.; Ma, C.; Xue, M. Preparation of CdS–TiO$_2$/Fe$_3$O$_4$ photocatalyst and its photocatalytic properties. *J. Sol-Gel Sci. Technol.* **2013**, *66*, 231–237. [CrossRef]

41. Avgouropoulos, G.; Ioannides, T.; Matralis, H. Influence of the preparation method on the performance of CuO–CeO$_2$ catalysts for the selective oxidation of CO. *Appl. Catal. B Environ.* **2005**, *56*, 87–93. [CrossRef]

42. Kumar, R.V.; Diamant, Y.; Gedanken, A. Sonochemical Synthesis and Characterization of Nanometer-Size Transition Metal Oxides from Metal Acetates. *Chem. Mater.* **2000**, *12*, 2301–2305. [CrossRef]

43. Zhang, H.; Zheng, L.; Ouyang, X.; Ni, Y. Carbon doping of Ti$_{0.91}$Co$_{0.03}$La$_{0.06}$O$_2$ nanoparticles for enhancing room-temperature ferromagnetism using carboxymethyl cellulose as carbon source. *Ceram. Int.* **2018**, *44*, 15754–15763. [CrossRef]

44. Li, J.; Liu, C.; Liu, Y. Au/graphene hydrogel: Synthesis, characterization and its use for catalytic reduction of 4-nitrophenol. *J. Mater. Chem.* **2012**, *22*, 8426–8430. [CrossRef]

45. Bordbar, M.; Negahdar, N.; Nasrollahzadeh, M. Melissa Officinalis L. leaf extract assisted green synthesis of CuO/ZnO nanocomposite for the reduction of 4-nitrophenol and Rhodamine B. *Sep. Purif. Technol.* **2018**, *191*, 295–300. [CrossRef]

46. You, J.-G.; Shanmugam, C.; Liu, Y.-W.; Yu, C.-J.; Tseng, W.-L. Boosting catalytic activity of metal nanoparticles for 4-nitrophenol reduction: Modification of metal naoparticles with poly(diallyldimethylammonium chloride). *J. Hazard. Mater.* **2017**, *324*, 420–427. [CrossRef] [PubMed]

47. Nabil, B.; Ahmida, E.A.; Christine, C.; Julien, V.; Abdelkrim, A. Polyfunctional cotton fabrics with catalytic activity and antibacterial capacity. *Chem. Eng. J.* **2018**, *351*, 328–339. [CrossRef]

48. Bouazizi, N.; El achari, A.; Campagne, C.; Vieillard, J.; Azzouz, A. Copper oxide coated polyester fabrics with enhanced catalytic properties towards the reduction of 4-nitrophenol. *J. Mater. Sci. Mater. Electron.* **2018**, *29*, 10802–10813. [CrossRef]

49. Islam, M.T.; Dominguez, N.; Ahsan, M.A.; Dominguez-Cisneros, H.; Zuniga, P.; Alvarez, P.J.J.; Noveron, J.C. Sodium rhodizonate induced formation of gold nanoparticles supported on cellulose fibers for catalytic reduction of 4-nitrophenol and organic dyes. *J. Environ. Chem. Eng.* **2017**, *5*, 4185–4193. [CrossRef]

50. Gan, Z.; Zhao, A.; Zhang, M.; Tao, W.; Guo, H.; Gao, Q.; Mao, R.; Liu, E. Controlled synthesis of Au-loaded Fe$_3$O$_4$@C composite microspheres with superior SERS detection and catalytic degradation abilities for organic dyes. *Dalton Trans.* **2013**, *42*, 8597–8605. [CrossRef]

51. Afshar, S.; Samari Jahromi, H.; Jafari, N.; Ahmadi, Z.; Hakamizadeh, M. Degradation of malachite green oxalate by UV and visible lights irradiation using Pt/TiO$_2$/SiO$_2$ nanophotocatalyst. *Sci. Iran.* **2011**, *18*, 772–779. [CrossRef]

52. Ganapuram, B.R.; Alle, M.; Dadigala, R.; Dasari, A.; Maragoni, V.; Guttena, V. Catalytic reduction of methylene blue and Congo red dyes using green synthesized gold nanoparticles capped by salmalia malabarica gum. *Int. Nano Lett.* **2015**, *5*, 215–222. [CrossRef]

53. Sallam, S.A.; El-Subruiti, G.M.; Eltaweil, A.S. Facile Synthesis of Ag–γ-Fe$_2$O$_3$ Superior Nanocomposite for Catalytic Reduction of Nitroaromatic Compounds and Catalytic Degradation of Methyl Orange. *Catal. Lett.* **2018**, *148*, 3701–3714. [CrossRef]

54. Momeni, M.M.; Mirhosseini, M.; Mohammadi, N. Efficient photo catalytic degradation of methyl orange over Ag–CuO nanostructures grown on copper foil under visible light irradiation. *J. Mater. Sci. Mater. Electron.* **2016**, *27*, 6542–6551. [CrossRef]

55. Alzahrani, S.A.; Malik, M.A.; Al-Thabaiti, S.A.; Khan, Z. Seedless synthesis and efficient recyclable catalytic activity of Ag@Fe nanocomposites towards methyl orange. *Appl. Nanosci.* **2018**, *8*, 255–271. [CrossRef]

Article

Advancing the Use of High-Performance Graphene-Based Multimodal Polymer Nanocomposite at Scale

Ibrahim A. Ahmad [1], Krzysztof K. K. Koziol [2], Suleyman Deveci [3], Hyun-Kyung Kim [1,4,*] and Ramachandran Vasant Kumar [1,*]

[1] Department of Materials Science and Metallurgy, University of Cambridge, 27 Charles Babbage Rd, Cambridge CB3 0FS, UK; iaiaa2@cam.ac.uk

[2] Enhanced Composites and Structures Centre, School of Aerospace, Transport and Manufacturing, Cranfield University, Cranfield MK43 0AL, UK; k.koziol@cranfield.ac.uk

[3] Innovation Centre, Borouge Pte. Ltd., PO Box 6951, Abu Dhabi, UAE; suleyman.deveci@borouge.com

[4] Gwangju Bio/Energy R&D Center, Korea Institute of Energy Research (KIER), 270-25 Samso-ro, Buk-gu, Gwangju 61003, Korea

* Correspondence: hkk28@cam.ac.uk (H.-K.K.); rvk10@cam.ac.uk (R.V.K.); +44 (0)1223 334327

Received: 16 October 2018; Accepted: 14 November 2018; Published: 17 November 2018

Abstract: The production of an innovative, high-performance graphene-based polymer nanocomposite using cost-effective techniques was pursued in this study. Well-dispersed and uniformly distributed graphene platelets within a polymer matrix, with strong interfacial bonding between the platelets and the matrix, provided an optimal nanocomposite system for industrial interest. This study reports on the reinforcement of high molecular weight multimodal-high-density polyethylene reinforced by a microwave-induced plasma graphene, using melt intercalation. The tailored process included designing a suitable screw configuration, paired with coordinating extruder conditions and blending techniques. This enabled the polymer to sufficiently degrade, predominantly through thermomechanical-degradation, as well as thermo-oxidative degradation, which subsequently created a suitable medium for the graphene sheets to disperse readily and distribute evenly within the polymer matrix. Different microscopy techniques were employed to prove the effectiveness. This was then qualitatively assessed by Raman spectroscopy, X-ray diffraction, rheology, mechanical testing, density measurements, thermal expansion, and thermogravimetric analysis, confirming both the originality as well as the effectiveness of the process.

Keywords: graphene; multimodal-high density polyethylene; melt extrusion; polymer; nanocomposite, polymer degradation; dispersion and distribution of graphene

---

1. Introduction

Multimodal high-density polyethylene (HDPE) is an engineered thermoplastic semi-crystalline polymer, which is widely used in automotive, films, pressure pipes and fittings, bottles, tubes, and cables jacketing [1–5]. It is a hybrid of at least two distinct polyethylene components, wherein each constituent has a different density and different molecular weight fractions [1–5]. This allows flexibility in engineering its microstructure to meet the desired balance of properties for concrete practical applications. Nevertheless, multimodal HDPE can be further improved, for example, with the addition of fillers or reinforcements, in order to overcome deficiencies in their mechanical or thermal properties [5–18]. It is feasible that a substantial benefit could be attained by strengthening the multimodal-hydrophobic polymers with graphene (g), deriving new and unique nanocomposite properties [5–22]. However, a proper dispersion and distribution of graphene platelets within the nonpolar polyolefin matrix is still a major challenge [6,15–22]. Irreversible agglomerates are formed through the van der Waals forces between the 2D-platelets, as the large surface area of the graphene platelets leads to the creation of interfacial regions, causing them to spontaneously restack themselves [4,6,15–21]. This creates defects which behave as voids, introducing degradation into the polymer matrices. Though numerous methods for producing polymer-graphene nanocomposites have recently emerged, each method is limited by its compatibility with only certain types of graphene and polymers, requiring chemical modifications on both constituents of the nanocomposite [6,15–21]. The fabrication techniques can also alter the pristineness of graphene by introducing structural defects on the graphene basal plane [15–22]. In addition, the production utilizes large amounts of solvent and supplementary chemicals, which incurs higher costs, as well as raising environmental, health, and safety concerns [15–22].

It is therefore the topic of this study to introduce a more cost-effective, optimal way of fabricating a nanocomposite of high molecular weight multimodal-HDPE matrix, reinforced with a bottom-up graphene. These polymers are indeed widely used in a long-term application in an extreme environment, which includes hydrostatic, thermal, and environmental stresses [5,21,23]. Herein we report a novel method for the preparation of high-performance polymer-graphene nanocomposite (PE-g) via melt intercalation, using a co-rotating intermeshing twin-screw extruder. Depicted in Figure 1 is a simple schematic diagram of the fabrication method followed in the present study.

In the present work, we attempted to degrade the polymer to a sufficient level, through thermo-oxidative, as well as thermo-mechanical degradation during the melt extrusion process. This created a compatible medium for the graphene to disperse and distribute thoroughly within the polymer matrix. The polymer is consequently able to interact physically or chemically with the residual oxygen functional groups at the graphene surface which contains almost 5% oxygens, or through the short molecules introduced by thermo-mechanical degradation, with defective sp^3 functional group on either the surface, or at the edge of the graphene sheets. Accordingly, a better stress transfer can potentially be achieved through the strong interfacial bonding created between graphene platelets and the polymer matrix. Achieving a thorough dispersion and distribution of graphene within the multimodal-HDPE, by melt intercalation, via co-rotating intermeshing twin-screw extruder, has never yet been reported according to the authors' knowledge. The results of this research provide greater insight into different melt intercalation factors, affecting the multimodal HDPE-graphene nanocomposite performance and criterion for effectively producing the next generation of black multimodal-polyethylene compounds for use in high-pressure pipes, automotive, and energy cable applications [5–21,23].

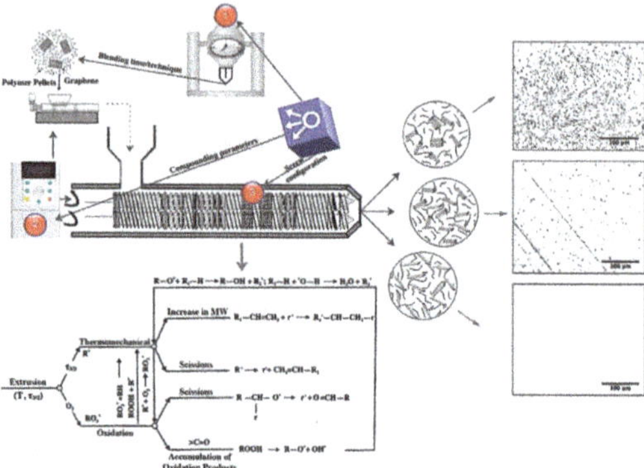

Figure 1. Simple schematic representation of the method followed in the present study. Gol'dberg–Zaikov model represents the general reaction mechanisms of all the thermo-mechanical and thermo-oxidative degradations that can occur during the melt extrusion of a polyethylene [24,25]. R represents the side chain of any hydrocarbon functional groups, r is the very short side chain of any hydrocarbon functional groups, MW is the molecular weight, denotes an active free radical site, and $\tau_{3/2}$ is the shear stress in x and y directions.

2. Experimental

2.1. Materials

Unstabilized high-density polyethylene powders, produced with Ziegler Natta catalyst via proprietary Borstar process, (Borouge, United Arab Emirates), with a melt flow rate of about 7.5 g/10 min (190 °C, 21.6 kg), Mw = 280 kg/mole, Mn = 8.49 kg/mole, Mw/Mn = 33, and a density of 950 kg/m^3 were used in this study. The multimodal high-density polyethylene matrix used in the present study was engineered specifically for nanocomposite applications. According to the production process technology, the melt flow rate (MFR) of the polymer gives an indication that the split MFR ratio between the reactors is significantly high [1–4]. The antioxidants' masterbatch containing Irganox 1010 and Irgafos 168 were added to the polymers at 0.5 wt.% for optimum stabilization during processing. Graphene was supplied by the FGV Cambridge Nanosystems Ltd. (Cambridge, United Kingdom), with ≥95% carbon purity, bulk density of 0.0266 g/ml, thickness < 1.0 nm, and flake size range of 150–500 nm. Carbon Black powder was provided by Orion Engineered Carbons GmbH (Frankfurt am Main, Germany), with ≥92 cc/100g oil absorption number, ash content of 0.10%, sulphur content of 0.10%, tint strength of 103%, average primary particle size of 20 nm, and a density of 1.7-1.9 g/cm^3 at 20 °C.

2.2. Nanocomposite Preparation

The graphene-based multimodal-HDPE nanocomposites (PE-g) were prepared via melt intercalation using a Coperion ZSK 18 twin extruder, with a screw diameter of 18 mm and a barrel length of 720 mm (L/D = 40). The screw rotation speed (rpm) was 600 min-1, barrel temperature profile was in the range of 170-240 °C (see Figure 2), and feed rate was between 1-2 kg/hr. Both the graphene and dry polyethylene powders were fed separately into the extruder via a spiral flow screw Brabender ISC-CM plus feeder. The nanofillers were fed at 0.1, 0.5, 1, 2, and 5 wt.% loadings. In order to prevent the polymer from severe degradation, an antioxidant masterbatch was simultaneously added through a side feeder, with the total loading of 0.5 wt.%. The extruded pellets were subsequently compression molded to about 0.4 mm thickness, following ISO 293 under 5 MPa, at a temperature of 200 °C. This was undertaken via a compression molding platen press (Dr. Collin P 400 M, Ebersberg, Germany), for an overall programming cycle of 32 min, at a heating and cooling rate of 15 °C/min. The specimens were successively conditioned at 23 ± 2 °C and 50 ± 5%, for at least 48 h, prior to being tested.

A schematic of a modular twin-extrusion screw configuration used in the present study is given in Figure 2. There are four main types of screw elements generally used in co-rotating twin screw extruders; forward and back flow convening elements (unboxed), kneading elements for dispersive mixing purposes (yellow and blue boxes), and toothed mixing elements for distributive mixing purposes (orange box) [26–30]. The screw consisted of 30% of 2-flighted right-handed normal and wide kneading elements (yellow box), with a 45° staggering angle, 9% of 16-flighted right-handed mixing elements (orange box), 6.7% of the 2-flighted left-handed narrow kneading elements (blue box) distributed over each dispersive segment. These percentages were based on the ratio of the mixing elements length to the total length of the screw shaft (720 mm).

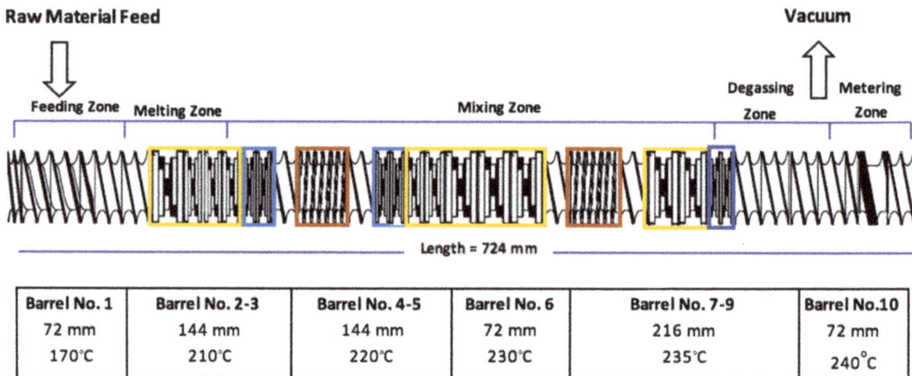

Barrel No. 1	Barrel No. 2-3	Barrel No. 4-5	Barrel No. 6	Barrel No. 7-9	Barrel No.10
72 mm	144 mm	144 mm	72 mm	216 mm	72 mm
170°C	210°C	220°C	230°C	235°C	240°C

Figure 2. Modular extrusion screw configuration based upon individual barrel sections and screw elements. The color boxes show the position of the dispersive and distributive kneading elements along the screw shaft (length of 724 mm).

Since graphene has the ability to shield the polymer from heating and becoming completely molten by enhancing the thermal stability in the feeding and melting zones [17,29–33], the nanocomposite constituents were simultaneously fed from separate feeders into the extruder to prevent graphene platelets from stabilizing around the multimodal-HDPE powders. A long dispersive segment was incorporated in the melting zone to increase the fusion rate of the polymer prior to entering the homogenization zone. The left-handed narrow kneading elements (blue boxes) were placed on each dispersive segment to melt the polymer entirely in the melting zone, and increase the residence time at each dispersive segment. It induces a distribution mixing rather than dispersive (shearing) mixing, especially as its pitch length is very narrow [26,27]. The two distributive elements (orange boxes) were placed between long dispersive segments in order to keep the nanocomposite constituents under continuous high-pressure, and to cause the dispersed (sheared) graphene sheets to instantaneously be pushed away. One of the distributive elements (orange box) was placed between left-handed narrow kneading elements to increase the residence time in a narrower axial length, at the beginning of the homogenization zone, by generating a reverse flow with the use of advancing discs which tend to compress the fluid. This modular assembly build enabled the polymer to degrade to a sufficient level in the targeted zone, under combined elongation and shear forces, prior to entering the homogenization zone. Resultantly, melting the polymer could be completed at the first kneading segment in the melting zones, preventing the graphene platelets from moving smoothly and re-connecting together through the van der Waals's interactions.

2.3. Characterization

2.3.1. Polarized Light Microscope (PLM)

Optical microscopy analyses were conducted on ZIESS Axio scope.A1 HAL 100/HBO 100, operated with an AxioCam MRc 5 camera, and AxioVision software. Film samples were sectioned to a thickness of 15 μm, using a fully automated rotary microtome Leica RM2265 (Leica microsystems, Wetzlar, Germany).

2.3.2. Transmission Electron Microscope (TEM)

Transmission electron microscopy (TEM) was performed using a Hitachi HT7700, at an accelerating voltage of 120 kV. Film samples were cryo-sectioned to a thickness of ~80 nm at -125 °C, using a Leica EM UC7/FC7 Cryo-Ultra-microtome.

2.3.3. Density Measurement

Density measurement was performed with an analytical balance, equipped with a density measurement kit (Metter Toledo XP205, Zurich, Switzerland), following an ASTM D792-method B, based on the Archimedes' principle where the weight of the sample immersed in an n-dodecane fluid decreases by an amount equal to the displacement of the liquid weight.

2.3.4. Raman Spectroscopy

Raman measurements were carried out using Renishaw inVia confocal Raman microscope with 633 nm and 532 nm lasers. With the exception of the deformation test, all the Raman data was collected using He-Ne ion laser with a wavelength of 633 nm (red, 1.96 eV). However, an Nd-YAG laser with a wavelength of 532 nm (green, 2.33 eV) was used to evaluate the stress-transfer along the interfacial surface between the polymer matrix and reinforcement. The dog-bone specimens, having been prepared for tensile testing, were deformed in a three-point bending rig. The strain (ε_f) was measured by calculating the deflection of the beam at the mid-span (δ), following the equation $\varepsilon_f = 6\,\delta\,t/L^2$, where t is the thickness of the beam specimen and L is the span between the supports [6].

2.3.5. X-ray Diffraction (XRD)

X-ray diffraction (XRD) measurements were performed using a Bruker D8-Advance diffractometer equipped with Cu Kα1 radiation ($\lambda = 1.54060$ Å). The diffraction patterns were recorded with a step size of 0.15087°, and dwell time of 5s.

2.3.6. Scanning Electron Microscope (SEM)

A scanning electron microscope (SEM) was utilized to assess the cryofractured cross-section surface of the uncoated polymers, using an FEI QUANT 250 FEG SEM, at an accelerating voltage of 2 kV. Samples were notched prior to being submerged in liquid nitrogen for ten minutes.

2.3.7. Tensile Testing

Tensile properties of the nanocomposites such as tensile modulus, stress and strain at both yield and break, as well as other aspects of the tensile stress-strain curve were studied on the stamped ISO 527-2 dog-bone specimens, type 1B. This was achieved using die punch equipment (Elastocon EP 02, Sweden) on the compression molded samples. These were measured with the Zwick/Roell Universal Testing Machine (UTM)–Z050, using a load cell of 2.5 kN, grip-to-grip separation of 115 mm, a gauge length of 50 mm, and a cross-head speed of 50 mm/min. A contact type extensometer (Zwick MultiXtense, Germany) was used to measure the strain of the specimen. The results were based on a minimum of 6 specimens.

2.3.8. Thermomechanical Analyzer (TMA)

Q400 Thermomechanical analyzer (TMA) was employed to determine the bending properties of the nanocomposites with dynamic mode through the 3-point bending test. Samples were cooled to -150 °C at the cooling rate of 3 °C/min for 10 min, and subsequently heated to 150 °C at the heating rate of 3 °C/min. The modulate force was 0.01 N at a frequency of 1 H_z in an ambient gas atmosphere (50 mL/min), using a wedge-shaped quartz probe. The dimension of the samples was $10 \times 3.4 \times 0.5$ mm^3. The same equipment was employed for studying the coefficient of thermal expansion (CTE) via a flat-tipped standard expansion probe in an ambient gas atmosphere (50 mL/min). Flat samples with dimensions of $6 \times 6 \times 4$ mm^3 were heated at 3 °C/min from room temperature to 100 °C. They were then held for 10 min, then cooled to 0 °C at the same rates of 3 °C/min, then held for 10 min, and subsequently heated at 3 °C/min to 120 °C, under a constant load of 0.05 N. The displacement was reset to zero at the start of the last sequence where the measurement started.

2.3.9. Rheology Analysis

Simple qualitative characterization related to an indirect measurement of molecular weight and processability of the polymer was studied upon the melt-flow rate (MFR). Pellets of 3-8 g were charged into a cylinder at 190 °C under the load of 21.6 kg, which was achieved using the Melt Indexer MI-4 manufactured by GÖETTFERT Werkstoff–Prüfmaschinen GmbH (Buchen, Germany), following ISO 1133 procedure B. The rheological behavior of the samples was studied using stress-controlled rotational rheometer, an Anton Paar Physica MCR 301 with CTD450 heating unit, at 190 °C under a nitrogen atmosphere. The compression molded sample, weighing 1.5 g, 25 mm in diameter, and 1.5 mm thick was conditioned at 40 °C for 48 h. The sample was then placed onto a 25 mm parallel plate fixture and trimmed to a thickness of 1.2 mm by slowly lowering the upper plate. Dynamic frequency sweep was conducted from 500 to 0.0154 rad/s at 5% strain. The reason for starting from the maximum frequency was to avoid sample degradation under high temperature and low angular speed. The polydispersity index (PDI) was measured as follows [33]:

$$PDI = \frac{100\,000}{G'(\omega_{COP})}, \quad \omega_{COP} = \omega(G' = G'') \tag{1}$$

where G' is the storage shear modulus, G'' is the loss shear modulus, ω is the angular frequency, and ω_{COP} is the crossover frequency obtained from the intersection of storage modulus and loss modulus in a log-log scale of a frequency sweep test.

2.3.10. Thermogravimetric Analysis (TGA)

Thermogravimetric analysis (TGA) was carried out with Q500 TGA (TA instruments, New Castle, USA) with a heating rate of 10 °C /min from room temperature to 1000 °C in a nitrogen atmosphere. High-resolution (Hi-Res)-dynamic mode was performed with a sensitivity of 2.00 and a resolution of 4 °C.

3. Results and Discussion

A twin-screw extrusion system with a modular screw configuration was utilized in this study to fabricate a polymer nanocomposite with well-dispersed and uniformly distributed graphene flakes. The screw configuration was optimized after several trials, starting with a screw design employed for compounding such polymers with a high-volume fraction of nanoparticles (carbon black) or nanoclays (talc). A combination of different elements was utilized and arranged according to mixing requirements and material properties to be attained [26–32]. A schematic of the optimal screw configuration for the studied nanocomposite model is given in Figure 2.

3.1. Morphology of Graphene Sheets in a Nanocomposite Matrix

The tailored process included designing a suitable screw configuration, paired with coordinating extruder conditions and blending techniques. This subsequently created a suitable medium for the graphene platelets to disperse readily, and distribute thoroughly within the multimodal-HDPE matrix, as demonstrated in Figure 3a,c. The mean particle size of the detected graphene particles and %area fraction (200 × 200 μm^2) was around 0.5 μm^2 and 0.0063, respectively. Graphene monolayers are transparent under an optical microscope, opacity of 2.3 ± 0.1%, while the optical loss become greater in the wrinkled and overlapped samples [34,35]. L. J. Cote et al. [35] found that the average light scattering from the wrinkled region was about 3.7 times that of the overlapped areas. For the nanocomposite produced using the pre-existing commercial approach, however, the mean particle size of the graphene agglomerates was calculated to be 4.12 μm^2, with maximum particle size of around 4.7 μm^2, and a %area fraction of 79.4 (see Figure 3b,d). The %area fraction and mean particle size were calculated based on transmission electron microscope (TEM) and light microscopy analysis, graphene particles of less than 0.05 μm^2 or 500 nm were excluded from the calculations, i.e. the average lateral size of graphene platelets ranges between 150–500 nm. A decrease in the %area fraction means a better distribution and fewer agglomerates.

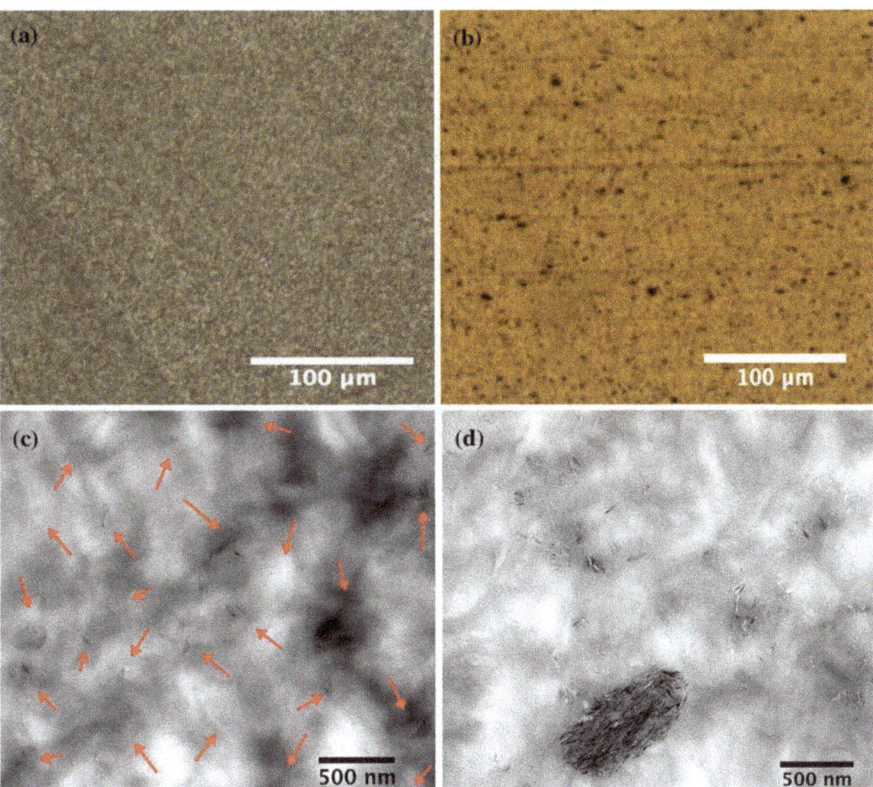

Figure 3. (**a,b**) Light microscopy images and (**c,d**) TEM images show the dispersion and distribution of 1 wt.% loading of graphene platelets within the multimodal-HDPE matrix (PE-g-1%). Images for the similar nanocomposite produced by a pre-existing processing protocol (**right**), were compared with PE-g-1% produced in this study (**left**). The TEM and light microscopy images were taken at 10k and 20x, respectively.

3.2. Dispersion and Distribution of Graphene Platelets Within the Nananocomposite Matrix

Figure 4a shows the X-ray diffraction (XRD) patterns of the neat multimodal-HDPE, graphene powder, and PE-g-1% samples. The diffraction peak (002) appeared in the XRD pattern of graphene at $2\theta = 26.07°$ and exhibited a broad band with a corresponding d-spacing of 0.3414 nm (Bragg's law), and average thickness of 1.941 nm (Scherrer's equation), whereas the weak diffraction peak (100) was observed at $2\theta = 42.89°$ [36,37]. This indicates the sample flakes consisted of 5-6 graphene layers, which is consistent with the TEM images shown in Figure S4. The weak intensity of these two characteristic diffraction peaks is due to the 2D nature of graphene, especially those with very few layers [38]. Interestingly, the XRD pattern of the PE-g-1% is similar to that of a neat multimodal-HDPE matrix, only showing the crystalline diffraction peaks {(110) and 200)} of the neat multimodal-HDPE matrix. Clearly the XRD results demonstrated that the graphene platelets almost exfoliated into individual sheets and dispersed well in the polymer matrix after the extrusion processing [36–42].

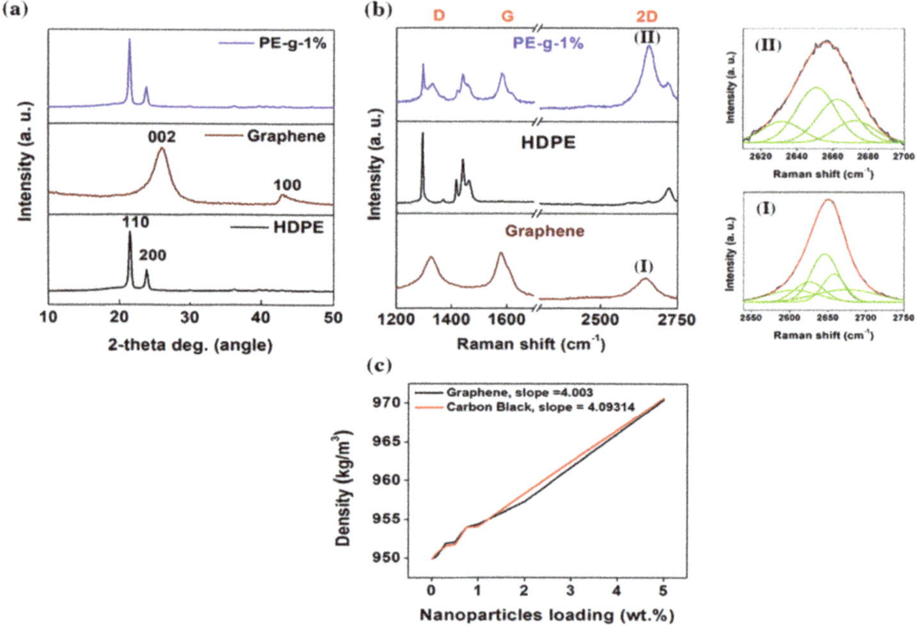

Figure 4. Dispersion and distribution of graphene platelets within the polymer matrix. (**a**) XRD patterns of the neat multimodal-HDPE, graphene powder, and PE-g-1%. (**b**) Overlaid Raman spectrum of the neat multimodal-HDPE, graphene powder, and PE-g-1%. The measured 2D Raman bands with 1.96 eV laser energy of graphene (**I**) and PE-g-1% (**II**) are fitted with four and five Lorentzians, respectively. (**c**) Density measurement of graphene/multimodal-HDPE (PE-g), and carbon black/multimodal-HDPE (PE-CB) nanocomposites as a function of nanofiller loading (0.1, 0.5, 1, 2, and 5 wt.%).

Overlay Raman spectra of the neat multimodal-HDPE, graphene powder, and PE-g-1% samples, are shown in Figure 4b. The three intense peaks appeared in the Raman spectra of graphene at 1327 cm^{-1}, 1577.5 cm^{-1}, and 2646 cm^{-1}, representing the characteristic D-band, G-band, and 2D-band peaks, respectively [43–45]. The G-band arises from the bond stretching of the sp^2 carbon atoms (chains or rings), while the breathing modes of the sp^2 carbon atoms in a hexagon ring gives rise to the D-band [43–45]. The D-band, therefore, requires a defect to be activated (by disorder or at the edge) [43–45]. It originates from one iTO phonon mode around the *K* point by double resonance, whereas the overtone of the D′ and D-bands gives rise to 2D′ and 2D-bands [43–45]. The 2D (or 2D′) peak does not require a defect for its activation, because it originates from the two iTO phonons with opposite momentum near Brillouin zone [43–45]. The position, full width at high maximum (FWHM), intensity ratio (I_{2D}/I_G), and Lorentzian fittings of the 2D peak provide good correlation with the number of layers of graphene in a flake sample [43–47]. In Figure 4bI, the 2D-band of graphene sample is fitted by five Lorentzians, with an overall FWHM of 65.52 cm^{-1}. Z. Lin et al. [46] and E. Dervishi et al. [47] in fact used up to five Lorentzian peaks to fit the 2D-band of their few-layer graphene produced by a bottom up approach. For PE-g-1%, the three prominent characteristic D, G, and 2D peaks associated with graphene were observed at 1327 cm^{-1}, 1582.5 cm^{-1}, and 2658 cm^{-1}, respectively. The 2D-band of the nanocomposite shown in Figure 4bII, is red-shifted from 2646 cm^{-1} to 2658 cm^{-1}, fitted by four Lorentzians, with an overall FWHM decreased from 65.52 cm^{-1} to ~53 cm^{-1}. Four fitted Lorentzians, each with a FWHM of ~24 cm^{-1}, most likely arose from the asymmetry between the valence and conduction bands present in the bilayer graphene [43–47]. Besides, the increase of the (I_{2D}/I_G) from 0.98 for graphene to 1.55 for PE-g-1%, reveals the reduction of the graphene layers [43].

Overall, these results indicate that the graphene platelets are indeed dispersed (thinned) through the melt extrusion process.

Figure 4c depicts the density (ρ_c) of the graphene/multimodal-HDPE (PE-g) and carbon black/multimodal-HDPE (PE-CB) nanocomposites at nanofiller loadings of 0.1, 0.5, 1, 2, and 5 wt.%. The PE-CB sample is a commercial grade, produced based on the same polymer matrix, but reinforced by a carbon black with a density of 1700-1900 kg·m^{-3}. A monolayer graphene is made up of covalently-bonded sp^2-hybridised carbon atoms, densely packed in a honeycomb lattice [21,22]. Therefore, the density of a defect-free monolayer graphene, with a thickness of 0.142 nm, is estimated to be around 2175 kg·m^{-3} [48]. On the other hand, carbon black is composed of primary particles that are permanently fused together through the covalent bonds, into an aggregate structure [49]. Each primary particle is made up of imperfect crystallites of turbostratic graphite structure, which are twisted into each other throughout the aggregates [49]. Accordingly, the graphene used in this study is likely to have a density closer to the carbon black density than a monolayer graphene, i.e., defective surface structure through the oxygen-containing functional groups (see Supplementary Materials Figure S3). As is evident from Figure 4c, the density of the multimodal-HDPE matrix ($\rho_m = 950$ kg·m^{-3}) increased linearly with the addition of the nanofillers, i.e., densities of both nanocomposites increased by the same amount. The slope values are calculated at 4.003 for PE-g and 4.093 for PE-CB, suggesting that the graphene platelets were homogenously dispersed and distributed throughout the polymer matrix. An increase in the nanocomposite density is attributed to the high density of the reinforcements (ρ_r) employed to reinforce the polymer matrix, according to the equation of the form ($\rho_c = 1/(W_r/\rho_r) + (W_m/]\rho_m)$), where W_r and W_m are the weight fractions of reinforcement and matrix, respectively [4]. With a greater incorporation of high-density reinforcement, a higher nanocomposite density is obtained. In the case of agglomeration however, most of the graphene platelets will be lost in the accumulation, thereby the increase in the nanocomposite density remains relatively small.

3.3. Interfacial Adhesion Strength between Graphene and Polymer Matrix

The interfacial adhesion strength between graphene sheets and a polymer matrix can be explored through the microscopic examination of cryofractured cross-sectional surfaces. Shown in Figure 5a is a SEM image of the neat multimodal-HDPE surface exposed by cryofracture. The SEM micrograph exhibits fibrils with various extents of surface fibrillation in the draw direction. The occurrence of fibrils may suggest that the fracture was due to chain slippage or scission in crystalline (long fibrils) and amorphous (short fibrils) regions [50]. Contrastingly, graphene was shown to have a significant effect on the microstructure of the adjacent polymer as evident by changes to the fibrous morphology of the PE-g-1% shown in Figure 5b. The SEM micrograph of the nanocomposite exhibits a number of graphene platelets protruding out of the fracture surface of the polymer matrix, i.e. embedded and strongly tied to the matrix.

These flakes are well dispersed and evenly distributed within the multimodal-HDPE matrix, which may have formed a continually interconnected network structure throughout the matrix [50–54]. Interestingly, the fractured surface of the nanocomposite become rough, compared to that of the unfilled multimodal-HDPE. Conceptually, the fracture toughness is quantified by the amount of the energy absorbed per unit crack extension [52]. Therefore, the significant change in the breaking (crack propagation) mechanism accordingly suggests that the strong interfacial bonding between the polymer matrix and graphene platelets likely split the material into cavities and molecular bundles under large loading [50]. The facilitated stress transfer along the large interfacial area between the reinforcement and matrix is expected to potentially display mechanical reinforcement [50–53]. In Figure 5c, the storage modulus measured by dynamic thermomechanical analysis (DTMA) increased by 75%, 84%, and 118% at -100 °C, -50 °C, and 23.5 °C, respectively. The tensile modulus increased by $\geq$35%, from 835 $\pm$ 13 MPa for neat multimodal-HDPE, to 1135 $\pm$ 17 MPa for PE-g-1%, as shown in Figure 5d. Moreover, the maximum tensile strain increased by 11%, from 615 $\pm$ 43 % for neat multimodal (extruded), to $680 \pm 31\%$ for the PE-g-1%. This increase in the tensile strain was possibly preceded by a prolonged exposure of the neat polymer to a high temperature in the extruder, under a combined high shear and elongation forces [55]. Thus, graphene has most likely acted as an antioxidant and protected the polymer from excessive thermo-oxidative degradation [55,56]. The maximum tensile strain of the nanocomposite is therefore compared to a non-extruded multimodal-HDPE for verification. Interestingly, the tensile strain decreased from >800% for neat multimodal-HDPE (non-extruded) to only $680 \pm 31\%$ for the PE-g-1%. This latter subject will be discussed in greater detail later in this study. Nevertheless, this indicates that graphene reinforced the polymer through the heat transfer from the polymer matrix to graphene platelets along the interface.

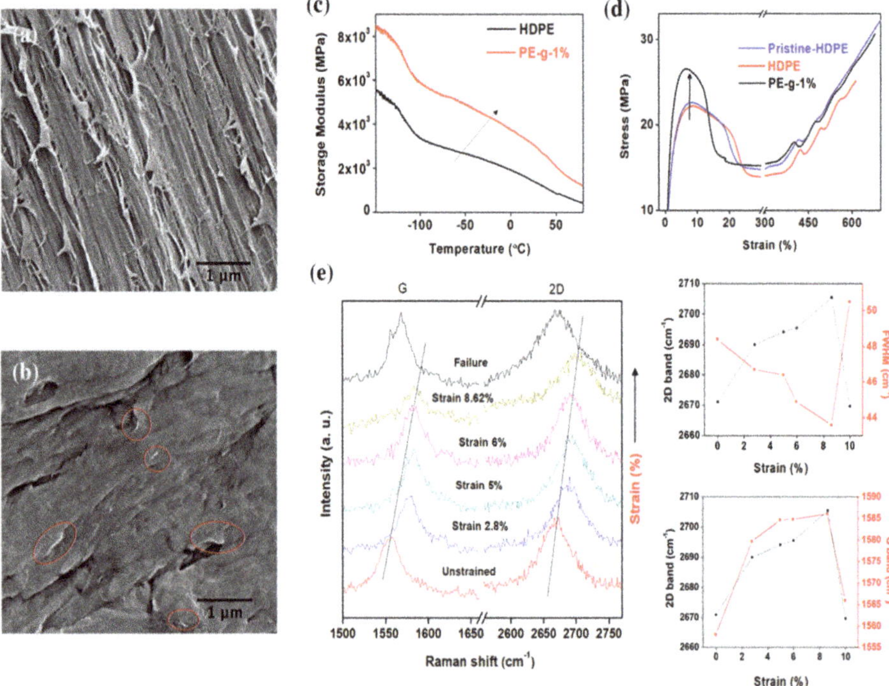

Figure 5. Assessment of the interfacial adhesion strength between graphene sheets and polymer matrix. SEM images of a cross-section fracture surface from (**a**) neat multimodal-HDPE and (**b**) PE-g-1%. (**c**) Dynamic-thermomechanical analysis (DTMA) of the neat multimodal-HDPE and PE-g-1%. (**d**) Tensile stress-strain curves for the pristine multimodal-HDPE (non-extruded), neat multimodal-HDPE (extruded), and PE-g-1%. The pristine polymer is the powder polyethylene. (**e**) Shift with strain of the 2D and G Raman bands of the graphene during deformation upon PE-g-1% nanocomposite (laser excitation energy 2.33 eV). The corresponding 2D and G Raman shifts as a function of applied strain are shown in the two graphs on the right.

The interfacial adhesion strength between the polymer matrix and graphene platelets was further investigated by the stress-induced Raman band shifts [57–60]. In Figure 5e, the 2D and G Raman bands of graphene in a nanocomposite shifted to higher wavenumbers as a function of applied strain, suggesting that the graphene platelets went into biaxial compression as reported in the literature [57,58]. Beyond ~9% strain however, these two bands reverted closer to that of the unstrained peak positions, due to relaxation of the graphene sheets upon debonding between the nanocomposite constituents [57–60]. The 2D and G Raman bands have significantly downshifted after ~9% strain, by ~34 cm^{-1} and 28 cm^{-1}, respectively. Surprisingly, the 9% strain is around the yield point as can be seen in the stress-strain curve shown in Figure 5d. Overall, the results show that a strong interfacial bonding is created between graphene sheets and polymer matrix [57–60].

3.4. Rheology and Thermal Stability Performance of a Nanocomposite

Figure 6a-b shows the rheological behaviour of neat multimodal-HDPE and PE-g-1%. Shown in Figure 6a is the pseudoplastic, non-Newtonian behavior of the viscoelastic polymer. The influence of graphene on the viscoelastic response of the polymer is revealed from the change in the absolute values of the storage (G') and loss (G'') moduli, as well as their frequency dependence [33,61].

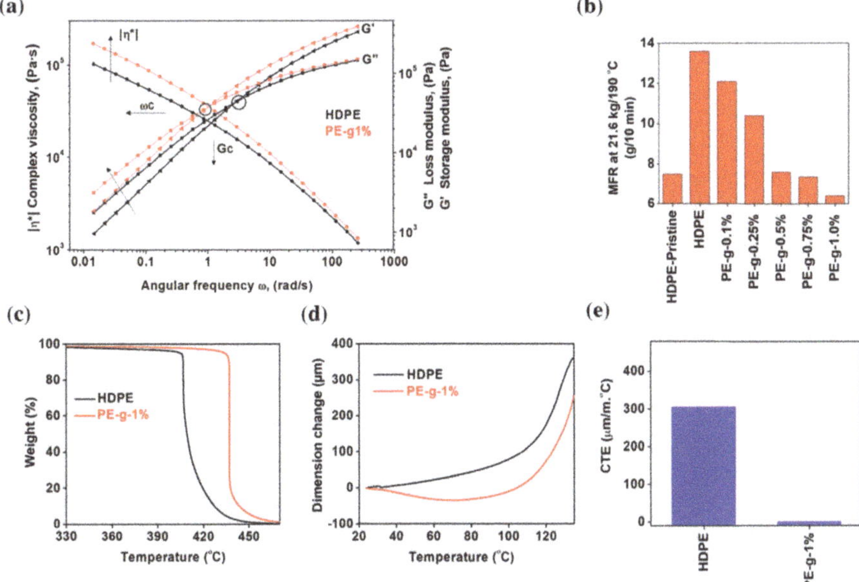

Figure 6. Thermal stability performance and rheological behaviors. (**a**) Dynamic frequency sweep measurements performed at 190 °C. ω_C is the crossover frequency point and G_C is the crossover modulus point in a log–log scale. (**b**) Melt flow rate (MFR) measurements of PE-g nanocomposites as a function of graphene loading (0.1, 0.25, 0.5, 0.75, and 1.0 wt.%). (**c**) Thermogravimetric thermograms performed in N_2 atmosphere. (**d**) Dimensional change as a function of temperature. (**e**) Coefficient of thermal expansion (CTE) measured at temperature difference range of 30-105 °C.

At a high-shear rate, both materials exhibited thinning behavior, which resulted in a decrease of extensional viscosity. However, the incorporation of 1 wt.% graphene increased the melt viscosity of the nanocomposite, though the relative increase gradually lessened at high-shear rate. The presence of graphene has considerably increased the pseudoplasticity at a low shear rate region. At the angular frequency (ω) of 0.0154 rad/s, the complex viscosity increased from 0.13 MPa·s for neat multimodal-HDPE to 0.24 MPa·s for the PE-g-1%. Furthermore, the loss and storage moduli of the neat polymer increased by a value of 92% and 77% with 1 wt.% loading of graphene, respectively. The greater amount of storage and loss moduli of PE-g-1% suggests that the formation of a strong interfacial bonding between the polymer matrix and the high-modulus graphene reduced the loss tangent, the nanocomposite accordingly became more elastic [33,61–63]. This is in addition to the thorough dispersion and distribution of the nanofillers, which led to a decrease in the degree of the chain mobility of the polymers, and thus suppressed the shear flow of the polyethylene macromolecular chains [4]. In addition, the crossover modulus point (G_C) and crossover frequency point (ω_C) have decreased from 0.044 MPa·s and 4.5 rad/s for the neat multimodal-HDPE to 0.035 MPa·s and 1.2 rad/s for PE-g-1%, respectively. The shift of ω_C to lower region indicates that the nanocomposite exhibited higher average molecular mass and/or the entangled molecules were induced by the three-dimensional network of graphene platelets within the matrix [33]. However, the G_C shifting to lower values after the reinforcement indicates that the polymer exhibited broader molecular weight distribution, which is evident from the increase in the polydispersity index (PDI) by 18.4%, i.e. the larger the PDI, the broader the molecular weight distribution [33]. The shift of G_c upon the addition of the reinforcement possibly arose also from exposing the neat polymer to high temperature, under a combination of high shear and elongation forces, for a prolonged period of time. In Figure 6b, the melt flow rate (MFR) of the extruded multimodal-HDPE decreased gradually from 13.6 g/min to only 6.42 g/min with 1 wt.% graphene loading at 21.6 kg/190 °C, whilst the MFR of the pristine multimodal-HDPE (non-extruded) was only was only 7.5 g/min. The lower the MFR, the higher the molecular weight or the viscosity. This indicates that graphene acted as a thermal barrier and enhanced the thermal stability of the polymer through the strong interface bonding.

The synergistic effect advantages of graphene are further investigated by thermal expansion and thermogravimetric analyses (TGA). As shown from the TGA thermograms in Figure 6c, the onset degradation temperature of PE-g-1% increased significantly by more 31 °C. The onset temperature at 5% mass loss (T5%) of neat multimodal-HDPE increased from 405 °C to 434.2 °C upon 1 wt.% loading of graphene (see Table 1). The reinforced polymers exhibited a greater melt strength during thermoforming such that the sagging resistance of the nanocomposite has been improved. The large aspect ratio of graphene with a platelet structure likely offered a larger interfacial surface with the polymer matrix which in turn slowed the diffusion of the decomposition products from a continuous network-structured protective layer created in the nanocomposite. It would seem as though graphene acted as an antioxidant and consequently protected the polymer from excessive thermal degradation [64–69]. The polymer could therefore be extruded in aggressive conditions, for example with a screw configuration of 37% of dispersive elements at a very low feed rate. This also implies that the amount of thermo-mechanical and thermo-oxidative degradations achieved was sufficient enough to produce an efficient reinforcement. This was achieved through the formation of a strong interfacial adhesion bond between the polymer matrix and graphene platelets, which has accordingly enhanced the thermal stability of the polymer.

Table 1. TGA data of the neat multimodal-HDPE and its nanocomposite.

Sample	Tonset, (°C)	T5%, (°C)	T30%, (°C)	T50%, (°C)	T80%, (°C)
Neat multimodal-HDPE	400	405	406.8	409	419.3
PE-g-1%	433.4	435	437	437	437.2

T5%, T30%, T50%, and T80%, are the onset temperatures at 5%, 30%, 50%, and 80% mass loss, respectively.

The coefficient of thermal expansion (CTE) of the PE-g-1% was calculated to be $0.55 \times 10^{-6}\,°\text{C}^{-1}$ over the temperature range of 30–103 °C, as shown in Figure 6d-e. The CTE started to become positive after 100 °C confirming that graphene sheets were well bonded with the polymer matrix, and suggests the continuous interconnected network structure formed in the polymer matrix hindered the reorientation of the polymer chains. Mounet et al. [70], Zakharchenko et al. [71], Yoon et al. [72], Bao et al. [73], and others found that graphene has negative thermal expansion at low temperatures. Mounet et al. [70] used a first-principles calculation, and estimated the CTE of graphene remains negative up to 2500 K. Zakharchenko et al. [71] found that the transition from negative to positive CTE occurs at ~900 K. The CTE of a single-layer graphene measured by Yoon et al. [72] via temperature-dependent Raman spectroscopy remained negative in the temperature range of 200–400 K. Therefore, it is not yet clear at what exact point the CTE changes from negative to positive.

4. Conclusions

A high-performance novel graphene-based multimodal polyethylene nanocomposite was produced directly without any particular treatment to the composite constituents. Different characterization techniques such as TEM, SEM, optical microscopes, Raman spectroscopy, Rheology, density measurements, TGA, and CTE were employed in order to demonstrate the novelty of the fabrication method. Microscopic analysis showed that graphene platelets were homogenously dispersed and distributed within the polymer matrix. The electronic and optical microscopies showed that the graphene was dispersed with an average size of less than 450 nm, and distributed with a %area fraction of only 0.0063. The adhesion strength between the graphene sheets and polymer matrix examined by microscopic examination of the cryofractured surfaces of the nanocomposite, mechanical testing and stress-induced Raman band shifts, revealed a strong interfacial bonding attained through thermo-mechanical and thermo-oxidative degradation to a controlled level during the melt extrusion process. The cryofractured surface of the nanocomposite became rough with fibrils almost entirely absent. Deflection of the nanocomposite led the characteristic 2D and G Raman peaks of graphene to shift significantly towards high wavenumbers. This was confirmed further by the mechanical testing where the storage modulus of the polyethylene reinforced with 1 wt.% of graphene increased up to 118% at room temperature.

The thermal performance of the nanocomposite was investigated via rheology testing, TGA and thermal expansion analysis. Loading a polymer with 1 wt.% graphene has resulted in a significant shift of the crossover frequency point and crossover modulus point to the lower regions. The homogenous dispersion and distribution of graphene platelets within the polymer matrix, as well as the strong adhesion bonding, led to the formation of an interconnected 3D network in the polymer which accordingly restricted the movement and expansion of the polymer chains movements. This has resulted in a significant increase in the onset degradation temperature by more 31 °C as a consequence of a thermal barrier or synergetic effects induced by graphene. The nanocomposites exhibited almost zero thermal expansion below 100 °C. The results of this study change the idea of there being difficulty in using melt extrusion for producing well dispersed and disturbed graphene-based hydrophobic polymer nanocomposites. Furthermore, considering there is no need to chemically treat graphene or polymer, or in fact change anything in the existing plants, such a result can significantly attract the industry.

Supplementary Materials: The following are available online at http://www.mdpi.com/2079-4991/8/11/947/s1, Figure S1: Type and use of each element used in the present study, Figure S2: The cross-sectional area of a two-flighted screw element, Figure S3: XPS wide range spectrum of graphene powder (CAE=50eV). A C1s (CAE=50eV) narrow scan peak fit is also presented in the inset, showing the components and the degree of oxygen species. They were fitted by the Gaussian-Lorentzian (GL60) functions. Figure S4: TEM images of the as-received graphene powders at different magnifications.

Author Contributions: Conceptualization, I.A.A.; methodology, I.A.A.; validation, I.A.A.; formal analysis, I.A.A.; investigation, I.A.A.; resources, I.A.A.; writing—original draft preparation, I.A.A.; writing—review and editing, I.A.A., S.D., and R.V.K.; visualization, I.A.A., and S.D., and H-K.K.; supervision, R.V.K., K.K.K.K, and H-K.K.; project administration, I.A.A.; funding acquisition, I.A.A.", please turn to the CRediT taxonomy for the term explanation.

Funding: This research was funded by Abu Dhabi National Oil Company (ADNOC) and Borouge Pte. Ltd., grant code LJGK GAAA, and the APC was funded by ADNOC.

Acknowledgments: This work was conducted under the framework of the Research and Development Program of the Korea Institute of Energy Research (KIER) (B8-2432-04) and ADNOC.

Conflicts of Interest: The authors declare no conflict of interest.

References

1. Chatzidoukas, C.; Kanellopoulos, V.; Kiparissides, C. On the production of polyolefins with bimodal molecular weight and copolymer composition distributions in catalytic gas-phase fluidized-bed reactors. *Macromol. Theor. Simul.* **2007**, *16*, 755–769. [CrossRef]

2. DesLauriers, P.J.; McDaniel, M.P.; Rohlfing, D.C.; Krishnaswamy, R.K.; Secora, S.J.; Benham, E.A.; Maeger, P.L.; Wolfe, A.R.; Sukhadia, A.M.; Beaulieu, B.B. A comparative study of multimodal vs. bimodal polyethylene pipe resins for PE-100 applications. *Polym. Eng. Sci.* **2005**, *45*, 1203–1213. [CrossRef]

3. Chen, K.; Tian, Z.; Luo, N.; Liu, B. Modeling and simulation of borstar bimodal polyethylene process based on a rigorous PC-SAFT equation of state model. *Ind. Eng. Chem. Res.* **2014**, *53*, 19905–19915. [CrossRef]

4. Young, R.J.; Lovell, P.A. *Introduction to Polymers*, 3rd ed.; CRC Press: New York, NY, USA, 2011.

5. *Polyolefin Market Analysis and Segment Forecast to 2025*, 1st ed.; Grand View Research, Inc.: San Francisco, CA, USA, 2018.

6. Bhattacharya, M. Polymer nanocomposites-A comparison between carbon nanotubes, graphene, and clay as nanofillers. *Materials* **2016**, *9*, 262. [CrossRef] [PubMed]

7. Deveci, S.; Fang, D. Correlation of molecular parameters, strain hardening modulus and cyclic fatigue test performances of polyethylene materials for pressure pipe applications. *Polym. Test.* **2017**, *62*, 246–253. [CrossRef]

8. Yi, X.S.; Wu, G.; Ma, D. Property balancing for polyethylene-based carbon black-filled conductive composites. *J. Appl. Polym. Sci.* **1998**, *67*, 131–138. [CrossRef]

9. Michael, T. *Additives for Polyolefins: Getting the Most out of Polypropylene, Polyethylene and TPO*, 2nd ed.; William Andrew: Oxford, UK, 2009.

10. Xanthos, M. Part One: Polymers and fillers. In *Functional Fillers for Plastics*, Ed. ed; Wiley-VCH Verlag GmbH & Co. KGaA: Weinheim, Germany, 2010.

11. Hawkins, W.L.; Hansen, R.H.; Matreyek, W.; Winslow, F.H. The effect of carbon black on thermal antioxidants for polyethylene. *J. Appl. Polym. Sci.* **1959**, *1*, 37–42. [CrossRef]

12. Deveci, S.; Antony, N.; Eryigit, B. Effect of carbon black distribution on the properties of polyethylene pipes—Part 1: Degradation of post yield mechanical properties and fracture surface analyses. *Polym. Degrad. Stab.* **2018**, *148*, 75–85. [CrossRef]

13. Pircheraghi, G.; Sarafpour, A.; Rashedi, R.; Afzali, K.; Adibfar, M. Correlation between rheological and mechanical properties of black PE100 compounds—Effect of carbon black masterbatch. *Express Polym. Lett.* **2017**, *11*, 622–634. [CrossRef]

14. *The Global Market for Carbon Black Report*, 5st ed.; Future Markets, Inc.: Edinburgh, Scotland, 2017.

15. Salavagione, H.J.; Martínez, G.; Ellis, G. Recent advances in the covalent modification of graphene with polymers. *Macromol. Rapid Commun.* **2011**, *32*, 1771–1789. [CrossRef] [PubMed]

16. Potts, J.R.; Dreyer, D.R.; Bielawski, C.W.; Ruoff, R.S. Graphene-based polymer nanocomposites. *Polymer* **2011**, *52*, 5–25. [CrossRef]

17. Kuilla, T.; Bhadra, S.; Yao, D.; Hoon, N.; Bose, S.; Hee, J. Recent advances in graphene based polymer composites. *Prog. Polym. Sci* **2010**, *35*, 350–1375. [CrossRef]

18. Zurutuza, A.; Marinelli, C. Challenges and opportunities in graphene commercialization. *Nat. Nanotechnol.* **2014**, *9*, 730–734. [CrossRef] [PubMed]

19. Huang, X.; Yin, Z.; Wu, S.; Qi, X.; He, Q.; Zhang, Q.; Yan, Q.; Boey, F.; Zhang, H. Graphene-based materials: Synthesis, characterization, properties, and applications. *Small* **2011**, *7*, 1876–1902. [CrossRef] [PubMed]

20. Tripathi, S.N.; Rao, G.S.S.; Mathur, A.B.; Jasra, R. Polyolefin/graphene nanocomposites: A review. *RSC Adv.* **2017**, *7*, 23615–23632. [CrossRef]

21. Phiri, J.; Gane, P.; Maloney, T.C. General overview of graphene: Production, properties and application in polymer composites. *Mater. Sci. Eng. B* **2017**, *215*, 9–28. [CrossRef]

22. Ferrari, A.C.; Bonaccorso, F.; Fal'ko, F.; Novoselov, K.S.; Roche, S.; Bøggild, P.; Borini, S.; Koppens, F.H.L.; Palermo, V.; Pugno, N.; et al. Science and technology roadmap for graphene, related two-dimensional crystals, and hybrid systems. *Nanoscale* **2015**, *7*, 4598–4810. [CrossRef] [PubMed]

23. Vasile, C.; Pascu, M. *Practical Guide to Polyethylene*; Rapra Technology Limited: Shropshire, UK, 2005.

24. El'darov, E.G.; Mamedov, F.V.; Gol'dberg, V.M.; Zaikov, G.E. A kinetic model of polymer degradation during extrusion. *Polym. Degrad. Stab.* **1996**, *51*, 271–279. [CrossRef]

25. Gol'dberg, V.M.; Zaikov, G.E. Kinetics of mechanical degradation in melts under model conditions and during processing of polymers-A review. *Polym. Degrad. Stab.* **1987**, *19*, 221–250. [CrossRef]

26. Schonfeld, S. *Compounding of Filled Polymers with the Co-rotating Twin Screw Extruder ZSK*; Coperion: Stuttgart, Germany, 2013.

27. Rauwendaal, C. *Polymer Extrusion*, 5th ed.; Hanser Gardner Publications: Munich, Germany, 2014.

28. Sakai, T. Screw extrusion technology—Past, present and future. *Polimery* **2013**, *58*, 847–857. [CrossRef]

29. Chiruvella, R.V.; Jaluria, Y.; Karwe, M.V.; Sernas, V. Transport in a twin-screw extruder for the processing of polymers. *Polym. Eng. Sci.* **1996**, *36*, 1531–1540. [CrossRef]

30. Montiel, R.; Patiño-Herrera, R.; Gonzalez-Calderón, J.A.; Pérez, E. Novel twin screw co-extrusion-electrospinning apparatus. *Am. J. Biomed. Eng.* **2016**, *6*, 19–24.

31. Yuan, B.; Bao, C.; Song, L.; Hong, N.; Liew, K.M.; Hu, Y. Preparation of functionalized graphene oxide/polypropylene nanocomposite with significantly improved thermal stability and studies on the crystallization behavior and mechanical properties. *Chem. Eng.* **2014**, *237*, 411–420. [CrossRef]

32. Yang, H.H.; Manas-Zloczower, I. Flow field analysis of the kneading disc region in a co-rotating twin screw extruder. *Polym. Eng. Sci.* **1992**, *32*, 1411–1417. [CrossRef]

33. Mezger, T.G. *The Rheology Handbook*, 4th ed.; Vincentz Network: Hanover, Germany, 2014.

34. Nair, R.R.; Blake, P.; Grigorenko, A.N.; Novoselov, K.S.; Booth, T.J.; Stauber, T.; Peres, N.M.R.; Geim, A.K. Fine structure constant defines visual transparency of graphene. *Science* **2008**, *320*, 1308. [CrossRef] [PubMed]

35. Cote, L.J.; Kim, J.; Zhang, Z.; Sun, C.; Huang, J. Tunable assembly of graphene oxide surfactant sheets: Wrinkles, overlaps and impacts on thin film properties. *Soft Matter.* **2010**, *6*, 6096–6101. [CrossRef]

36. Stobinski, L.; Lesiak, B.; Malolepszy, A.; Mazurkiewicz, M.; Mierzwa, B.; Zemek, J.; Jiricek, P.; Bieloshapka, I. Graphene oxide and reduced graphene oxide studied by the XRD, TEM and electron spectroscopy methods. *J. Electron. Spectrosc. Relat. Phenomena* **2014**, *195*, 145–154. [CrossRef]

37. Biris, A.S.; Pruneanu, S.M.; Pogacean, F.; Lazar, M.D.; Borodi, G.; Ardelean, S.; Dervishi, E.; Watanabe, F. Few-layer graphene sheets with embedded gold nanoparticles for electrochemical analysis of adenine. *Int. J. Nanomed.* **2013**, *8*, 1429–1438. [CrossRef] [PubMed]

38. Kuila, T.; Bose, S.; Mishra, A.K.; Khanra, P.; Kim, N.H.; Lee, J.H. Effect of functionalized graphene on the physical properties of linear low density polyethylene nanocomposites. *Polym. Test.* **2012**, *31*, 31–38. [CrossRef]

39. Khan, Q.A.; Shaur, A.; Khan, T.A.; Joya, Y.F.; Awan, M.S. Characterization of reduced graphene oxide produced through a modified Hoffman method. *Cogent. Chem.* **2017**, *3*, 1298980. [CrossRef]

40. Song, P.; Cao, Z.; Cai, Y.; Zhao, L.; Fang, Z.; Fu, S. Fabrication of exfoliated graphene-based polypropylene nanocomposites with enhanced mechanical and thermal properties. *Polymer* **2011**, *52*, 4001–4010. [CrossRef]

41. Patwary, F.; Mittal, V. Degradable polyethylene nanocomposites with silica, silicate and thermally reduced graphene using oxo-degradable pro-oxidant. *Heliyon* **2015**, *1*, e00050. [CrossRef] [PubMed]

42. Wei, P.; Bai, S. Fabrication of a high-density polyethylene/graphene composite with high exfoliation and high mechanical performance via solid-state shear milling. *RSC Adv.* **2015**, *5*, 93697–93705. [CrossRef]

43. Malard, L.M.; Pimenta, M.A.; Dresselhaus, G.; Dresselhaus, M.S. Raman spectroscopy in graphene. *Phys. Rep.* **2009**, *473*, 51–87. [CrossRef]

44. Ferrari, A.C. Raman spectroscopy of graphene and graphite: Disorder, electron–phonon coupling, doping and nonadiabatic effects. *Solid State Commun.* **2007**, *143*, 47–57. [CrossRef]

45. Ferrari, A.C.; Basko, D.M. Raman spectroscopy as a versatile tool for studying the properties of graphene. *Nat. Nanotechnol.* **2013**, *8*, 235. [CrossRef] [PubMed]

46. Lin, Z.; Ye, X.; Han, J.; Chen, Q.; Fan, P.; Zhang, H.; Xie, D.; Zhu, H.; Zhong, M. Precise Control of the Number of Layers of Graphene by Picosecond Laser Thinning. *Sci. Rep* **2015**, *5*, 11662. [CrossRef] [PubMed]

47. Dervishi, E.; Li, Z.; Watanabe, F.; Biswas, A.; Xu, Y.; Biris, A.R.; Saini, V.; Biris, A.S. Large-scale graphene production by RF-cCVD method. *Chem. Commun.* **2009**, *27*, 4061–4063. [CrossRef] [PubMed]

48. Choi, W.; Lee, J.W. *Graphene: Synthesis and Applications*, 1st ed.; CRC Press: Boca Raton, FL, USA, 2012.

49. Albers, P.; Maier, M.; Reisinger, M.; Hannebauer, B.; Weinand, R. Physical boundaries within aggregates-Differences between amorphous, para-crystalline, and crystalline Structures. *Cryst. Res. Technol.* **2015**, *50*, 846–865. [CrossRef]

50. Istrate, O.M.; Paton, K.R.; Khan, U.; O'Neill, A.; Bell, A.P.; Coleman, J.N. Reinforcement in melt-processed polymer–graphene composites at extremely low graphene loading level. *Carbon* **2014**, *78*, 243–249. [CrossRef]

51. Wang, J.; Bai, J.; Zhang, Y.; Fang, H.; Wang, Z. Shear-induced enhancements of crystallization kinetics and morphological transformation for long chain branched polylactides with different branching degrees. *Sci. Rep.* **2016**, *6*, 26560. [CrossRef] [PubMed]

52. Phillips, D.C.; Tetelman, A.S. The fracture toughness of fibre composites. *Composites* **1972**, *3*, 216–223. [CrossRef]

53. Panzavolta, S.; Bracci, B.; Gualandi, C.; Letizia, M.; Treossi, E.; Kouroupis-agalou, K.; Rubini, K.; Bosia, F.; Brely, L.; Pugno, N.M.; et al. Structural reinforcement and failure analysis in composite nanofibers of graphene oxide and gelatin. *Carbon* **2014**, *78*, 566–577. [CrossRef]

54. Rafiee, M.A.; Rafiee, J.; Srivastava, I.; Wang, Z.; Song, H.; Yu, Z.; Koratkar, N. Fracture and Fatigue in Graphene Nanocomposites. *Small* **2010**, *6*, 179–183. [CrossRef] [PubMed]

55. Butler, T.I. The Influence of Extruder Residence Time. *J. Plast. Film Sheeting* **1990**, *6*, 247–259. [CrossRef]

56. Qiu, Y.; Wang, Z.; Owens, A.C.E.; Kulaots, I.; Chen, Y.; Kane, A.B.; Hurt, R.H. Antioxidant chemistry of graphene-based materials and its role in oxidation protection technology. *Nanoscale* **2014**, *6*, 11744–11755. [CrossRef] [PubMed]

57. Proctor, J.E.; Gregoryanz, E.; Novoselov, K.S.; Lotya, M.; Coleman, J.N.; Halsall, M.P. High-pressure Raman spectroscopy of graphene. *Phys. Rev. B* **2009**, *80*, 73408. [CrossRef]

58. Srivastava, I.; Mehta, R.J.; Yu, Z.Z.; Schadler, L.; Koratkar, N. Raman study of interfacial load transfer in graphene nanocomposites. *Appl. Phys. Lett.* **2011**, *98*, 63102. [CrossRef]

59. Shin, Y.; Lozada, H.M.; Sambricio, J.L.; Grigorieva, I.V.; Geim, A.K.; Casiraghi, C. Raman spectroscopy of highly pressurized graphene membranes. *Appl. Phys. Lett.* **2016**, *108*, 221907. [CrossRef]

60. Gong, L.; Young, R.J.; Kinloch, I.A.; Riaz, I.; Jalil, R.; Novoselov, K.S. Optimizing the Reinforcement of Polymer-Based Nanocomposites by Graphene. *ACS Nano* **2012**, *6*, 2086–2095. [CrossRef] [PubMed]

61. Zhang, Q.; Rastogi, S.; Chen, D.; Lippits, D.; Lemstra, P.J. Low percolation threshold in single-walled carbon nanotube/high density polyethylene composites prepared by melt processing technique. *Carbon* **2006**, *44*, 778–785. [CrossRef]

62. Adhikari, A.R.; Lozano, K.; Chipara, M. Non-isothermal crystallization kinetics of polyethylene/carbon nanofiber composites. *J. Compos. Mater.* **2012**, *46*, 823–832. [CrossRef]

63. Jiasheng, Q.; Pingsheng, H. Non-isothermal crystallization of HDPE/nano-SiO$_2$ composite. *J. Mater. Sci.* **2003**, *38*, 2299–2304. [CrossRef]

64. Foroozan, T.; Soto, F.A.; Yurkiv, V.; Sharifi, A.S.; Deivanayagam, R.; Huang, Z.; Rojaee, R.; Mashayek, F.; Balbuena, P.B.; Shahbazian, Y.R. Synergistic Effect of Graphene Oxide for Impeding the Dendritic Plating of Li. *Adv. Funct. Mater.* **2018**, *28*, 1705917. [CrossRef]

65. Sarabia, R.R.; Ramos, F.G.; Martin, G.I.; Weisenberger, M.C. Synergistic effect of graphene oxide and wet-chemical hydrazine/deionized water solution treatment on the thermoelectric properties of PEDOT:PSS sprayed films. *Synth. Met.* **2016**, *222*, 330–337. [CrossRef]

66. Achaby, M.E.; Arrakhiz, F.E.; Vaudreuil, S.; el Kacem Qaiss, A.; Bousmina, M.; Fassi, F.O. Mechanical, thermal, and rheological properties of graphene-based polypropylene nanocomposites prepared by melt mixing. *Polym. Compos.* **2012**, *33*, 733–744. [CrossRef]

67. Kashiwagi, T.; Du, F.; Douglas, J.F.; Winey, K.I.; Harris, R.H.; Shields, J.R. Nanoparticle networks reduce the flammability of polymer nanocomposites. *Nat. Mater.* **2005**, *4*, 928–933.

68. Kashiwagi, T. Polymer combustion and flammability—Role of the condensed phase. *Symp. Combust.* **1994**, *25*, 1423–1437. [CrossRef]

69. Andersson, T.; Wesslén, B.; Sandström, J. Degradation of low density polyethylene during extrusion. I. Volatile compounds in smoke from extruded films. *J. Appl. Polym. Sci.* **2002**, *86*, 1580–1586. [CrossRef]

70. Mounet, N.; Marzari, N. First-principles determination of the structural, vibrational and thermodynamic properties of diamond, graphite, and derivatives. *Phys. Rev. B* **2005**, *71*, 205214. [CrossRef]

71. Zakharchenko, K.V.; Katsnelson, M.I.; Fasolino, A. Finite Temperature Lattice Properties of Graphene beyond the Quasiharmonic Approximation. *Phys. Rev. Lett.* **2009**, *102*, 46808. [CrossRef] [PubMed]

72. Yoon, D.; Son, Y.W.; Cheong, H. Negative Thermal Expansion Coefficient of Graphene Measured by Raman Spectroscopy. *Nano Lett.* **2011**, *11*, 3227–3231. [CrossRef] [PubMed]

73. Bao, W.; Miao, F.; Chen, Z.; Zhang, H.; Jang, W.; Dames, C.; Lau, C.N. Controlled ripple texturing of suspended graphene and ultrathin graphite membranes. *Nat. Nanotechnol.* **2009**, *4*, 562–566. [CrossRef] [PubMed]

 *nanomaterials* 

Article

Microwave-Assisted Rapid Synthesis of Reduced Graphene Oxide-Based Gum Tragacanth Hydrogel Nanocomposite for Heavy Metal Ions Adsorption

Bhawna Sharma [1], Sourbh Thakur [1,2,*], Djalal Trache [3], Hamed Yazdani Nezhad [4] and Vijay Kumar Thakur [5,6,*]

[1]   School of Chemistry, Faculty of Sciences, Shoolini University, Solan, Himachal Pradesh 173229, India; sharmabhanusln@gmail.com

[2]   Center for Computational Materials Science, Institute of Physics, Slovak Academy of Sciences, 84511 Bratislava, Slovakia

[3]   UER Chimie Appliquée, Ecole Militaire Polytechnique, Bordj El-Bahri, Algiers 16046, Algeria; djalaltrache@gmail.com

[4]   Department of Mechanical Engineering and Aeronautics, City University of London, London EC1V0HB, UK; hamed.yazdani@city.ac.uk

[5]   Biorefining and Advanced Materials Research Center, Scotland's Rural College (SRUC), Kings Buildings, West Mains Road, Edinburgh EH9 3JG, UK

[6]   Department of Mechanical Engineering, School of Engineering, Shiv Nadar University, Uttar Pradesh 201314, India

*   Correspondence: thakursourbh@gmail.com or sourbh.thakur@savba.sk (S.T.); vijay.thakur@sruc.ac.uk (V.K.T.)

Received: 28 June 2020; Accepted: 12 August 2020; Published: 18 August 2020

Abstract: Reduced graphene oxide (RGO) was synthesized in this research via Tour's method for the use of filler in the hydrogel matrix. The copolymerization of N,N-dimethylacrylamide (DMA) onto the gum tragacanth (GT) was carried out to develop gum tragacanth-cl-N,N-dimethylacrylamide (GT-cl-poly(DMA)) hydrogel using N,N'-methylenebisacrylamide (NMBA) and potassium persulfate (KPS) as cross-linker and initiator correspondingly. The various GT-cl-poly(DMA) hydrogel synthesis parameters were optimized to achieve maximum swelling of GT-cl-poly(DMA) hydrogel. The optimized GT-cl-poly(DMA) hydrogel was then filled with RGO to form reduced graphene oxide incorporated gum tragacanth-cl-N,N-dimethylacrylamide (GT-cl-poly(DMA)/RGO) hydrogel composite. The synthesized samples were used for competent adsorption of Hg^{2+} and Cr^{6+} ions. Fourier transform infrared, X-ray powder diffraction, field emission scanning electron microscopy, energy-dispersive X-ray spectroscopy were used to characterize the gum tragacanth-cl-N,N-dimethylacrylamide hydrogel and reduced graphene oxide incorporated gum tragacanth-cl-N,N-dimethylacrylamide hydrogel composite. The experiments of adsorption-desorption cycles for Hg^{2+} and Cr^{6+} ions were carried out to perform the reusability of gum tragacanth-cl-N,N-dimethylacrylamide hydrogel and reduced graphene oxide incorporated gum tragacanth-cl-N,N-dimethylacrylamide hydrogel composite. From these two samples, reduced graphene oxide incorporated gum tragacanth-cl-N,N-dimethylacrylamide exhibited high adsorption ability. The Hg^{2+} and Cr^{6+} ions adsorption by gum tragacanth-cl-N,N-dimethylacrylamide and reduced graphene oxide incorporated gum tragacanth-cl-N,N-dimethylacrylamide were best suited for pseudo-second-order kinetics and Langmuir isotherm. The reported maximum Hg^{2+} and Cr^{6+} ions adsorption capacities were 666.6 mg g^{-1} and 473.9 mg g^{-1} respectively.

Keywords: reduced graphene oxide; gum tragacanth; hydrogel; hydrogel composite; mercury ion; chromium ion; reusability

---

1. Introduction

Continuous industrialization leads to the excessive release of toxic pollutants into various sources of water. The heavy metal poisoning of water has now become a pandemic concern due to its dangerous impacts to human health because these pollutants are non-degradable, poisonous, cancer-causing agent and are hard to separate from water [1]. For metal uptake from water, several approaches have been established, among them, treatment via adsorption is the most appealing one. Adsorption technology is widely used for removing pollutants due to its simple operation, cost and easy implementation [2]. Specifically, hydrogels based on biopolymer have now become very useful in adsorptive wastewater treatment [3]. Gum tragacanth based hydrogel is highly adsorptive because of the presence of hydroxyl (–OH) and carboxyl (–COOH) groups [4,5]. It is a renewable, cost-effective and environmentally friendly polysaccharide that can be easily polymerized to form cross-linked structures [6–8].

Gum tragacanth is commonly found in the sap of different legumes in the Middle East. The biological source of gum tragacanth is a plant named *Astragalus gummifer*. It is a complex mixture of polysaccharides including bassorin and tragacanthin units. When mixed with water, gum tragacanth produces a colloidal hydrosol. The bassorin unit can (composed of 60–70% of the compound) swells to form a gel [9]. Mallakpour et al. reported the glutaraldehyde cross-linked gum tragacanth/CaCO$_3$ hydrogel composite as an adsorbent for the abstraction of Pb^{2+} ion [10]. Moghaddam et al. synthesized methoxyl gum tragacanth-glutamic acid/polyacrylamide hydrogel via electron beam radiations as an adsorbent for trapping uranium ions from toxic uranium solution [11].

The adsorption and stability of hydrogel can be improved by using reduced graphene oxide as filler in the hydrogel matrix. Reduced graphene oxide (RGO) can result in high C/O with better mechanical strength [12]. The reduced graphene oxide is partially decorated with an oxygen-rich functional group that acts as active sites for interaction. The high RGO surface, large porosity and defect sites are the features that help pollutants adsorption [13]. Sahraei et al. reported adsorption of Cr^{6+} metal using chitosan/reduced-graphene oxide/montmorillonite composite hydrogel. The composite hydrogel showed maximum Cr^{6+} absorption of 87.03 mg g^{-1} [14]. Zhuang et al. synthesized molybdenum disulfide/RGO hydrogel as an adsorbent for mercury ions removal [15].

The synthesized gum tragacanth-cl-*N,N*-dimethylacrylamide (GT-cl-poly(DMA)) and reduced graphene oxide incorporated gum tragacanth-cl-*N,N*-dimethylacrylamide (GT-cl-poly(DMA)/RGO) hydrogel composite were efficient in adsorption of Hg^{2+} and Cr^{6+} as compared to previously reported adsorbents in the literature (Table 1). This was due to the perfect combination of reduced graphene oxide (RGO), gum tragacanth (GT), and *N,N*-dimethylacrylamide (DMA) led to the presence of many –OH, –NH$_2$ and –COOH hydrophilic groups. The gum tragacanth-cl-*N,N*-dimethylacrylamide hydrogel and reduced graphene oxide incorporated gum tragacanth-cl-*N,N*-dimethylacrylamide hydrogel composite exhibited the highest removal capacity of 625 mg g^{-1} and 666.6 mg g^{-1} respectively for Hg^{2+}. Similarly, for Cr^{6+}, removal capacities were 401.6 mg g^{-1} and 473.9 mg g^{-1} respectively.

The previously reported works (Table 1) have not comprehensively considered the factors responsible for the high adsorption capability of the adsorbent. In this work, we achieved a better adsorption capacity of 666.6 mg g^{-1} and 473.9 mg g^{-1} for mercury and chromium ions within less time using a low adsorbent dose. Specifically, prepared reduced graphene oxide incorporated gum tragacanth cross-linked poly *N,N*-dimethylacrylamide hydrogel composite shows a very high adsorption percentage of 99% for mercury metal ion under optimal conditions (adsorbent dose = 0.035 g and time = 270 min, T = 25 °C, the concentration of mercury solution = 20 ppm) which means it is highly efficient for mercury adsorption. Also, compared to recently reported studies, we are able to synthesize our adsorbents in very short period (90 s) with high swelling percentage (Table 2) using microwave radiations. This is one of the key points where our synthesis part shows novelty. Hence, we developed the simple and fast synthetic route for the preparation of efficient, sustainable and eco-friendly graphene oxide incorporated gum tragacanth-cl-*N,N*-dimethylacrylamide hydrogel with high adsorption rate for heavy metal ions.

Table 1. Comparison of different adsorbents with gum tragacanth-cl-*N,N*-dimethylacrylamide (GT-cl-poly(DMA)) hydrogel and reduced graphene oxide incorporated gum tragacanth-cl-*N,N*-dimethylacrylamide (GT-cl-poly(DMA)/RGO) hydrogel composite for adsorption of Hg^{2+} and Cr^{6+} metal ions.

Serial Number	Adsorbent	Hg^{2+} Adsorption Capacity (mg g^{-1})	Cr^{6+} Adsorption Capacity (mg g^{-1})	References
1.	Poly(allylamine-co-methacrylamide-co-dimethylthiourea)	198.23	-	[16]
2.	Sulfhydryl-functional paramagnetic solid-phase adsorbent	51.32	-	[17]
3.	Diethylenetriaminepentaacetic acid-modified cellulose adsorbent	476.2	-	[18]
4.	Cross-linked magnetic chitosan-phenylthiourea resin	135.5	-	[19]
5.	Carboxyl methylcellulose and chitosan-derived nanostructured	-	347.0	[20]
6.	Carboxymethyl cellulose–stabilized sulfidated nano zerovalent iron	-	355.9	[21]
7.	Fungal strain (Rhizopus sp.)	-	9.95	[22]
8.	Surfactant-modified Auricularia auricula spent substrate	-	21.74	[23]
9.	GT-cl-poly(DMA) hydrogel and GT-cl-poly(DMA)/RGO hydrogel composite	636.94 and 666.66	-	Present work
10.	GT-cl-poly(DMA) hydrogel and GT-cl-poly(DMA)/RGO hydrogel composite	-	416.66 and 476.19	Present work

Table 2. Comparative analysis for swelling percentage of hydrogels.

Serial Number	Sample	Synthetic Route	Time for Synthesis (s)	Swelling Percentage (%)	References
1.	Carboxymethyl cellulose-cl-poly(lactic acid-co-itaconic acid) hydrogel	Microwave assisted method	90 s	332%	[24]
2.	Tragacanth gum-g-poly(itaconic acid) hydrogel	Microwave-assisted method	220 s	800%	[25]
3.	Chitosan-polyethylene glycol hydrogel membrane	Microwave assisted method	120 s	96.4%	[26]
4.	IPN [(GcA-coll)-cl-poly (AAm-ip-AA)]	Microwave assisted method	150 s	382.1%	[27]
5.	GT-cl-poly(DMA) hydrogel	Microwave assisted method	90 s	957.2%	Present work
6.	GT-cl-poly(DMA)/RGO hydrogel composite	Microwave assisted method	90 s	971.9%	Present work

In this work, we developed first-time gum tragacanth-cl-*N,N*-dimethylacrylamide hydrogel and reduced graphene oxide incorporated gum tragacanth-cl-*N,N*-dimethylacrylamide hydrogel composite for adsorption of Hg^{2+} and Cr^{6+}. The RGO was synthesized from graphite and incorporated in GT-cl-poly(DMA) hydrogel matrix to increase the adsorption efficiency. The sorption study was explained by kinetic and isotherm models. The effect of pH, adsorbent dose and RGO loading on adsorption were performed. The gum tragacanth-cl-*N,N*-dimethylacrylamide hydrogel was systematically designed based on swelling. The adsorbed samples were desorbed successfully by using 0.1 M HNO$_3$ and used further for adsorption experiments.

2. Materials and Methods

2.1. Materials

Gum tragacanth (GT), (molecular weight = 8.4×10^5 g mol^{-1}), potassium persulfate (KPS), *N,N'*-methylenebisacrylamide (NMBA), *N,N*-dimethylacrylamide (DMA) were purchased from Sigma-Aldrich (Sigma Aldrich Chemicals Pvt. Ltd., Delhi, India). Graphite powder, H_3PO_4, $KMnO_4$, H_2O_2 and concentrated H_2SO_4 (98 wt %) were obtained from LOBA-Chemie (Loba Chemie Pvt. Ltd., Tarapur, Maharashtra, India). Hg^{2+} and Cr^{6+} ions solutions were obtained by dissolving mercury chloride (Sigma Aldrich Chemicals Pvt. Ltd., Delhi, India) and potassium chromate (Sigma Aldrich Chemicals Pvt. Ltd., Delhi, India) reagents in distilled water.

2.2. Synthesis of Reduced Graphene Oxide (RGO)

For the synthesis, a mixture of 50 mL concentrated H_2SO_4 and 5.5 mL of H_3PO_4 was taken in a beaker. Then, 1.0 g of graphite powder and $KMnO_4$ (4.0 g) were added to the mixture after maintaining 10 °C under magnetic stirring in ice bath. The mixture was heated to 35 °C governed by a water bath. After 2 h of stirring, the reaction mixture was sonicated 10 times with the help of ultrasonicator. To stop the reaction, deionized water (200 mL) and 1 M NaOH solution were added dropwise to maintain the pH = 6 of solution mixture. The reaction process was then followed with the addition of H_2O_2 (15 mL) which led to a change in suspension color to yellow. The mixture was kept overnight. Thereafter, the solution of ascorbic acid (0.227 mol L^{-1}) was added dropwise under magnetic stirring at 95 °C. In this step, the color of the solution changes slowly from greenish-yellow to black. Finally, the mixture was allowed to settle down for 1 h. Black precipitates were formed which were then centrifuged and washed several times using ethanol.

2.3. Synthesis of Gum Tragacanth-cl-N,N-dimethylacrylamide (GT-cl-poly(DMA)) Hydrogel

Microwave-assisted copolymerization method was used in the synthesis of gum tragacanth-cl-*N,N*-dimethylacrylamide hydrogel [5]. In a typical reaction, 0.5 g of gum tragacanth (GT) (in 11 mL of deionized water) was taken in 50 mL beaker, stirred until GT was uniformly mixed with distilled water. After this, KPS (10×10^{-1} mol L^{-1}) and NMBA (5.8×10^{-1} mol L^{-1}) were added into the GT solution. Magnetic stirring was continued to get a homogenous mixture and then 4.4×10^{-1} mol L^{-1} of DMA was added in this mixture. The solution mixture was placed under microwave radiations for 90 s to generate active radicals needed for the initiation of the polymerization reaction. The prepared gel was washed using acetone and dried inside the preheated (50 °C) hot air oven for 24 h.

2.4. Synthesis of Reduced Graphene Oxide Incorporated Gum Tragacanth-cl-N,N-dimethylacrylamide (GT-cl-poly(DMA)/RGO) Hydrogel Composite

For the preparation of reduced graphene oxide incorporated gum tragacanth-cl-*N,N*-dimethylacrylamide hydrogel composite, 0.5 g of GT was added in a solution of RGO (0.005–0.025 g in 11 mL of deionized water). Physical agitation was applied until the mixture becomes homogenously uniform. Thereafter, the mixture was treated similarly following the procedure in Section 2.3. The incorporation of RGO was confirmed physically by monitoring the color change from light orange to black, as presented in Scheme 1. The optimized quantities that are used in the preparation of hydrogels are given in Table 3.

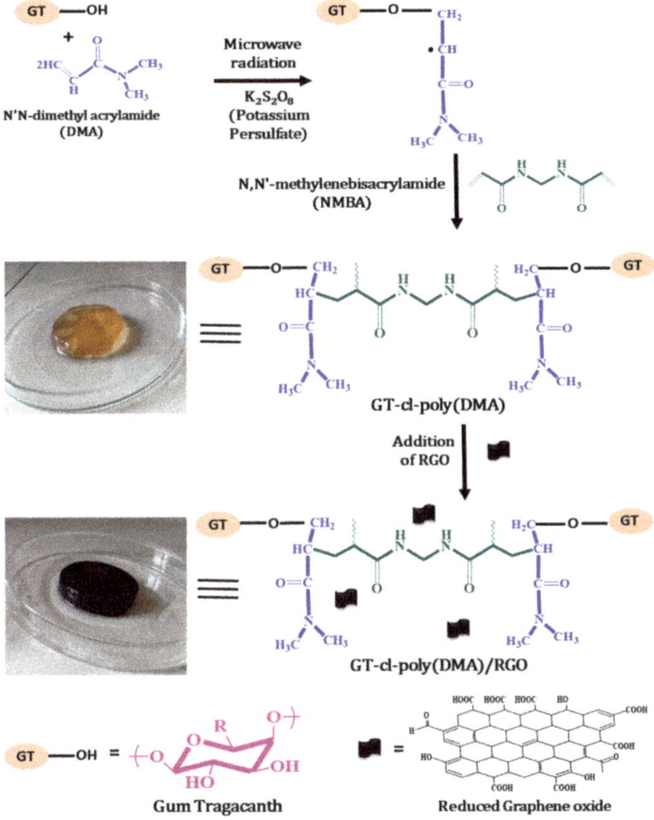

Scheme 1. General scheme for the synthesis of gum tragacanth-cl-*N,N*-dimethylacrylamide hydrogel and reduced graphene oxide incorporated gum tragacanth-cl-*N,N*-dimethylacrylamide hydrogel composite.

Table 3. Optimized quantities used for the preparation of hydrogels.

Serial Number	Sample Name	GT (g)	KPS (g)	DMA (mL)	NMBA (g)	Solvent (mL)	RGO (g)	Swelling %
1.	GT-cl-poly(DMA) hydrogel	0.500	0.030	0.5	0.030	11	-	957.2%
2.	GT-cl-poly(DMA)/RGO hydrogel composite	0.500	0.030	0.5	0.030	11	0.020	971.9%

2.5. Characterization

Fourier transform infrared spectra of GT, RGO, gum tragacanth-cl-*N,N*-dimethylacrylamide hydrogel and reduced graphene oxide incorporated gum tragacanth-cl-*N,N*-dimethylacrylamide hydrogel composite were measured through L1600312 spectrum TWOLITA/ZnSe FTIR spectrophotometer (Agilent Technologies, Santa Clara, CA, USA). Field emission scanning electron microscopy images were collected at different resolutions from a Nova Nano SEM-450 FESEM (JFEI, USA (S.E.A.), Hillsboro, ORE, USA). X-ray powder diffraction was collected from a SmartLab 9 kW rotating anode x-ray diffractometer (Rigaku Corporation, Tokyo, Japan).

2.6. Swelling Study

Various parameters such as initiator concentration (KPS) solvent volume, time, cross-linker concentration (NMBA), microwave power, monomer concentration (DMA) and amount of RGO were optimized to obtain the maximum swelling percentage. The swelling percentages of gum tragacanth-cl-*N*,*N*-dimethylacrylamide and reduced graphene oxide incorporated gum tragacanth-cl-*N*,*N*-dimethylacrylamide hydrogels in deionized water were examined for a fixed period of 16 h. The pre-weighed dry piece of the sample was added in 50 mL of distilled water for 16 h. Then, swelled hydrogel was weighed. The swelling percentage was calculated using Equation (1) [28]:

$$Swelling\ \% = \frac{W_s - W_d}{W_d} \times 100 \tag{1}$$

where, W_s = weight of swelled gum tragacanth-cl-*N*,*N*-dimethylacrylamide hydrogel, W_d = weight of dry gum tragacanth-cl-*N*,*N*-dimethylacrylamide hydrogel.

2.7. Adsorption of Hg^{2+} and Cr^{6+}

The batch adsorption analyses were performed in 150 mL beaker using adsorption shaker (200 rpm) at pH of 5.5 and 3.5 for Hg^{2+} and Cr^{6+} removal respectively. For more illustration, 0.010–0.070 g of adsorbents were used in Hg^{2+} and Cr^{6+} ions solution (50 mL, 20 ppm) at 25 °C for fixed period. After adsorption, the mixture was filtered to determine the concentration of Hg^{2+} and Cr^{6+} ions using 1,3-diphenylcarbazide method [14]. The concentration of adsorbed Hg^{2+} and Cr^{6+} was calculated by evaluating the absorbance of heavy metal ion solution using UV-Vis spectrophotometer at 532 nm and 370 nm respectively. The amount of adsorbed Hg^{2+} and Cr^{6+} was calculated by Equation (2) [28]:

$$q_e = \frac{(C_o - C_e)V}{M} \tag{2}$$

where q_e = equilibrium gum tragacanth-cl-*N*,*N*-dimethylacrylamide and reduced graphene oxide incorporated gum tragacanth-cl-*N*,*N*-dimethylacrylamide adsorption capacity, C_o = initial Hg^{2+} and Cr^{6+} concentration (mg L^{-1}), C_e = equilibrium Hg^{2+} and Cr^{6+} concentration (mg L^{-1}), V = volume (L) of Hg^{2+} and Cr^{6+} ion solution, M = weight (g) of gum tragacanth-cl-*N*,*N*-dimethylacrylamide and reduced graphene oxide incorporated gum tragacanth-cl-*N*,*N*-dimethylacrylamide hydrogels. Adsorption-desorption analyses were conducted by using 0.1 M HNO$_3$. We have optimized the HNO$_3$ concentration (0.02 M, 0.04 M, 0.06 M, 0.08 M, 0.1 M, 0.12 M) for desorption experiments. The maximum adsorption-desorption rate was found at optimized 0.1 M HNO$_3$. The Hg^{2+} and Cr^{6+} loaded GT-cl-poly(DMA) hydrogel and GT-cl-poly(DMA)/RGO hydrogel composite were desorbed by using 100 mL of 0.1 M HNO$_3$ followed by neutralization with 0.1 M NaOH. Finally, the desorbed adsorbent was washed by distilled water and dried at room temperature for further adsorption of Hg^{2+} and Cr^{6+}.

3. Results

3.1. Mechanism for Synthesis of Gum Tragacanth-cl-N,N-dimethylacrylamide Hydrogel

In this work, gum tragacanth-cl-*N*,*N*-dimethylacrylamide hydrogel and reduced graphene oxide incorporated gum tragacanth-cl-*N*,*N*-dimethylacrylamide hydrogel composite were prepared by radical copolymerization of DMA and GT in the presence radical initiator (KPS) and the cross-linking action of NMBA. Under microwave radiations, KPS was decomposed and radical ions were generated [25]. These primary free radicals led to the generation of DMA monomer radical (through addition reaction with KPS) and GT alkoxy radical (through abstraction of hydrogen by KPS). The grafting of DMA radical and GT alkoxy radical was carried through radical copolymerization reaction (Scheme 1). The crosslinker NMBA led to the cross-linkages between different chains to facilitate the construction of three-dimensional gum tragacanth-cl-*N,N*-dimethylacrylamide hydrogel polymeric network. Finally, the dispersion of RGO led to the generation of reduced graphene oxide incorporated gum tragacanth-cl-*N,N*-dimethylacrylamide hydrogel composite.

3.2. Optimization of Swelling for Reduced Graphene Oxide Incorporated Gum Tragacanth-cl-N, N-dimethylacrylamide Hydrogel Composite

3.2.1. Initiator (KPS) Concentration

The swelling percentage of GT-cl-poly(DMA) hydrogel was affected by the concentration of KPS and results are shown in Figure 1a. The maximum swelling of 565.5% was observed at 10×10^{-1} mol L^{-1} of KPS. Below this concentration ($<10 \times 10^{-1}$ mol L^{-1}), the swelling percentage was lower due to inadequate initiator, which was unable to produce appropriate active sites on GT-cl-poly(DMA).

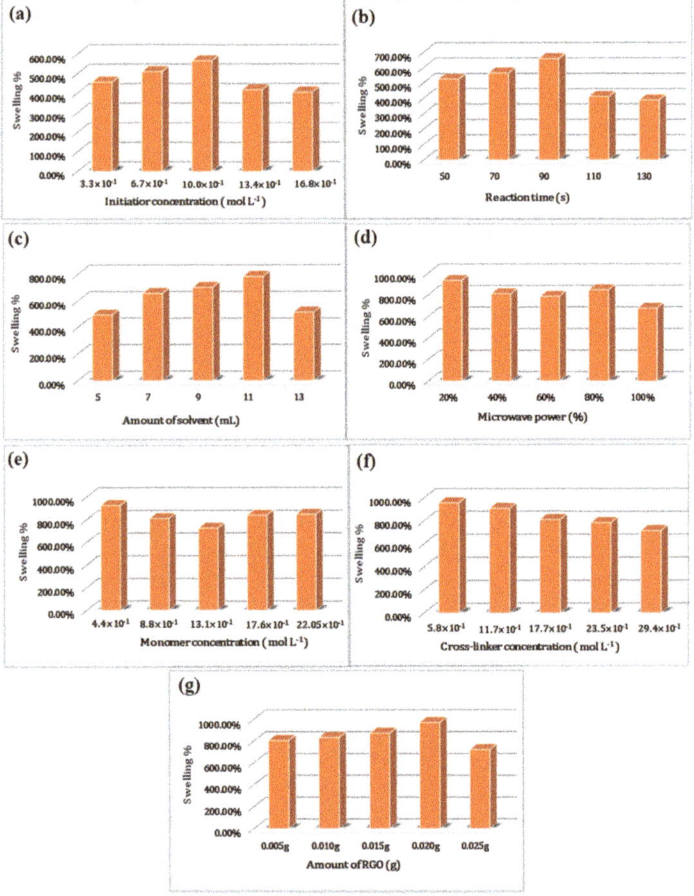

Figure 1. Effect of (**a**) initiator (KPS) concentration, (**b**) reaction time (s), (**c**) amount of solvent (mL), (**d**) microwave power (%), (**e**) monomer (DMA) concentration, (**f**) cross-linker (NMBA) concentration and (**g**) amount of reduced graphene oxide (RGO) on gum tragacanth-cl-*N,N*-dimethylacrylamide hydrogel swelling.

3.2.2. Reaction Time

The GT-cl-poly(DMA) hydrogel showed the highest swelling percentage (657.8%) at 90 s (Figure 1b). The swelling percentage was decreased from 90 s to 130 s, this might be due to the formation of excess branched chains that could inhibit the expansion of the polymer.

3.2.3. Solvent

The solvent volume was varied from 5 mL to 13 mL in the formation of GT-cl-poly(DMA) hydrogel (Figure 1c). The maximum swelling (784.4%) was obtained at solvent volume of 11 mL. At higher solvent volume beyond 11 mL, the swelling percentage of GT-cl-poly(DMA) hydrogel was decreased, the excess solvent volume lowered the concentration of KPS, DMA and NMBA resulted in poor degree of polymerization.

3.2.4. Microwave Power

The swelling percentage was maximum at 20% of microwave power for GT-cl-poly(DMA) hydrogel (Figure 1d). The swelling percentage was lower at microwave power above 20%. This was due to the formation of excess radical led to increase the homo-polymerization rate reaction. Hence, the microwave power was kept at 20% for further synthesis reaction of GT-cl-poly(DMA) hydrogel.

3.2.5. Monomer (DMA) Concentration

The swelling percentage was affected by concentration of monomer (DMA) used in the development of GT-cl-poly(DMA) hydrogel. The maximum swelling percentage was observed at 4.4×10^{-1} mol L^{-1} of DMA concentration (Figure 1e). The increase in the concentration of DMA monomer from 4.4×10^{-1} mol L^{-1} to 22.05×10^{-1} mol L^{-1} led to decrease the swelling percentage due to the self cross-linking of DMA.

3.2.6. Cross-Linker (NMBA)

Figure 1f demonstrates the effect of NMBA concentration (5.8×10^{-1}–29.4×10^{-1} mol L^{-1}) on the swelling percentage of GT-cl-poly(DMA) hydrogel. The recorded highest swelling percentage was 957.2% at NMBA concentration of 5.8×10^{-1} mol L^{-1}. Beyond 5.8×10^{-1} mol L^{-1}, the excess network was developed due to the more cross-linking points. Thus, the excess network led to decrease the available pores and swelling percentage of GT-cl-poly(DMA) hydrogel.

3.2.7. RGO Loading

Figure 1g shows the influence of RGO loading on the swelling of GT-cl-poly(DMA) hydrogel. The rise in the amount of RGO from 0.005 g to 0.020 g was attributed to an increase in swelling percentage. This was due to an increase in hydrophilic group and surface area of GT-cl-poly(DMA) on the incorporation of RGO. Any further increase in RGO loading (> 0.020 g) was found to decrease the swelling percentage of GT-cl-poly(DMA) hydrogel. This might be attributed to the increased cross-linking density of composite hydrogel networks and the aggregations of excessive RGO in the hydrogel matrix.

3.3. FTIR

The FTIR graphs of samples are presented in Figure 2. In spectrum of RGO (Figure 2a), broadband of nearly 3358 cm^{-1} can be attributed to –OH stretching mode. The peak at 1421 cm^{-1} corresponds to the carboxylic acid and peak at 1625 cm^{-1} belongs to the –C=C group in the aromatic rings. The peak at 1128 cm^{-1} is due to –C–O stretching in the C–OH functional groups of RGO [29]. The shifting of –C–O stretching from 1128 cm^{-1} to 1125 cm^{-1} in GT-cl-poly(DMA)/RGO hydrogel composite is related to the successful incorporation of RGO in GT-cl-poly(DMA) hydrogel. The bands at 1638 cm^{-1} and 1748 cm^{-1} correspond to asymmetric stretching of the carboxylate group and asymmetric stretching of

C=O in galacturonic acid respectively [30], peak at 1142 cm^{-1} ascribed to antisymmetric vibrations of C–O–C linkage in glycosidic groups [31]. The asymmetric stretching of C=O shows shifting of peaks from 1748 cm^{-1} to 1750 cm^{-1} after the crosslinking of poly(DMA). The GT-cl-poly(DMA)/RGO hydrogel composite (Figure 2a) shows shift related to asymmetric stretching of C=O from 1750 cm^{-1} to 1758 cm^{-1} suggesting interaction between RGO and GT-cl-poly(DMA) hydrogel. The peaks at 1608 cm^{-1} and 1410 cm^{-1} in GT-cl-poly(DMA) hydrogel ascribed to stretching vibrations of poly(DMA) amide group [32]. These stretching vibrations of poly(DMA) show peak shifting from 1608 cm^{-1} to 1612 cm^{-1} and from 1410 cm^{-1} to 1403 cm^{-1} in GT-cl-poly(DMA)/RGO hydrogel composite which confirms the changes in the structure of poly(DMA) after RGO incorporation. In GT-cl-poly(DMA)/RGO, the broadband of O–H stretching vibration shifted from 3403 cm^{-1} to 3382 cm^{-1} which may be attributed to the RGO interaction with GT-cl-poly(DMA) through intermolecular hydrogen bonds. The peaks intensity of GT-cl-poly(DMA)/RGO hydrogel composite is slightly lower than the GT-cl-poly(DMA) hydrogel, which also confirms the RGO dispersion in GT-cl-poly(DMA)/RGO hydrogel composite. The absorption band at 2910 cm^{-1} was attributed to stretching vibrations of aliphatic C–H [10]. Also, peaks at 1048 cm^{-1} and 1052 cm^{-1} in the spectra of hydrogels correspond to the –C–O bending. After the adsorption of Hg^{2+} and Cr^{6+} on GT-cl-poly(DMA) hydrogel and GT-cl-poly(DMA)/RGO hydrogel composite, peak due to carboxylate groups was shifted from 1612 cm^{-1} to 1621 cm^{-1} and the intensity of the peaks decreases (Figure 2b). The peaks at 1410 cm^{-1} and 1048 cm^{-1} were shifted to 1403 cm^{-1} and 1061 cm^{-1} respectively, which was probably due to the interactions of metal ions to the active site of adsorbent. The peaks intensity of Hg^{2+} loaded GT-cl-poly(DMA) hydrogel and GT-cl-poly(DMA)/RGO hydrogel composite was lower than the Cr^{6+} loaded GT-cl-poly(DMA) hydrogel and GT-cl-poly(DMA)/RGO hydrogel composite, which supports higher Hg^{2+} adsorption than Cr^{6+} adsorption.

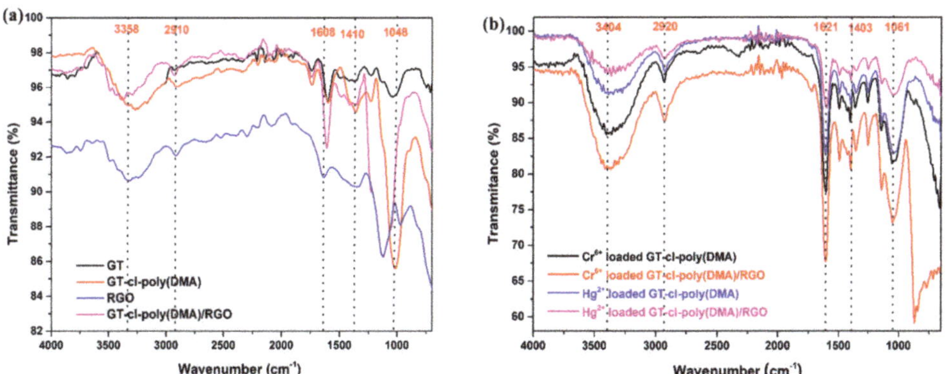

Figure 2. Fourier transform infrared of (**a**) gum tragacanth (GT), gum tragacanth-cl-*N,N*-dimethylacrylamide (GT-cl-poly(DMA)) hydrogel, RGO and reduced graphene oxide incorporated gum tragacanth-cl-*N,N*-dimethylacrylamide (GT-cl-poly(DMA)/RGO) hydrogel composite, (**b**) Hg^{2+} and Cr^{6+} loaded GT-cl-poly(DMA) hydrogel and GT-cl-poly(DMA)/RGO hydrogel composite.

3.4. XRD

The XRD pattern of GT, gum tragacanth-cl-*N,N*-dimethylacrylamide hydrogel (GT-cl-poly(DMA)), RGO and reduced graphene oxide incorporated gum tragacanth-cl-*N,N*-dimethylacrylamide hydrogel composite (GT-cl-poly(DMA)/RGO) is shown in Figure 3. The RGO formation was confirmed by the characteristic peak at $2\theta = 24.3°$ [33]. On applying Bragg's law, the calculated interlayer spacing of RGO was 0.367 nm. Another peak of RGO at $2\theta = 43.6°$ corresponded to the fingermark of graphite indicating the regeneration of graphitic onto RGO [34]. According to Scherrer's formula the calculated particle size of RGO at $2\theta = 24.3°$ was 0.894 nm. In the case of GT, the diffraction peak occurred at 2θ = 22.9° and 26.2°, exhibited semi-crystalline in nature [35]. The slight shifting of the peak at 2θ = 26.6° in XRD pattern of GT-cl-poly(DMA) hydrogel confirms the crosslinking of DMA with polysaccharide by destroying semi-crystalline structure into the amorphous structure. The broad peak in GT-cl-poly(DMA)/RGO hydrogel composite indicates the poor ordered arrangement of RGO in GT-cl-poly(DMA) hydrogel supported by SEM morphology.

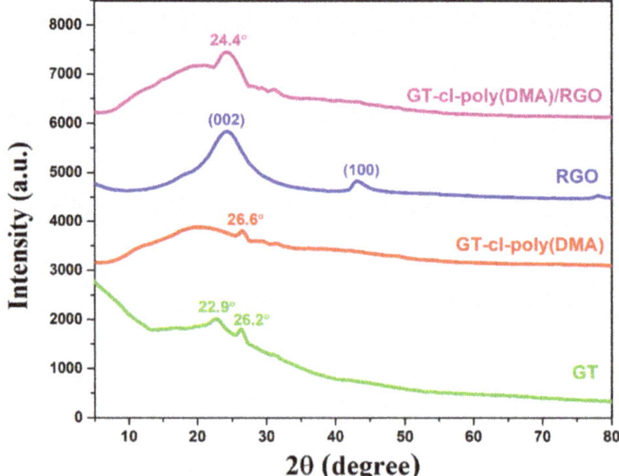

Figure 3. X-ray diffraction pattern for gum tragacanth (GT), gum tragacanth-cl-*N*,*N*-dimethylacrylamide (GT-cl-poly(DMA)) hydrogel, RGO and reduced graphene oxide incorporated gum tragacanth-cl-*N*,*N*-dimethylacrylamide (GT-cl-poly(DMA)/RGO) hydrogel composite.

3.5. SEM

Microscopic images of RGO, gum tragacanth-cl-*N*,*N*-dimethylacrylamide hydrogel and reduced graphene oxide incorporated gum tragacanth-cl-*N*,*N*-dimethylacrylamide hydrogel composite are shown in Figure 4. In RGO, Figure 4a shows the aggregated wrinkled structure, which means particles were closely associated. The RGO morphology showed the formation of agglomerated RGO with estimated average grain size of 20–25 nm. Figure 4b shows the distribution of certain bulges on a quite smooth, porous and compact surface of gum tragacanth-cl-*N*,*N*-dimethylacrylamide hydrogel. After the incorporation of RGO, reduced graphene oxide incorporated gum tragacanth-cl-*N*,*N*-dimethylacrylamide hydrogel composite showed (Figure 4c) the rough and irregular surface with reduced size which was beneficial in fast adsorption of Hg^{2+} and Cr^{6+}.

3.6. EDS

The elemental distribution of carbon, oxygen, nitrogen, mercury and chromium in hydrogel matrix was evaluated by EDS analysis and the spectra for Hg^{2+} adsorbed GT-cl-DMA hydrogel, Hg^{2+} adsorbed GT-cl-poly(DMA)/RGO hydrogel composite, Cr^{6+} adsorbed GT-cl-poly(DMA) and Cr^{6+} adsorbed GT-cl-poly(DMA)/RGO are presented in Figure 5. It is evident from the elemental analysis that Hg^{2+} and Cr^{6+} ions were successfully adsorbed by GT-cl-poly(DMA) and GT-cl-poly(DMA)/RGO. Importantly, the weight percentages of Hg^{2+} and Cr^{6+} ions were higher in the case of Hg^{2+} adsorbed GT-cl-poly(DMA)/RGO (6.68%) (Figure 5b) and Cr^{6+} adsorbed GT-cl-poly(DMA)/RGO (0.86%) (Figure 5d) than the Hg^{2+} adsorbed GT-cl-poly(DMA) (1.20%) (Figure 5a) and Cr^{6+} adsorbed GT-cl-poly(DMA) (0.46%) respectively (Figure 5c). Hence, GT-cl-poly(DMA)/RGO hydrogel composite showed better adsorption capability than GT-cl-poly(DMA) hydrogel. Also, the weight percentage of carbon is higher the in case of GT-cl-poly(DMA)/RGO hydrogel composite than the GT-cl-poly(DMA) hydrogel which confirmed the successful dispersion of RGO in GT-cl-poly(DMA) hydrogel matrix.

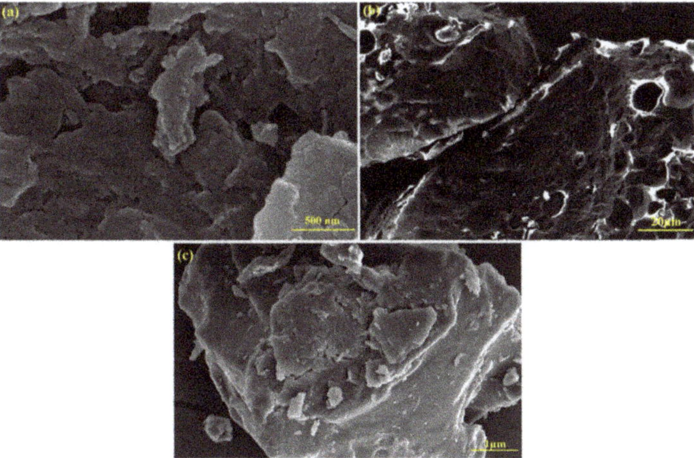

Figure 4. Scanning electron microscope images of (**a**) RGO, (**b**) gum tragacanth-cl-*N,N*-dimethylacrylamide hydrogel and (**c**) reduced graphene oxide incorporated gum tragacanth-cl-*N,N*-dimethylacrylamide hydrogel composite.

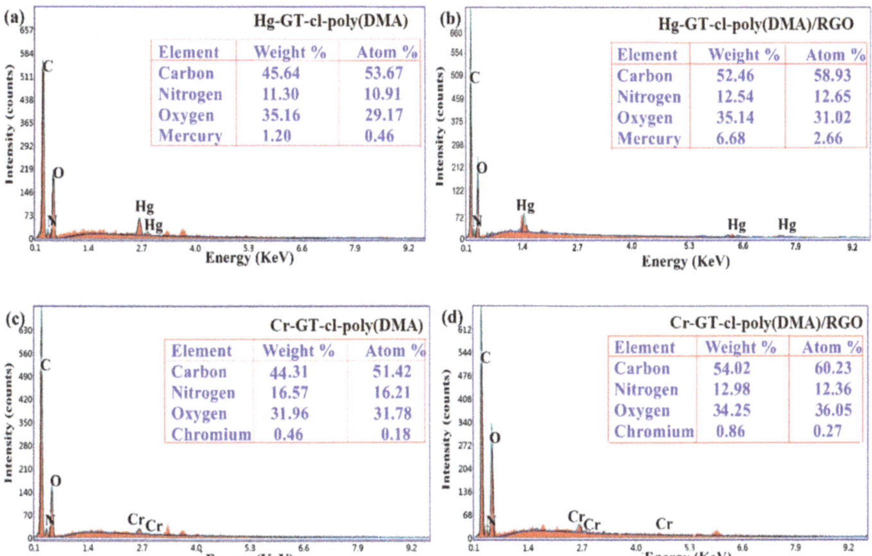

Element	Weight %	Atom %
Carbon	45.64	53.67
Nitrogen	11.30	10.91
Oxygen	35.16	29.17
Mercury	1.20	0.46

Element	Weight %	Atom %
Carbon	52.46	58.93
Nitrogen	12.54	12.65
Oxygen	35.14	31.02
Mercury	6.68	2.66

Element	Weight %	Atom %
Carbon	44.31	51.42
Nitrogen	16.57	16.21
Oxygen	31.96	31.78
Chromium	0.46	0.18

Element	Weight %	Atom %
Carbon	54.02	60.23
Nitrogen	12.98	12.36
Oxygen	34.25	36.05
Chromium	0.86	0.27

Figure 5. Energy dispersive X-ray spectroscopy and elemental weight percentage for (**a**) Hg^{2+} adsorbed gum tragacanth-cl-*N,N*-dimethylacrylamide (GT-cl-poly(DMA)) hydrogel, (**b**) Hg^{2+} adsorbed reduced graphene oxide incorporated gum tragacanth-cl-*N,N*-dimethylacrylamide (GT-cl-poly(DMA)/RGO) hydrogel composite, (**c**) Cr^{6+} adsorbed gum tragacanth-cl-*N,N*-dimethylacrylamide (GT-cl-poly(DMA)) hydrogel and (**d**) Cr^{6+} adsorbed reduced graphene oxide incorporated gum tragacanth-cl-*N,N*-dimethylacrylamide (GT-cl-poly(DMA)/RGO) hydrogel composite.

3.7. Application of Gum Tragacanth-cl-N,N-dimethylacrylamide (GT-cl-poly(DMA)) Hydrogel and Reduced Graphene Oxide Incorporated Gum Tragacanth-cl-N,N-dimethylacrylamide (GT-cl-poly(DMA)/RGO) Hydrogel Composite for Removal of Hg^{2+} and Cr^{6+}

3.7.1. Influence of RGO Loading on the Removal of Hg^{2+} and Cr^{6+}

In the GT-cl-poly(DMA) hydrogel matrix, different quantities of RGO (0.005 g, 0.01 g, 0.015 g, 0.02 g and 0.025 g) were incorporated to study the effect RGO loading on metal ions removal. The adsorption percentages for without RGO were 70.6% and 20.4% for Hg^{2+} and Cr^{6+} respectively. The adsorption percentages for Hg^{2+} (Figure 6a) and Cr^{6+} (Figure 6b) ions were enhanced on raising the concentration of RGO from 0.005 g to 0.020 g. The RGO contains carboxylic groups which boost interactions with metal ions resulting in high percentage adsorption [30]. The removal efficiency of Hg^{2+} and Cr^{6+} was 90.7% and 38.4% at RGO loading of 0.020 g. The development of tough three-dimensional networks was responsible for the decrease in adsorption percentage at higher RGO loading (>0.020). Therefore, 0.020 g was the optimized dose of RGO in the formation of GT-cl-poly(DMA)/RGO for removal of Hg^{2+} and Cr^{6+}.

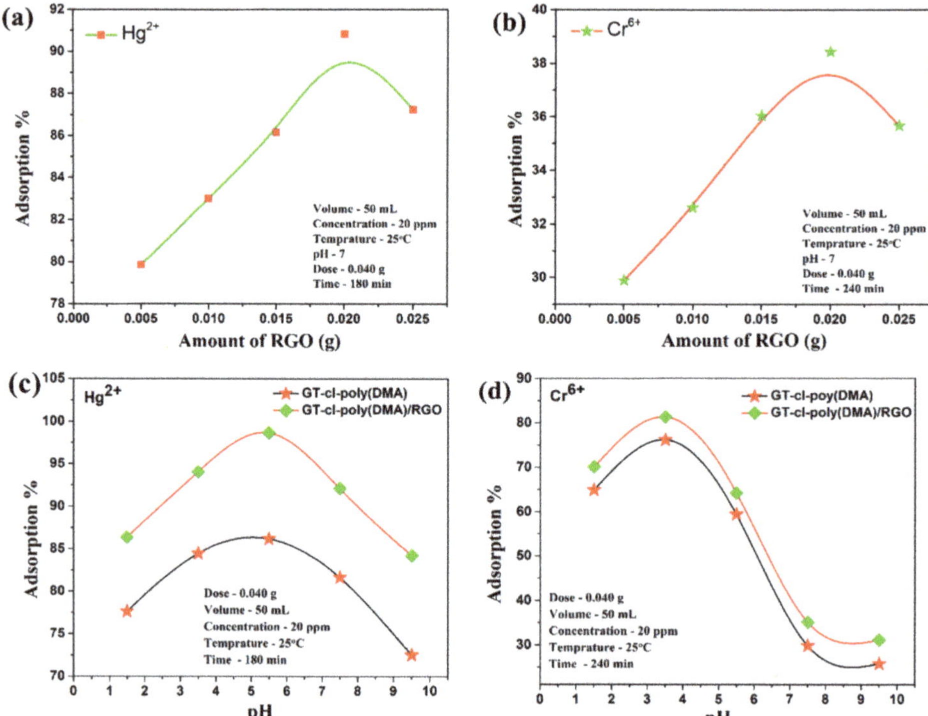

Figure 6. *Cont.*

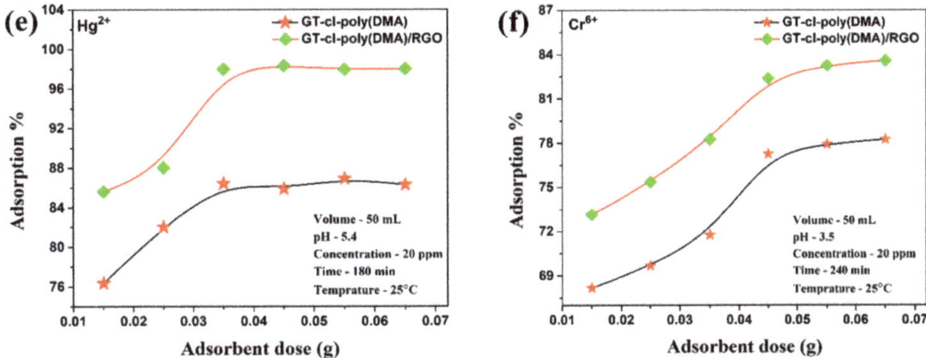

Figure 6. Influence of (**a**) RGO loading on Hg^{2+} adsorption (**b**) RGO loading on Cr^{6+} adsorption, (**c**) pH on adsorption of Hg^{2+}, (**d**) pH on adsorption of Cr^{6+}, (**e**) adsorbent dose on adsorption of Hg^{2+}, (**f**) adsorbent dose on adsorption of Cr^{6+}. Adsorbent: gum tragacanth-cl-*N,N*-dimethylacrylamide (GT-cl-poly(DMA)) hydrogel and reduced graphene oxide incorporated gum tragacanth-cl-*N,N*-dimethylacrylamide (GT-cl-poly(DMA)/RGO) hydrogel composite.

3.7.2. Influence of pH on Removal of Hg^{2+} and Cr^{6+} by GT-cl-poly(DMA) Hydrogel and GT-cl-poly (DMA)/RGO Hydrogel Composite

The impact of pH on adsorption percentage for Hg^{2+} (Figure 6c) and Cr^{6+} (Figure 6d) by gum tragacanth-cl-*N,N*-dimethylacrylamide (GT-cl-poly(DMA)) hydrogel and reduced graphene oxide incorporated gum tragacanth-cl-*N,N*-dimethylacrylamide (GT-cl-poly(DMA)/RGO) hydrogel composite are shown in Figure 6. The removal percentage of Hg^{2+} was first increased from pH 1.5 (77.6% for GT-cl-poly(DMA), 86.3% for GT-cl-poly(DMA)/RGO) to 5.5 (86.1% for GT-cl-poly(DMA), 97.6% for GT-cl-poly(DMA)/RGO) and then decreased from pH 5.5 (86.1%, 97.6%) to 9.5 (72.5%, 84.2%). The reported highest removal efficiencies for Hg^{2+} were 86.1% and 97.6% by GT-cl-poly(DMA) hydrogel and GT-cl-poly(DMA)/RGO) hydrogel composite respectively at 5.5 pH. At low pH, the concentration of H^+ ions was high which could compete with Hg^{2+} on GT-cl-poly(DMA)/RGO surface resulting in poor binding of Hg^{2+} [36]. However, high pH was responsible for the decrease in the concentration of H^+ ions in the solution and improves the binding potential of Hg^{2+} ions to the surface of GT-cl-poly(DMA)/RGO) (Scheme 2a). Hence adsorption of Hg^{2+} was increased from pH 1.1 to 5.5. The dominant species were $Hg(OH)_2$ and $HgCl_4^{2-}$ [37] at pH above 5.5. The electrostatic repulsion among $Hg(OH)_2$ or $HgCl_4^{2-}$ [38] and negatively charged GT-cl-poly(DMA)/RGO) was responsible for low Hg^{2+} adsorption percentage at pH above 5.5.

In the case of Cr^{6+}, the recorded maximum adsorption percentages were 76.1% and 81.5% (Figure 6d) for GT-cl-poly(DMA) hydrogel and GT-cl-poly(DMA)/RGO hydrogel composite correspondingly at pH 3.5. The dominant Cr^{6+} species [39] are as: H_2CrO_4 (pH < 3.5), $HCrO_4^-$ (pH < 7), CrO_4^{2-} (pH > 7). The Cr^{6+} ions were exists in solution as negatively charged $HCrO_4^-$ at pH 3.5. Therefore, electrostatic attraction of $HCrO_4^-$ [40] took place at pH 3.5 with protonated positively charged group of GT-cl-poly(DMA)/RGO) (Scheme 2b). Hence, Cr^{6+} exhibited maximum adsorption percentage at 3.5 pH. At pH < 3.5, electrostatic attraction for adsorption was reduced due to the dominance of H_2CrO_4. Also, with increasing pH from 3.5 to 7, the protonated group on GT-cl-poly(DMA)/RGO decreases which reduces the electrostatic attraction. At pH > 7, electrostatic repulsion between dominant CrO_4^{2-} species [40] and deprotonated GT-cl-poly(DMA)/RGO was attributed to low Cr^{6+} adsorption.

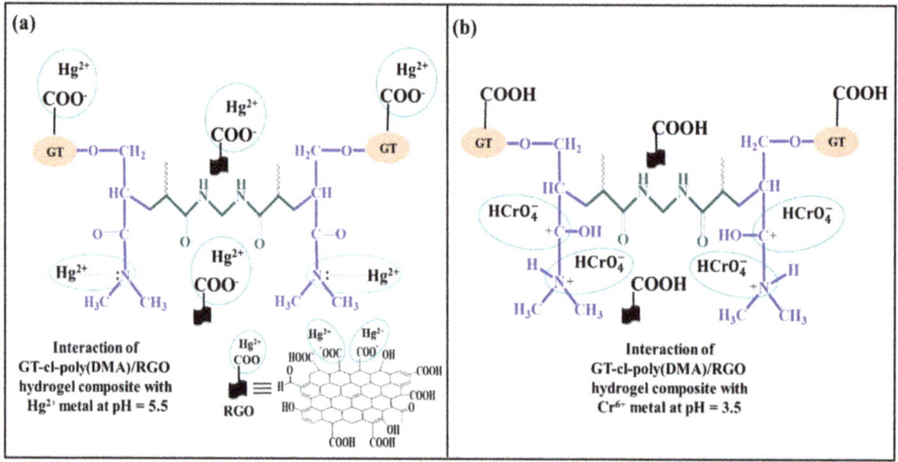

Scheme 2. Possible interactions of (**a**) Hg^{2+} at pH 5.5 and (**b**) Cr^{6+} at pH 3.5 with GT-cl-poly(DMA)/RGO hydrogel composite adsorbent.

3.7.3. Influence of GT-cl-poly(DMA) Hydrogel and GT-cl-poly(DMA)/RGO Hydrogel Composite Dose for Removal of Hg^{2+} and Cr^{6+}

The effect of adsorbents dosages (0.015 – 0.065 g) on the removal of metal ions are represented in Figure 6e,f. The adsorption percentage was increased by increasing the adsorbent dosage. This was due to the existence of more adsorption sites with enhanced dose of adsorbent. The removal efficiencies of Hg^{2+} were found to be 86.4% and 98.4% by GT-cl-poly(DMA) hydrogel and GT-cl-poly(DMA)/RGO hydrogel composite correspondingly at dose of 0.035 g. The reported Cr^{6+} ion removal percentages were 77.2% and 82.3% by using GT-cl-poly(DMA) hydrogel and GT-cl-poly(DMA)/RGO hydrogel composite respectively at optimized dose of 0.045 g. Thus, 0.035 g (for Hg^{2+}) and 0.045 g (for Cr^{6+}) were the ideal doses used for experiments.

3.8. Adsorption Kinetics

The pseudo first-order rate equation is given as:

$$\log\left(q_e - q_t\right) = \log q_e - \frac{K_1}{2.303}t \tag{3}$$

where, q_e and q_t are the adsorption capacity at equilibrium (mg g⁻¹) and time t respectively and K_1 is the pseudo first order kinetics rate constant.

The pseudo second-order rate equation is given as:

$$\frac{t}{q_t} = \frac{1}{K_2\, q_e^2} + \frac{t}{q_e} \tag{4}$$

where K_2 is the pseudo second-order kinetics rate constant.

The removal mechanism for Hg^{2+} and Cr^{6+} by GT-cl-poly(DMA) and GT-cl-poly(DMA)/RGO were solved by different kinetic models as given in Equations (3) and (4). The parameters (pseudo-first-order: R^2, K_1, q_e) were calculated from Figure 7a,b (Table 4). The parameters (pseudo-second-order: K_2, q_e) and correlation coefficient (R^2) were calculated from Figure 7c,d (Table 4). The higher R^2 values for the pseudo-second-order kinetic model supports Hg^{2+} and Cr^{6+} ions adsorption onto GT-cl-poly(DMA) hydrogel and GT-cl-poly(DMA)/RGO hydrogel composite through the pseudo-second-order kinetic model.

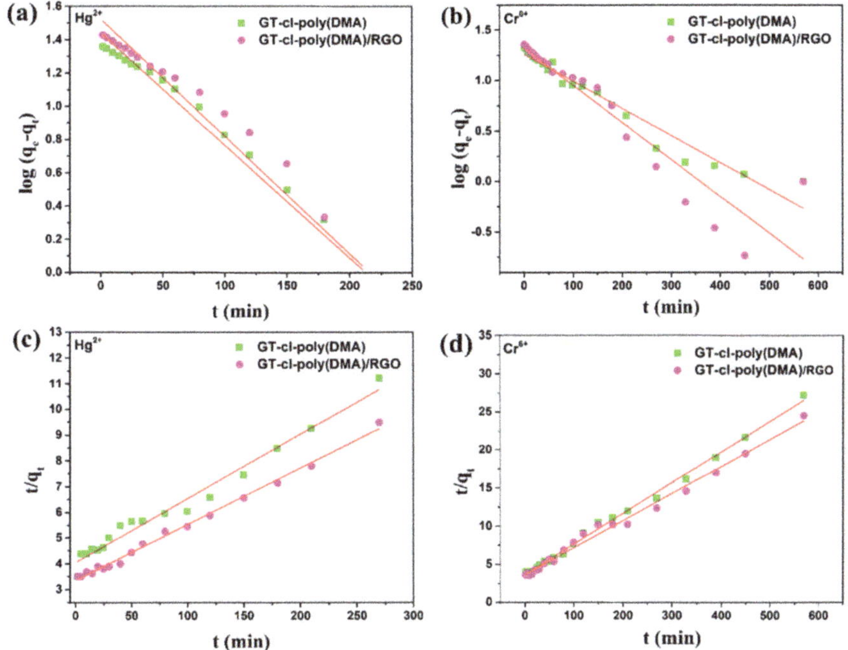

Figure 7. Pseudo first order for (**a**) Hg^{2+} and (**b**) Cr^{6+}, pseudo second order for (**c**) Hg^{2+} and (**d**) Cr^{6+}. (Experimental conditions for Hg^{2+}: adsorbent dose—0.035 g, pH—5.5, metal ion concentration—20 ppm, rpm = 200 and for Cr^{6+}: adsorbent dose—0.045 g, pH—3.5, metal ion concentration—20 ppm, rpm = 200).

Table 4. Kinetics model parameters for Hg^{2+} and Cr^{6+} removal by GT-cl-poly(DMA) hydrogel and GT-cl-poly(DMA)/RGO hydrogel composite.

Kinetic Model	Parameters	GT-cl-poly(DMA)		GT-cl-poly(DMA)/RGO	
		Hg^{2+}	Cr^{6+}	Hg^{2+}	Cr^{6+}
	R^2	0.946	0.931	0.909	0.863
Pseudo-first-order	q_e (cal)	25.7	19.0	33.5	21.0
kinetics	q_e (exp)	28.2	22.6	40.8	25.9
	k_1	0.011	0.008	0.016	0.006
	R^2	0.989	0.995	0.994	0.989
Pseudo-second-order	q_e (cal)	29.4	25	45.8	28.3
kinetics	q_e (exp)	28.2	22.6	40.8	25.9
	k_2	5.90	4.42	1.42	3.44

3.9. Adsorption Isotherms

The Langmuir model is expressed according to Equation (5) as:

$$\frac{C_e}{q_e} = \frac{1}{q_m b} + \frac{C_e}{q_m} \tag{5}$$

where C_e is the equilibrium concentration of metal ions solution, q_e is the amount of equilibrium adsorbed metal ions, q_m is maximum adsorption capacity and b is the Langmuir isotherm constant. The separation factor R_L of Langmuir isotherm was examined by using Equation (6) as:

$$R_L = \frac{1}{1 + b \times C_o} \tag{6}$$

where C_o represent the initial concentration of metal ions. The R_L values show the nature of adsorption i.e. irreversible ($R_L = 0$), linear ($R_L = 1$), unfavorable ($R_L > 1$) and favorable ($0 < R_L < 1$). The Freundlich isotherm model is given by Equation (7) as:

$$\ln q_e = \ln K_F + \frac{1}{n}\ln C_e \tag{7}$$

where K_F and n are Freundlich constants and indicate the adsorption capacity and adsorption intensity of adsorbent respectively.

The interaction between adsorbent (GT-cl-poly(DMA) hydrogel and GT-cl-poly(DMA)/RGO hydrogel composite) and adsorbate (Hg^{2+} and Cr^{6+}) was explained through isotherms model Equations (5) and (7). The Langmuir parameters were calculated from the graph between C_e/q_e and C_e (Figure 8a–d) and presented in Table 5. The Freundlich parameters were determined from the graph of $\ln q_e$ vs $\ln C_e$ (Figure 9a–d) and depicted in Table 5. For the Langmuir isotherm, the higher R^2 suggests that the Langmuir isotherm was best suited for the removal of Hg^{2+} and Cr^{6+} ions on GT-cl-poly(DMA) hydrogel and GT-cl-poly(DMA)/RGO hydrogel composite. For Hg^{2+}, GT-cl-poly(DMA) and GT-cl-poly(DMA)/RGO showed higher removal capacity of 625 mg g^{-1} and 666.6 mg g^{-1} respectively. Similarly, for Cr^{6+}, the maximum reported removal capacities were 401.6 mg g^{-1} and 473.9 mg g^{-1} by GT-cl-poly(DMA) hydrogel and GT-cl-poly(DMA)/RGO hydrogel composite respectively.

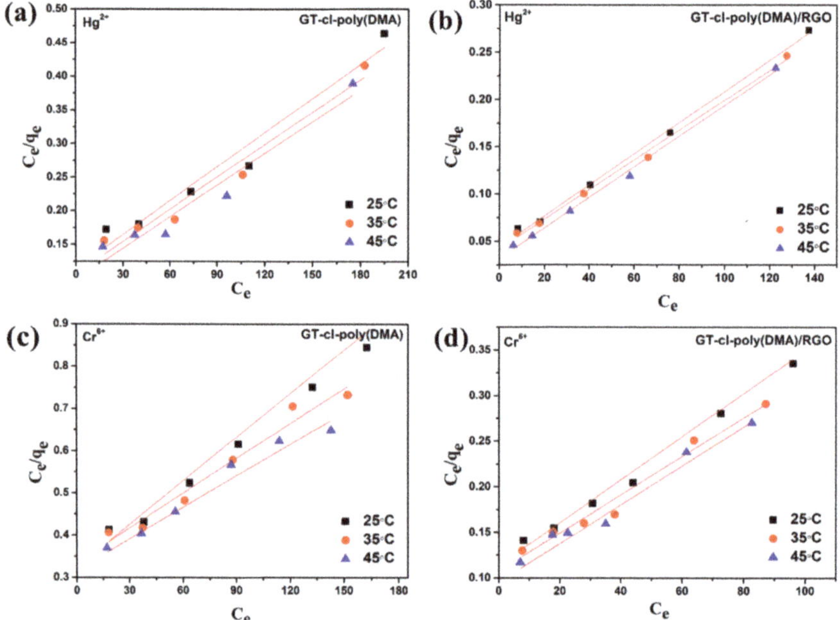

Figure 8. Langmuir isotherm model for Hg^{2+} adsorption by (**a**) gum tragacanth-cl-*N,N*-dimethylacrylamide (GT-cl-poly(DMA)) hydrogel (**b**) reduced graphene oxide incorporated gum tragacanth-cl-*N,N*-dimethylacrylamide (GT-cl-poly(DMA)/RGO) hydrogel composite, Langmuir isotherm model for Cr^{6+} adsorption by (**c**) gum tragacanth-cl-*N,N*-dimethylacrylamide (GT-cl-poly(DMA)) hydrogel (**d**) reduced graphene oxide incorporated gum tragacanth-cl-*N,N*-dimethylacrylamide (GT-cl-poly(DMA)/RGO) hydrogel composite. (Experimental conditions for Hg^{2+}: adsorbent dose—0.035 g, pH—5.5, metal ion concentration—20–300 ppm, rpm = 200 and for Cr^{6+}: adsorbent dose—0.045 g, pH—3.5, metal ion concentration—20–500 ppm, rpm = 200).

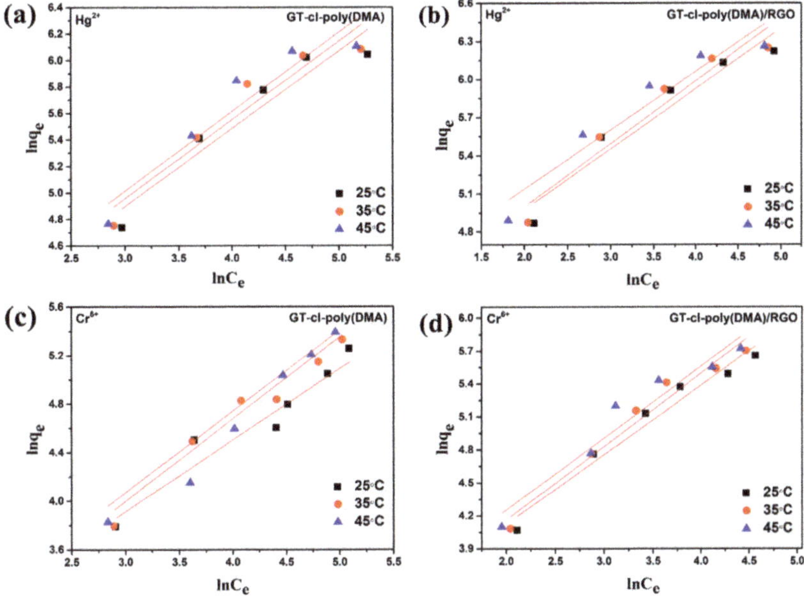

Figure 9. Freundlich isotherm model for Hg^{2+} adsorption by (**a**) gum tragacanth-cl-*N,N*-dimethylacrylamide (GT-cl-poly(DMA)) hydrogel (**b**) reduced graphene oxide incorporated gum tragacanth-cl-*N,N*-dimethylacrylamide (GT-cl-poly(DMA)/RGO) hydrogel composite, Freundlich isotherm model for Cr^{6+} adsorption by (**c**) gum tragacanth-cl-*N,N*-dimethylacrylamide (GT-cl-poly(DMA)) hydrogel (**d**) reduced graphene oxide incorporated gum tragacanth-cl-*N,N*-dimethylacrylamide (GT-cl-poly(DMA)/RGO) hydrogel composite, (Experimental conditions for Hg^{2+}: adsorbent dose—0.035 g, pH—5.5, metal ion concentration—20–300 ppm, rpm = 200 and for Cr^{6+}: adsorbent dose—0.045 g, pH—3.5, metal ion concentration—20–500 ppm, rpm = 200).

Table 5. Isotherm model parameters for Hg^{2+} and Cr^{6+} adsorption by GT-cl-poly(DMA) hydrogel and GT-cl-poly(DMA)/RGO hydrogel composite.

Isotherm Models	Temperature	Parameters	GT-cl-poly(DMA)		GT-cl-poly(DMA)/RGO	
			Hg^{2+}	Cr^{6+}	Hg^{2+}	Cr^{6+}
Langmuir	25 °C	q_m (mg g^{-1})	591.7	289.8	628.9	423.7
		b (L mg^{-1})	0.014	0.010	0.036	0.020
		R_L	0.405–0.121	0.839–0.370	0.217–0.052	0.856–0.339
		R^2	0.987	0.964	0.996	0.989
	35 °C	q_m (mg g^{-1})	621.1	362.3	662.2	467.2
		b (L mg^{-1})	0.015	0.008	0.035	0.019
		R_L	0.400–0.119	0.874–0.452	0.220–0.054	0.869–0.371
		R^2	0.960	0.962	0.994	0.989
	45 °C	q_m (mg g^{-1})	625	401.6	666.6	473.9
		b (L mg^{-1})	0.016	0.007	0.046	0.022
		R_L	0.385–0.112	0.887–0.438	0.167–0.039	0.864–0.353
		R^2	0.963	0.956	0.995	0.9623
Freundlich	25 °C	K_F (mg g^{-1})	1.144	0.765	1.393	1.05
		n	1.72	1.70	2.12	1.58
		R^2	0.889	0.898	0.905	0.955
	35 °C	K_F (mg g^{-1})	1.156	0.679	1.389	1.038
		N	1.69	1.47	2.04	1.52
		R^2	0.891	0.895	0.914	0.955
	45 °C	K_F (mg g^{-1})	1.168	0.726	1.438	1.080
		n	1.66	1.49	2.17	1.53
		R^2	0.868	0.876	0.898	0.924

3.10. Relationship between the Adsorption and Swelling

For the investigation of the correlation between the swelling of GT-cl-poly(DMA)/RGO hydrogel composite and adsorption of the Hg^{2+} and Cr^{6+} onto GT-cl-poly(DMA)/RGO hydrogel composite, the adsorption percentage was determined using aqueous metal solution (20 mg L^{-1}) and the swelling experiments were performed in distilled water. The relationship between the adsorption values versus swelling values is presented in Figure 10. It is clear from Figure 10 that the adsorption percentage is directly proportional to the swelling percentage of the adsorbent. The adsorption percentages for Hg^{2+} and Cr^{6+} were increased from 78.9% to 90.7% and 29.8% to 38.4% respectively when the swelling percentage of GT-cl-poly(DMA)/RGO rise from 834.6% to 971.9%.

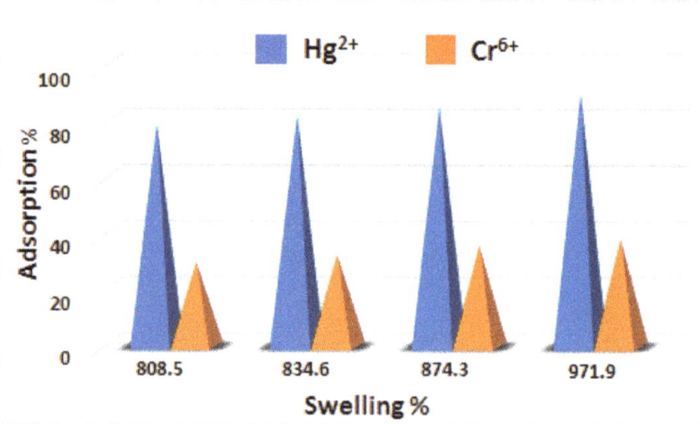

Figure 10. Adsorption percentage and swelling percentage of reduced graphene oxide incorporated gum tragacanth-cl-*N,N*-dimethylacrylamide (GT-cl-poly(DMA)/RGO) hydrogel composite.

3.11. Adsorption-Desorption Study

For an ideal adsorbent, ability for regeneration without considerable loss of removal percentage is of paramount importance. The five cycles of adsorption-desorption and their effects on percentage adsorption are presented in Figure 11. The Hg^{2+} ions adsorption percentages were 83.4% (1st cycle), 80.9% (2nd cycle), 78.4% (3rd cycle), 75% (4th cycle), 73.6% (5th cycle) and 94.1% (1st cycle), 92.7% (2nd cycle), 89.8% (3rd cycle), 87.9% (4th cycle), 85.5% (5th cycle) for GT-cl-poly(DMA) hydrogel and GT-cl-poly(DMA)/RGO hydrogel composite respectively (Figure 11a). For Cr^{6+}, GT-cl-poly(DMA) and GT-cl-poly(DMA)/RGO exhibited 77.2% (1st cycle), 74.7% (2nd cycle), 71% (3rd cycle), 68.4% (4th cycle), 66.3% (5th cycle) and 82.3% (1st cycle), 80.1% (2nd cycle), 78.5% (3rd cycle), 76.9% (4th cycle), 73.1% (5th cycle) respectively (Figure 11b). Hence, the synthesized GT-cl-poly(DMA) hydrogel and GT-cl-poly(DMA)/RGO hydrogel composite can be effectively reused for up to five cycles, which leads to reduction in cost.

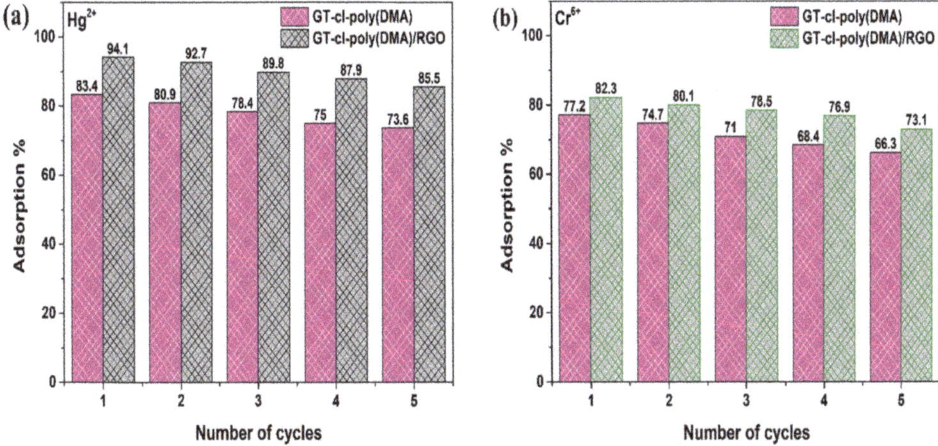

Figure 11. Recycling ability of gum tragacanth-cl-*N,N*-dimethylacrylamide (GT-cl-poly(DMA)) hydrogel and reduced graphene oxide incorporated gum tragacanth-cl-*N,N*-dimethylacrylamide (GT-cl-poly(DMA)/RGO) hydrogel composite for the removal of (**a**) Hg^{2+} and (**b**) Cr^{6+} metal ions up to five cycles.

4. Conclusions

We developed novel reduced graphene oxide incorporated gum tragacanth-cl-*N,N*-dimethylacrylamide (GT-cl-poly(DMA)/RGO) hydrogel composite as reusable adsorbent for Hg^{2+} and Cr^{6+} ions. The reported maximum swelling percentage was 971.9% for reduced graphene oxide incorporated gum tragacanth-cl-*N,N*-dimethylacrylamide hydrogel composite at optimized synthesis conditions (KPS concentration: 10.0×10^{-1} mol L^{-1}, solvent: 11 mL, reaction time: 90 s, microwave power: 20%, DMA concentration: 4.4×10^{-1} mol L^{-1}, NMBA concentration: 5.8×10^{-1} mol L^{-1} and amount of RGO: 0.020 g). The adsorption efficiencies of 99% and 82% were reported for Hg^{2+} and Cr^{6+} by using GT-cl-poly(DMA)/RGO hydrogel composite at optimized adsorption conditions (for Hg^{2+}, pH: 5.5, adsorbent dose: 0.035 g, RGO loading: 0.020 g, Hg^{2+} concentration: 20 ppm, Hg^{2+} volume: 50 mL, time: 270 min, temperature: 25 °C and for Cr^{6+}, pH: 3.5, adsorbent dose: 0.045 g, RGO loading: 0.020 g, Cr^{6+} concentration: 20 ppm, Cr^{6+} volume: 50 mL, time: 570 min, temperature: 25 °C). The Q_{max} of Hg^{2+} and Cr^{6+} onto reduced graphene oxide incorporated gum tragacanth-cl-*N,N*-dimethylacrylamide hydrogel composite were 666.6 mg g^{-1} and 473.9 mg g^{-1} correspondingly, which were higher than the Q_{max} of gum tragacanth-cl-*N,N*-dimethylacrylamide hydrogel (Hg^{2+} = 625 mg g^{-1}, Cr^{6+} = 401.6 mg g^{-1}). The Hg^{2+} and Cr^{6+} adsorption were better depicted through pseudo-second-order and Langmuir isotherm. The gum tragacanth-cl-*N,N*-dimethylacrylamide and reduced graphene oxide incorporated gum tragacanth-cl-*N,N*-dimethylacrylamide adsorbents can be effectively reused for up to five cycles for adsorption of Hg^{2+} and Cr^{6+} ions. Thus, developed adsorbents are highly efficient in heavy metal ion adsorption and can be exploited for environmental remediation application.

Author Contributions: Experiments, B.S.; original draft writing, B.S., S.T.; data analysis, B.S., S.T.; writing, review and editing, S.T., V.K.T., D.T., H.Y.N.; supervision, S.T., V.K.T. All authors have read and agreed to the published version of the manuscript.

Funding: This research received no external funding.

References

1. Siddiqui, E.; Pandey, J. Assessment of heavy metal pollution in water and surface sediment and evaluation of ecological risks associated with sediment contamination in the Ganga River: A basin-scale study. *Environ. Sci. Pollut. Res.* **2019**, *26*, 10926–10940. [CrossRef] [PubMed]
2. Thakur, S.; Sharma, B.; Verma, A.; Chaudhary, J.; Tamulevicius, S.; Thakur, V.K. Recent approaches in guar gum hydrogel synthesis for water purification. *Int. J. Polym. Anal. Chem.* **2018**, *23*, 621–632. [CrossRef]
3. Thakur, S.; Sharma, B.; Verma, A.; Chaudhary, J.; Tamulevicius, S.; Thakur, V.K. Recent progress in sodium alginate based sustainable hydrogels for environmental applications. *J. Clean. Prod.* **2018**, *198*, 143–159. [CrossRef]
4. Hussain, I.; Sayed, S.M.; Fu, G. Facile and cost-effective synthesis of glycogen-based conductive hydrogels with extremely flexible, excellent self-healing and tunable mechanical properties. *Int. J. Biol. Macromol.* **2018**, *118*, 1463–1469. [CrossRef]
5. Chaudhary, J.; Thakur, S.; Sharma, M.; Gupta, V.K.; Thakur, V.K. Development of Biodegradable Agar-Agar/Gelatin-Based Superabsorbent Hydrogel as an Efficient Moisture-Retaining Agent. *Biomolecules* **2020**, *10*, 939. [CrossRef]
6. Nazarzadeh, E.Z.; Makvandi, P.; Tay, F.R. Recent progress in the industrial and biomedical applications of tragacanth gum: A review. *Carbohydr. Polym.* **2019**, *212*, 450–467. [CrossRef] [PubMed]
7. Ates, B.; Koytepe, S.; Ulu, A.; Gurses, C.; Thakur, V.K. Chemistry, Structures, and Advanced Applications of Nanocomposites from Biorenewable Resources. *Chem. Rev.* **2020**. [CrossRef] [PubMed]
8. Mohammadinejad, R.; Maleki, H.; Larrañeta, E.; Fajardo, A.R.; Nik, A.B.; Shavandi, A.; Sheikhi, A.; Ghorbanpour, M.; Farokhi, M.; Govindh, P. Status and future scope of plant-based green hydrogels in biomedical engineering. *Appl. Mat. Today* **2019**, *16*, 213–246. [CrossRef]
9. Nejatian, M.; Abbasi, S.; Azarikia, F. Gum Tragacanth: Structure, characteristics and applications in foods. *Int. J. Biol. Macromol.* **2020**, *160*, 846–860. [CrossRef]
10. Mallakpour, S.; Abdolmaleki, A.; Tabesh, F. Ultrasonic-assisted manufacturing of new hydrogel nanocomposite biosorbent containing calcium carbonate nanoparticles and tragacanth gum for removal of heavy metal. *Ultrason. Sonochem.* **2018**, *41*, 572–581. [CrossRef]
11. Moghaddam, R.H.; Dadfarnia, S.; Shabani, A.M.H.; Tavakol, M. Synthesis of composite hydrogel of glutamic acid, gum tragacanth, and anionic polyacrylamide by electron beam irradiation for uranium (VI) removal from aqueous samples: Equilibrium, kinetics, and thermodynamic studies. *Carbohydr. Polym.* **2019**, *206*, 352–361. [CrossRef] [PubMed]
12. Guex, L.G.; Sacchi, B.; Peuvot, K.F.; Andersson, R.L.; Pourrahimi, A.M.; Ström, V.; Farris, S.; Olsson, R.T. Experimental review: Chemical reduction of graphene oxide (GO) to reduced graphene oxide (rGO) by aqueous chemistry. *Nanoscale* **2017**, *9*, 9562–9571. [CrossRef] [PubMed]
13. Peng, W.; Li, H.; Liu, Y.; Song, S. A review on heavy metal ions adsorption from water by graphene oxide and its composites. *J. Mol. Liq.* **2017**, *230*, 496–504. [CrossRef]
14. Yu, P.; Wang, H.-Q.; Bao, R.-Y.; Liu, Z.; Yang, W.; Xie, B.-H.; Yang, M.-B. Self-assembled sponge-like chitosan/reduced graphene oxide/montmorillonite composite hydrogels without cross-linking of chitosan for effective Cr (VI) sorption. *ACS Sustain. Chem. Eng.* **2017**, *5*, 1557–1566. [CrossRef]
15. Zhuang, Y.-T.; Zhang, X.; Wang, D.-H.; Yu, Y.-L.; Wang, J.-H. Three-dimensional molybdenum disulfide/graphene hydrogel with tunable heterointerfaces for high selective Hg (II) scavenging. *J. Colloid Interf. Sci.* **2018**, *514*, 715–722. [CrossRef]
16. Kim, M.Y.; Seo, H.; Lee, T.G. Removal of Hg (II) ions from aqueous solution by poly (allylamine-co-methacrylamide-co-dimethylthiourea). *J. Ind. Eng. Chem.* **2020**, *84*, 82–86. [CrossRef]
17. Zhang, M.; Ma, J.; Xiao, Y.; Zhang, C.; Wang, Q.; Zheng, W. Preparation sulfhydryl functionalized paramagnetic Ni0. 25Zn0. 75Fe2O4 microspheres for separating Pb (II) and Hg (II) ions from aqueous solution. *Colloids Surf. A* **2020**, *586*, 124205. [CrossRef]
18. Li, B.; Li, M.; Zhang, J.; Pan, Y.; Huang, Z.; Xiao, H. Adsorption of Hg (II) from aqueous solution by diethylenetriaminepentaacetic acid-modified cellulose. *Int. J. Biol. Macromol.* **2019**, *122*, 149–156. [CrossRef]
19. Monier, M.; Abdel-Latif, D.A. Preparation of cross-linked magnetic chitosan-phenylthiourea resin for adsorption of Hg (II), Cd (II) and Zn (II) ions from aqueous solutions. *J. Hazard. Mater.* **2012**, *209*, 240–249. [CrossRef]

20. Li, S.-S.; Wang, X.-L.; An, Q.-D.; Xiao, Z.-Y.; Zhai, S.-R.; Cui, L.; Li, Z.-C. Upon designing carboxyl methylcellulose and chitosan-derived nanostructured sorbents for efficient removal of Cd (II) and Cr (VI) from water. *Int. J. Biol. Macromol.* **2020**, *143*, 640–650. [CrossRef]

21. Zhao, L.; Zhao, Y.; Yang, B.; Teng, H. Application of Carboxymethyl Cellulose–Stabilized Sulfidated Nano Zerovalent Iron for Removal of Cr (VI) in Simulated Groundwater. *Water Air Soil Pollut.* **2019**, *230*, 113. [CrossRef]

22. Espinoza-Sánchez, M.A.; Arévalo-Niño, K.; Quintero-Zapata, I.; Castro-González, I.; Almaguer-Cantú, V. Cr (VI) adsorption from aqueous solution by fungal bioremediation based using Rhizopus sp. *J. Environ. Manage.* **2019**, *251*, 109595. [CrossRef] [PubMed]

23. Dong, L.; Jin, Y.; Song, T.; Liang, J.; Bai, X.; Yu, S.; Teng, C.; Wang, X.; Qu, J.; Huang, X. Removal of Cr (VI) by surfactant modified Auricularia auricula spent substrate: Biosorption condition and mechanism. *Environ. Sci. Pollut. Res.* **2017**, *24*, 17626–17641. [CrossRef] [PubMed]

24. Sood, S.; Gupta, V.K.; Agarwal, S.; Dev, K.; Pathania, D. Controlled release of antibiotic amoxicillin drug using carboxymethyl cellulose-cl-poly (lactic acid-co-itaconic acid) hydrogel. *Intern. J. Biol. Macromol.* **2017**, *101*, 612–620. [CrossRef]

25. Pathania, D.; Verma, C.; Negi, P.; Tyagi, I.; Asif, M.; Kumar, N.S.; Al-Ghurabi, E.H.; Agarwal, S.; Gupta, V.K. Novel nanohydrogel based on itaconic acid grafted tragacanth gum for controlled release of ampicillin. *Carbohydr. Polym.* **2018**, *196*, 262–271. [CrossRef]

26. Wang, Z.; Zhao, Z.; KhaN, N.R.; Hua, Z.; Huo, J.; Li, Y. Microwave assisted chitosan-polyethylene glycol hydrogel membrane synthesis of curcumin for open incision wound healing. *Pharmazie* **2020**, *75*, 118–123.

27. Kaur, S.; Jindal, R.; Kaur Bhatia, J. Synthesis and RSM-CCD optimization of microwave-induced green interpenetrating network hydrogel adsorbent based on gum copal for selective removal of malachite green from waste water. *Polym. Eng. Sci.* **2018**, *58*, 2293–2303. [CrossRef]

28. Verma, A.; Thakur, S.; Mamba, G.; Gupta, R.K.; Thakur, P.; Thakur, V.K. Graphite modified sodium alginate hydrogel composite for efficient removal of malachite green dye. *Inter. J. Biol. Macromol.* **2020**, *148*, 1130–1139. [CrossRef]

29. Makhado, E.; Pandey, S.; Ramontja, J. Microwave assisted synthesis of xanthan gum-cl-poly(acrylic acid) based-reduced graphene oxide hydrogel composite for adsorption of methylene blue and methyl violet from aqueous solution. *Int. J. Biol. Macromol.* **2018**, *119*, 255–269. [CrossRef]

30. Rahmani, Z.; Sahraei, R.; Ghaemy, M. Preparation of spherical porous hydrogel beads based on ion-crosslinked gum tragacanth and graphene oxide: Study of drug delivery behavior. *Carbohydr. Polym.* **2018**, *194*, 34–42. [CrossRef]

31. Martín-Alfonso, J.E.; Číková, E.; Omastová, M. Development and characterization of composite fibers based on tragacanth gum and polyvinylpyrrolidone. *Compos. B Eng.* **2019**, *169*, 79–87. [CrossRef]

32. Pandey, V.S.; Verma, S.K.; Yadav, M.; Behari, K. Guar gum-gN,N'-dimethylacrylamide: Synthesis, characterization and applications. *Carbohydr. Polym.* **2014**, *99*, 284–290. [CrossRef] [PubMed]

33. Zhao, M.; Tesfay Reda, A.; Zhang, D. Reduced Graphene Oxide/ZIF-67 Aerogel Composite Material for Uranium Adsorption in Aqueous Solutions. *ACS Omega* **2020**, *5*, 8012–8022. [CrossRef] [PubMed]

34. Wu, J.; Wei, Y.; Ding, H.; Wu, Z.; Yang, X.; Li, Z.; Huang, W.; Xie, X.; Tao, K.; Wang, X. Green Synthesis of 3D Chemically Functionalized Graphene Hydrogel for High-Performance NH_3 and NO_2 Detection at Room Temperature. *ACS Appl. Mater. Inter.* **2020**, *12*, 20623–20632. [CrossRef]

35. Mallakpour, S.; Tabesh, F. Tragacanth gum based hydrogel nanocomposites for the adsorption of methylene blue: Comparison of linear and non-linear forms of different adsorption isotherm and kinetics models. *Int. J. Biol. Macromol.* **2019**, *133*, 754–766. [CrossRef]

36. Khraisheh, M.A.M.; Al-Ghouti, M.A.; Allen, S.J.; Ahmad, M.N.M. The effect of pH, temperature, and molecular size on the removal of dyes from textile effluent using manganese oxides-modified diatomite. *Water Environ. Res.* **2004**, *76*, 2655–2663. [CrossRef]

37. Al-Ghouti, M.A.; Da'ana, D.; Abu-Dieyeh, M.; Khraisheh, M. Adsorptive removal of mercury from water by adsorbents derived from date pits. *Sci. Rep.* **2019**, *9*, 1–15. [CrossRef]

38. Mei, J.; Zhang, H.; Li, Z.; Ou, H. A novel tetraethylenepentamine crosslinked chitosan oligosaccharide hydrogel for total adsorption of Cr (VI). *Carbohydr. Polym.* **2019**, *224*, 115154. [CrossRef]

39. Li, Y.-S.; Li, T.-T.; Song, X.-F.; Yang, J.-Y.; Liu, G.; Qin, J.-T.; Dong, Z.-B.; Chen, H.-G.; Liu, Y. Enhanced adsorption-photocatalytic reduction removal for Cr (VI) based on functionalized TiO_2 with hydrophilic monomers by pre-radiation induced grafting-ring opening method. *Appl. Surf. Sci.* **2020**, *514*, 145789. [CrossRef]

40. Vilela, P.B.; Dalalibera, A.; Duminelli, E.C.; Becegato, V.A.; Paulino, A.T. Adsorption and removal of chromium (VI) contained in aqueous solutions using a chitosan-based hydrogel. *Environ. Sci. Pollut. Res.* **2019**, *26*, 28481–28489. [CrossRef]

 *nanomaterials*

Article

Synthesis and Characterization of Multi-Walled Carbon Nanotube/Graphene Nanoplatelet Hybrid Film for Flexible Strain Sensors

JianRen Huang [1,2], **Shiuh-Chuan Her** [1,*], **XiaoXiang Yang** [2,3] and **MaNan Zhi** [1,2]

[1] Department of Mechanical Engineering, Yuan Ze University, Chung-Li 320, Taiwan;
    chinafzhjr@gmail.com (J.R.H.); zhimnlucky@163.com (M.N.Z.)
[2] School of Mechanical Engineering and Automation, Fuzhou University, Fuzhou 350108, China;
    yangxx@fzu.edu.cn
[3] Vice President Office, Quanzhou Normal University, Quanzhou 362000, China
* Correspondence: mesch@saturn.yzu.edu.tw; Tel.: +886-3-463-8800

Received: 11 September 2018; Accepted: 2 October 2018; Published: 4 October 2018

Abstract: Graphene nanoplatelet (GNP) and multi-walled carbon nanotube (MWCNT) hybrid films were prepared with the aid of surfactant Triton X-100 and sonication through a vacuum filtration process. The influence of GNP content ranging from 0 to 50 wt.% on the mechanical and electrical properties was investigated using the tensile test and Hall effect measurement, respectively. It showed that the tensile strength of the hybrid film is decreasing with the increase of the GNP content while the electrical conductivity exhibits an opposite trend. The effectiveness of the MWCNT/GNP hybrid film as a strain sensor is presented. The specimen is subjected to a flexural loading, and the electrical resistance measured by a two-point probe method is found to be function of applied strain. Experimental results demonstrate that there are two different linear strain-sensing stages (0–0.2% and 0.2–1%) in the resistance of the hybrid film with applied strain. The strain sensitivity is increasing with the increase of the GNP content. In addition, the repeatability and stability of the strain sensitivity of the hybrid film were conformed through the cyclic loading–unloading tests. The MWCNT/GNP hybrid film shows promising application for strain sensing.

Keywords: graphene nanoplatelet; multi-walled carbon nanotube; hybrid film; vacuum filtration; strain sensing

---

1. Introduction

Since the discoveries of carbon nanotubes (CNT) by Iijima [1] and graphene nanoplatelets (GNP) by Novoselov et al. [2], they have received a great attention as raw materials for the development of nanomaterials due to their excellent thermal, electrical and mechanical properties, low density and high specific surface area [3]. Nowadays, enormous efforts have been devoted to the use of CNT and graphene in various applications, such as energy storage [4,5], field effect transistors (FETs) [6], electrodes [7,8] and sensors [9,10]. Thin films or paper-like materials consisting of GNP or CNT have drawn extensive attention and they are being widely employed for supercapacitors [11], pressure sensors [12], monitoring cure behavior of polymer composite [13], flexible temperature sensors [14], and as reinforcing fillers in polymers [15–17]. These free-standing thin films are cohesively bound by van der Waals interactions among entangled CNTs and GNPs. The main idea behind the fabrication of thin film is to utilize the excellent properties of individual GNP and CNT in macroscopic form. This thin film is advantageous to facilitate easier handling of GNPs and CNTs and to improve the capability of using GNPs and CNTs in industry [18]. These thin films are suitable for both lightweight structural and functional applications.

Flexible strain sensors have been highly desirable in applications such as electronic skin, structural health monitoring, and robot sensors in recent years. CNT and GNP are applicable for piezoresistive strain sensors and have been of great interest among researchers. Lu et al. [19] employed a flexible GNP/epoxy strain sensor to monitor the deformation and damage in structural composites. Liu et al. [20] reported a highly reliable strain sensor based on graphene composite film with layered structure. Moriche et al. [21] studied the strain monitoring mechanism of GNPs incorporated into epoxy matrix. Sanli et al. [22] investigated the piezoresistive performance of strain-sensitive MWCNT/epoxy nanocomposites. Wang et al. [23] utilized a CNT composite film as a strain sensor to monitor biaxial strain under tensile tests. Wang et al. [24] developed a new processing technique of MWCNT strain sensors with tunable strain gauge factors. Natarajan et al. [25] examined the efficiency and effectiveness in terms of piezoresistive properties of natural rubber composites based on MWCNT, carbon black and their mixtures (hybrid). The change of relative resistance, was found to be as much as $\sim$1300 at around 120% elongation. Boland et al. [26] incorporated graphene into a lightly cross-linked polysilicone, resulting in a change of its electromechanical properties substantially. These nanocomposites were sensitive electromechanical sensors with gauge factors >500 that can measure pulse, blood pressure, and even the impact associated with the footsteps of a small spider. Li. et al. [27] fabricated flexible and electrical conductive carbon cotton/polydimethylsiloxane composites by vacuum-assisted infusion for highly sensitive pressure sensor. The flexible pressure sensor exhibited a maximum sensitivity of $6.04\,KPa^{-1}$ in a wide working pressure up to 700 kPa. Samad et al. [28] developed a graphene foam/polydimethylsiloxane flexible sensor to sense both compressive and bending strains in the form of change in electrical resistance. They found that resistances can be increased to 120% and 52% of its original value by applying a 30% compressive strain and bending a sample to a radius of 1 mm, respectively. Samad et al. [29] fabricated freestanding, mechanically stable, and highly electrically conductive graphene foam with two-step facile, adaptable and scalable techniques. They demonstrated the capability of graphene foam as strain/pressure sensor for both high and low strains and pressures with tunable densities.

The potential applications of CNTs and GNPs are limited because CNTs are easy to entangle and agglomerate due to the large aspect ratio and GNPs also tend to restack due to van der Waals and strong interactions. One of the most efficient ways to avoid the agglomeration is to incorporate CNTs with GNPs to produce a nanocomposite material or a hybrid. CNTs can bridge adjacent graphene layers and retard graphene interlayer stacking, resulting in an increased contact surface area between GNPs. Hybrid CNT/graphene films are typically bonded by π–π interaction, which can induce functionalization due to the difference in geometry between the GNP and the CNT. Apart from the non-covalent interaction, covalent bonds and hydrogen bonds have also been used to construct hybrid graphene and CNT nanomaterials, which are confirmed to be of ultrahigh strength, modulus, electrical conductivity and thermal performance [17]. Most of the existing literature, as mentioned above, studied the sensing capability of the carbon nanomaterials such as MWCNT and GNP individually. Relatively few studies have been reported on electrical and mechanical properties of hybrid films. This work seeks to explore the piezoresistive behavior and strain-sensing ability of the MWCNT/GNP hybrid film. The synergistic effect of MWCNT and GNP on the electrical conductivity is presented. These are the novelty and originality of present work. In this study, MWCNT/GNP hybrid films were fabricated by vacuum filtration of mixed dispersion with varied MWCNT-to-GNP weight ratios. A series of hybrid films with different amounts of MWCNTs and GNPs were prepared. The loading of GNPs varied from 0 wt.% to 50 wt.% and that for MWCNTs was 100 wt.% to 50 wt.%, respectively. The effect of GNP content on mechanical properties, electrical conductivity and strain-sensing performance of hybrid films are investigated. A controllable strain sensitivity of the hybrid film can be achieved by varying the GNP content. It is important in understanding the MWCNT/GNP hybrid films so as to further improve their properties for end applications.

2. Experiments

In this study, two types of nanomaterials, a multi-walled carbon nanotube (MWCNT) and a graphene nanoplatelet (GNP), were used to fabricate the hybrid films with different weight ratios.

2.1. Materials

GNPs were purchased from UChees Co. (Taiwan, China) with 1~10 nm thickness, 0.5–20 µm lateral dimension and surface area of 400–700 m^2/g. MWCNTs, grown by CVD, were purchased from Conjutek Co. Taiwan with the diameter in the range of 10–50 nm, length of 100–200 µm, surface area of 400–700 m^2/g, and purity >98.5%. Both of GNPs and MWCNTs were used as received without any modification.

2.2. Film Preparation

MWCNTs and GNPs have strong tendencies to form bundles and aggregate together because of their high surface area and the strong van der Waals interaction. They are also hydrophobic and have poor solubility in aqueous solutions [30]. In this study, the MWCNT/GNP hybrid films were prepared with the aid of surfactant Triton X-100. The molecular structure of surfactant Triton X-100 (Big Sun Chemical Corp., New Taipei City, Taiwan) contains a hydrophilic polyethylene oxide group and hydrocarbon lipophilic (or hydrophobic group), which improves the dispersibility of MWCNTs and GNPs in aqueous solution [31]. The hybrid films with different weight ratios of MWCNT and GNP were prepared using the following process. The total mass of MWCNTs and GNPs was held constant at 0.16 g. Surfactant Triton X-100 with the weight of 5 g were dissolved in 500 ml deionized water and dispersed by a sonicate tip (Q700, Qsonica L.L.C., Newtown, CT, USA) for 30 min at 30 W. The sonicator was operated at pulse mode (10 s on and 20 s off). Then, a total mass 0.16 g of MWCNTs and GNPs with a desired weight ratio were added to the suspension and dispersed by a sonicate tip for 3h at 30 W. Upon completion of the dispersion process, the MWCNT/GNP suspension was filtered through a Polytetrafluoroethylene (PTFE) microporous membrane (pore size 0.45 µm, diameter 90 mm) by a vacuum filtration. The experimental setup of the vacuum filtration is shown in Figure 1. After filtration, the hybrid film was peeled off from the filter membrane and washed by a large amount of isopropyl alcohol to remove any residual surfactant. The film was dried in a vacuum oven preheated to 40 °C for 12h. The typical thickness of the hybrid film was 60–80 µm. Following the same process, a series of hybrid films with GNP weight percentage ranging from 0 wt.% to 50 wt.% were fabricated to investigate the effect of GNP on the mechanical and electrical properties. In this study, hybrid films with GNP weight percentages of 0%, 10%, 20%, 30%, 40% and 50% were denoted as GNP-0, GNP-10, GNP-20, GNP-30, GNP-40 and GNP-50, respectively.

Figure 1. Experimental setup of the vacuum filtration and as-prepared hybrid film.

2.3. Morphology

Field emission scanning electron microscope (JSM-7600F, Jeol Ltd., Tokyo, Japan) was performed to characterize the surface morphology and cross-section view of the hybrid film. An accelerating voltage of 10 kV and a working distance of 5–10 mm were adopted to generate the field emission scanning electron microscope (FESEM) images of the hybrid film. The samples were sputter-coated with a conductive gold layer before taken the image.

A typical MWCNT/GNP hybrid film is shown in Figure 1 It can be seen that the film is highly flexible, which can be rolled up or bear small radius bending without any damage or fracture. The surface morphology and cross-section view of the hybrid film with 0 wt.% (GNP-0), 20 wt.% (GNP-20) and 50 wt.% (GNP-50) of GNP are presented in Figure 2. The surface morphology of the film GNP-0 (0 wt.% GNP and 100 wt.% MWCNT) exhibits homogenous and densely packed mass of randomly oriented MWCNTs without any agglomeration, and this orientation gives rise to its isotropic properties as shown in Figure 2a. For the hybrid film GNP-20 (20 wt.% GNP and 80 wt.% MWCNT), most of the GNPs are covered by the MWCNTs as shown in Figure 2c. As the content of GNP increases, some of the GNPs can be observed on the top of MWCNTs as shown in Figure 2e for hybrid film GNP-50 (50 wt.% GNP and 50 wt.% MWCNT). GNPs and MWCNTs are uniformly dispersed and highly entangled with each other. From the cross-section SEM images shown in Figure 2b,d and f, MWCNTs and GNPs are successfully deposited to form densely packed film with layered structure. Similar layered structure with MWCNTs distributed between GNP sheets in the flexible GNP/MWCNT film using as a high performance supercapacitor was also reported by Lu et al. [11]. It can be attributed to the filtration-induced directional flow during the fabrication process. Under the vacuum filtration pressure, the 2D GNP tended to self-adjust their basal planes parallel to the filter membrane plane due to the large aspect ratio of GNP sheets, resulting in significant alignment of GNP sheets [32]. Clearly, GNP sheets served as the supporters to hold the MWCNTs in-between, generating a more compact and aligned structure of hybrid films. Graphene sheets uniformly spread on MWCNTs and alternately stacked layer structure are observed. With the increase of GNP content, the long and tortuous MWCNTs are embedded between the GNP layers, which can prevent the aggregation of GNPs. It appears that MWCNTs were preferentially oriented and bridged the gap between the GNP layers.

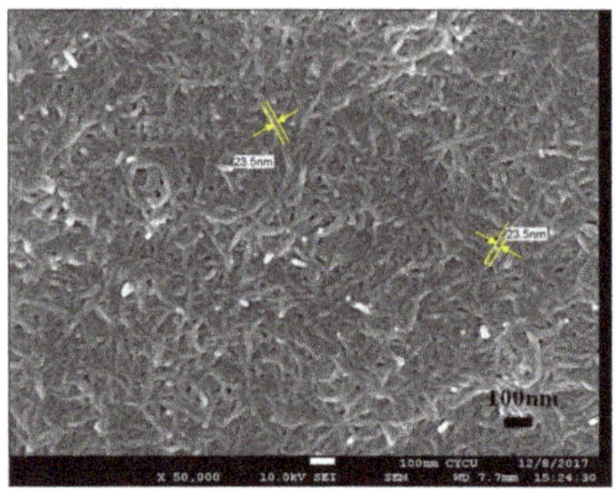

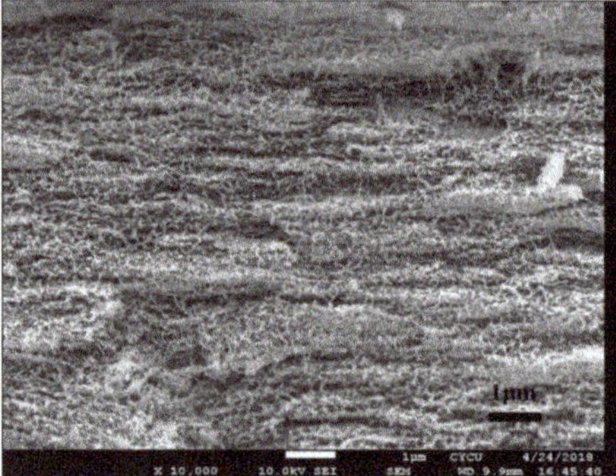

(a) GNP-0 surface morphology   **(b)** GNP-0 cross-section view

Figure 2. *Cont.*

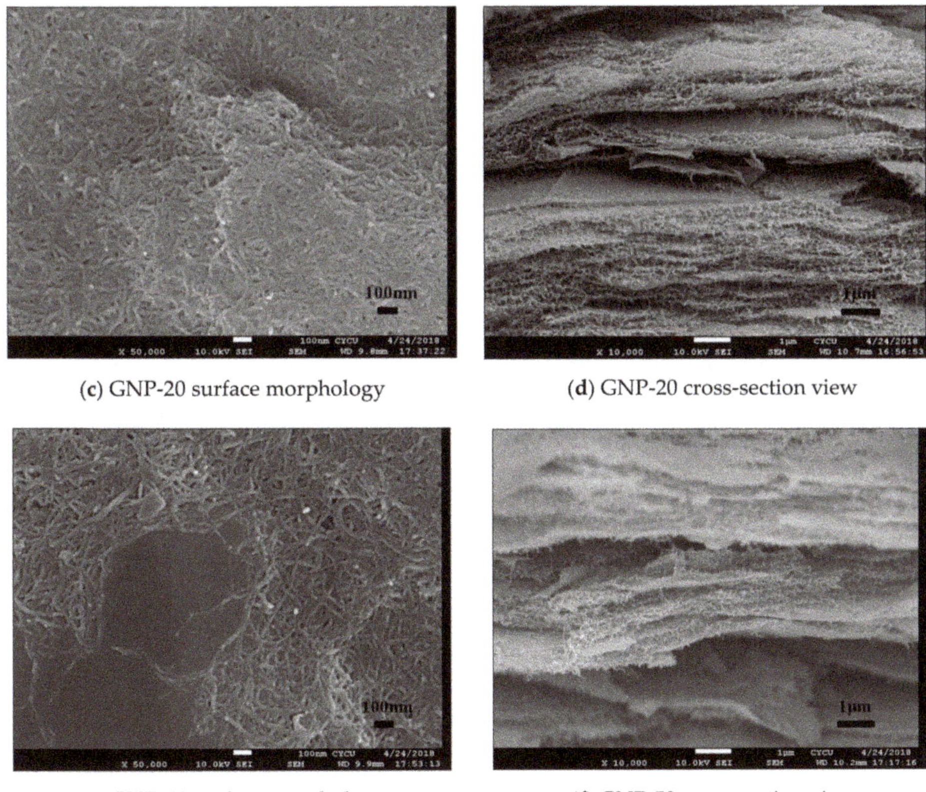

(**c**) GNP-20 surface morphology   (**d**) GNP-20 cross-section view

(**e**) GNP-50 surface morphology   (**f**) GNP-50 cross-section view

Figure 2. SEM images of hybrid films (**a**) surface morphology of GNP-0; (**b**) cross-section view of GNP-0; (**c**) surface morphology of GNP-20; (**d**) cross-section view of GNP-20; (**e**) surface morphology of GNP-50; (**f**) cross-section view of GNP-50.

3. Results and Discussions

3.1. Mechanical Properties

The mechanical properties of hybrid film were evaluated by uniaxial tensile testing. The specimens were cut into a rectangle strip with 30 mm in length and 10 mm in width. Tests were conducted using a universal testing machine with 200 N load cell at a constant cross-head speed of 0.5 mm/min. To reveal the reproducibility of the results, three samples were fabricated and tested for each hybrid film. The experimental results reported in this work are the averaged values.

Figure 3 plots the typical stress–strain curves of the hybrid films with different GNP contents ranging from 0 wt.% to 50 wt.%. The mechanical properties including the Young's modulus, tensile strength and fracture strain can be extracted from the stress–strain curve. Table 1 lists the tensile strength and fracture strain of the hybrid films. Based on published literature, the mechanical properties of CNT buckypaper, tensile strength of 2–94 MPa, Young's modulus of 2.1 MPa to 3.84 GPa, and fracture strain of 0.3–2% have been reported [18]. Present results are within the range of typical MWCNT buckypaper. It can be observed that both the tensile strength and fracture strain are decreasing with the increase of the GNP content as shown in the inset of Figure 3. Tensile strength and fracture strain of GNP-0 (0 wt.% GNP and 100 wt.% MWCNT) are 17 MPa and 8.2%, respectively, which are 105% and 86% higher than that of GNP-50 (50 wt.% GNP and 50 wt.% MWCNT). These results can be inferred from the SEM images that the MWCNT bundles exhibit not only strong Van der Waals and π–π interactions but also mechanical interlocking through entanglements and form a strong robust network [33]. However, the graphene sheets are mainly assembled by an in-plane contacting via Van der Waals forces without being strongly inter-connected. Thus, the tensile strength of the hybrid film is decreasing as the MWCNT content decreases.

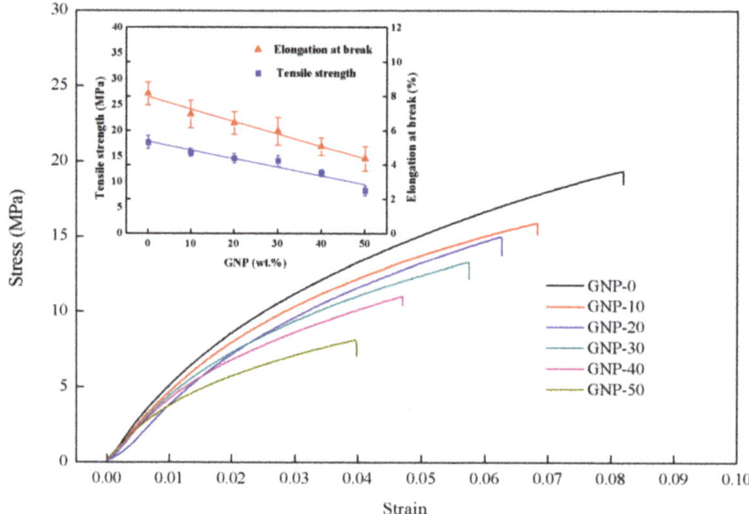

Figure 3. Stress-strain curves of the hybrid films with different weight percentage of graphene nanoplatelets (GNP).

Table 1. Mechanical properties of the hybrid film with different weight percentage of graphene nanoplatelets (GNP).

Hybrid Film	Tensile Strength (MPa)	Fracture Strain (%)
GNP-0	17 ± 1.3	8.2 ± 0.7
GNP-10	16 ± 0.8	7.0 ± 0.8
GNP-20	15 ± 1.0	6.5 ± 0.7
GNP-30	14 ± 1.1	6.0 ± 0.8
GNP-40	12 ± 0.6	5.1 ± 0.5
GNP-50	8.3 ± 0.9	4.4 ± 0.7

The enlarged stress–strain curves as shown in Figure 4 can be divided into three stages. In stage I (strain ranging from 0 to 0.2%), the wavy MWCNTs are first straightened upon tensile loading, causing little change in stress with linear stress-strain relationship [17]. In stage II (strain ranging from 0.2 to 1.0%), the deformation happens under fairly law stress and the joints between MWCNTs and GNPs inside the film are stretched resulting in a higher elastic modulus which is analogized to the disentanglement of polymer chain [18]. In stage III (strain >1.0%), with further stretching of the hybrid film, interlock between GNPs and MWCNTs gradually fails and the network becomes loosely, a non-linear stress-strain relationship is observed. At the initial stage of loading, significant straightening took place both in the GNPs [34] and MWCNTs leading to alignment along the tensile direction, after which the curves became almost linear at higher strains. Elastic moduli of the hybrid films in stages I and II of the tensile testing are shown in the inset of Figure 4. The elastic modulus of the hybrid film in stage II is higher than that of stage I by approximately 200 MPa. It appears that GNPs were easier to be straightened due to the slippage of the overlapped GNPs; i.e., it is more flexible than MWCNTs which were interlocked with each other. The tensile strength and Young's modulus of the hybrid film measured as a function of GNP content are shown in Figures 3 and 4, respectively. Both properties consistently increased with decreasing GNP content, indicating the dominance of MWCNT on mechanical properties of the hybrid film.

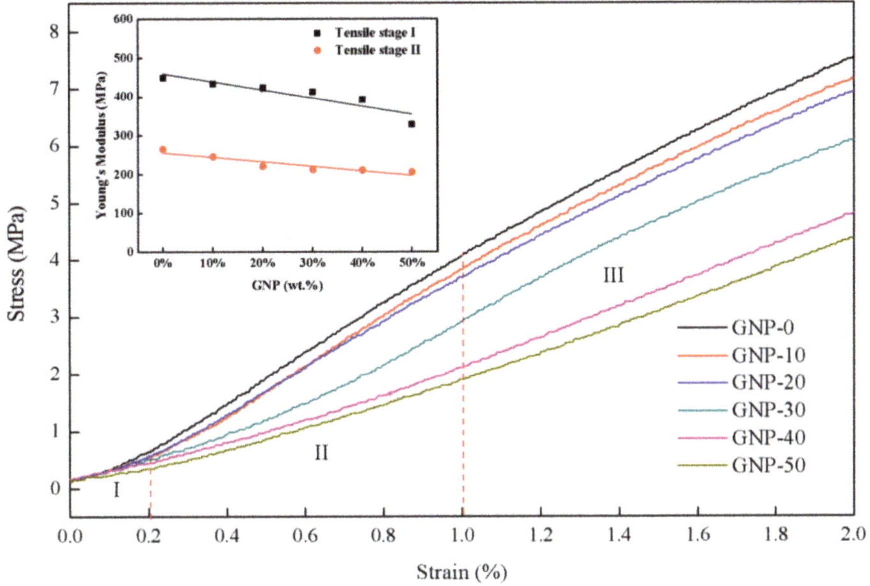

Figure 4. Enlarged stress –strain curve of MWCNT/GNP hybrid films.

3.2. Electrical Properties

The sheet resistivity of the hybrid film was measured by the Hall effect (Ecopia HMS-3000). To demonstrate the reproducibility of the experimental results, three specimens with square shape (10 × 10 mm) were cut from different locations of the hybrid film and tested. The average value was reported with standard deviation.

Table 2 lists the electrical resistivity and conductivity of the hybrid film with different weight percentages of GNP. It can be seen that the conductivity of the hybrid film is increasing with the increase of the content of GNP as shown in Figure 5. While both MWCNTs and GNPs are highly conductive, GNPs are more conductive for two reasons. Firstly, their two-dimensional nature results in a better connectivity and so a greater choice of conductive paths for electrons to flow through. Secondly, their planar nature allows them to pack more closely than MWCNTs, giving lower porosity [35]. Thus, GNP is the dominant factor on the electrical property of the hybrid film. A remarkable increase in the electrical conductivity from 47.72 S/cm to 192.60 S/cm was observed when the GNP content was increased from 0 to 50 wt.%. The conductivity of the hybrid film GNP-50 was enhanced by 304% in comparison with the GNP-0, due to the formation of 3D conductive networks [36]. The hybrid film exhibits a well-stacked layered structure throughout the cross section. The MWCNT network bridges the gap between the GNPs. Larger lateral dimension of GNP acts as a strong holder while MWCNT serves as a wire to connect GNP. The conductivity of the hybrid film depends on the conductive network formed by the MWCNTs and the inherent conductivity of GNPs. At a low weight fraction of GNP, MWCNT and GNP are not close-packed to form effective conductive pathways in the hybrid film. The overlap of MWCNTs introduces larger interfacial resistance that further decreases the conductivity of the hybrid film with too much MWCNTs in the grapheme layer. When the fractions of GNPs were increased, the percolated network of MWCNTs and GNPs was formed which provided efficiently conductive pathways for electron transfer in the hybrid film. The decrease in the sheet resistivity by incorporation of GNPs demonstrates that two-dimensional GNPs provide a more efficient percolating network than one-dimensional MWCNTs. Furthermore, GNP worked as strong holders with a large surface area to support contact between the MWCNT and GNP, resulting in a further reduction of the contact resistance. In the MWCNT-dominated hybrid film, a pronounced synergistic effect on conductivity can be observed. The electrical conductivity is related to both in-plane and through-thickness conduction of electrons. It is clear to see from Figure 2 that 1-D MWCNTs act as bridges to connect 2-D GNPs and provide additional channels for the electron transfer within the hybrid film. This leads to a decreased electrical resistance and may be considered as the major reason for the synergistic effect of the MWCNT and GNP hybrid films. In addition, high electrical conductivity of GNP in the basal plane enhances the synergistic effect on electrical conductivity.

Table 2. Electrical properties of hybrid film with different weight percentage of GNP.

GNP wt.%	Resistivity ($\Omega \cdot cm$)	Conductivity (S/cm)
GNP-0	$2.1 \times 10^{-2} \pm 1.4 \times 10^{-3}$	48 ± 3.0
GNP-10	$1.4 \times 10^{-2} \pm 7.0 \times 10^{-4}$	72 ± 3.6
GNP-20	$1.2 \times 10^{-2} \pm 2.9 \times 10^{-4}$	87 ± 2.1
GNP-30	$8.1 \times 10^{-2} \pm 2.4 \times 10^{-4}$	124 ± 3.7
GNP-40	$7.0 \times 10^{-3} \pm 2.0 \times 10^{-4}$	142 ± 4.0
GNP-50	$5.2 \times 10^{-3} \pm 3.0 \times 10^{-5}$	193 ± 1.1

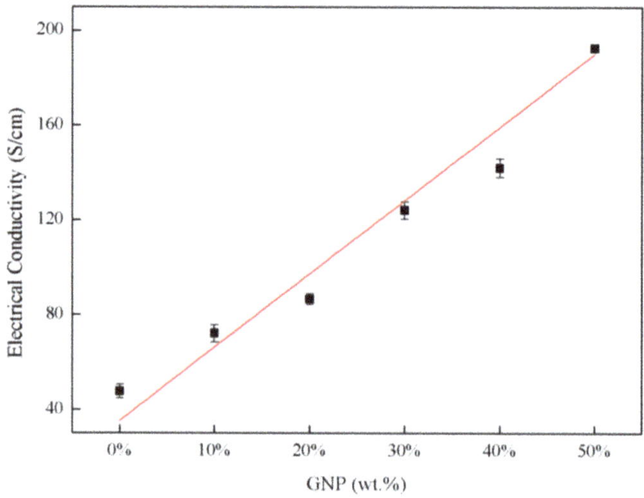

Figure 5. Electrical conductivity of MWCNT/GNP hybrid films with different weight percentages of GNP.

3.3. Self-Strain Sensing Properties

The prepared hybrid film was cut into a rectangular strip (30 × 10 mm) for evaluation of the piezoresistive response and sensing performance. The strain-monitoring capability of the hybrid film sensor was tested in a flexural test. Electrical resistance and mechanical strain during the test were measured simultaneously by a digital Multimeter (Keithley 2450) and strain gauge, respectively. The hybrid film sensor was attached to the center of an Al (Al6064-T6) test specimen (dimensions: 200 × 19 × 2 mm) using epoxy to make perfect bonding between the specimen and hybrid film sensor. When the load was applied on the Al specimen, the hybrid film bonded through high strength epoxy and the metallic strain gauge experienced the same strain. Two copper electrodes were adhered to the hybrid film sensor at a distance of 25 mm using silver paste to minimize the contact resistance. The electrical resistance in monitoring tests was measured by the two-point method due to the simplicity of the method regarding scalability to real applications [21]. Resistivity and strain data were recorded by the digital data acquisition system (cDAQ-9174 NI) through the Lab VIEW software. In this work, a four-point-bending test was conducted to study the piezoresistive behavior of the hybrid film sensor. The spans between the two inner points and two outer points are 60 mm and 120 mm, respectively. A schematic diagram and experimental setup of the four-point-bending test are shown in Figure 6.

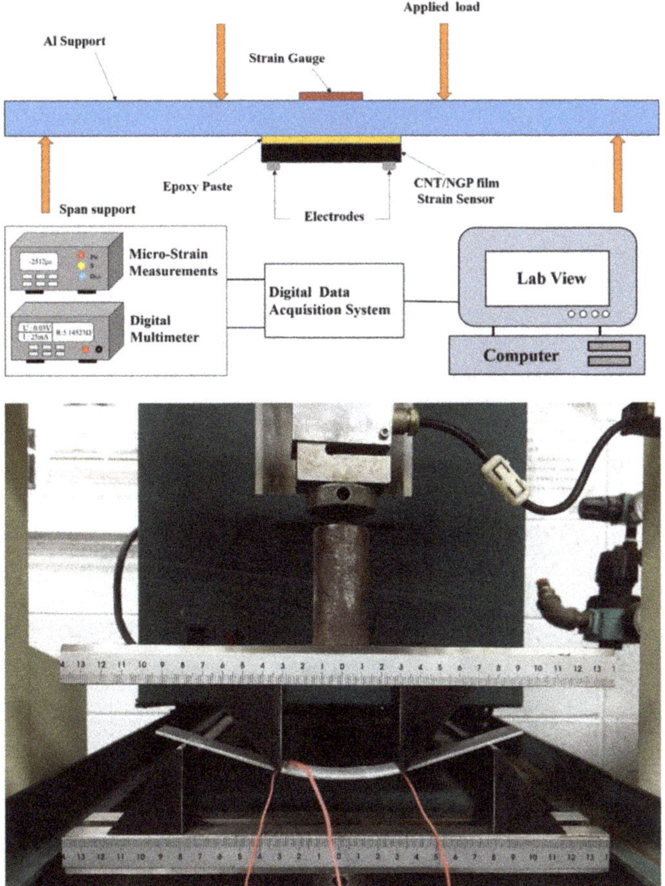

Figure 6. Schematic diagram and experimental setup of the four-point-bending test.

The addition of GNPs increases the conductivity of the hybrid film as described in Section 3.2. Four-point-bending tests were performed to monitor the electrical resistance change of the hybrid film with different GNP contents induced by the strain. Gauge factor is an important parameter which can be used to describe the sensitivity of the strain sensor. It is defined as the ratio of the normalized electrical resistance and strain induced in the sensor as follows.

$$\mathrm{GF} = \frac{\Delta R / R_0}{\varepsilon} \tag{1}$$

where ΔR is the resistance change with strain, R_0 is the initial resistance prior to straining, ε is the applied strain.

Representative normalized resistance–strain curves of the experimental results are plotted in Figure 7 for various GNP contents ranging from 0 to 50 wt.%. It can be observed that the normalized resistance behaves in positive piezoresistive trend, i.e., the normalized resistance change monotonic increases with the increase of the strain. Moreover, the normalized resistance of the hybrid film is increasing with the increase of GNP content. For the increase of the resistance curve, an evident change occurs around at the strain of 0.2%. The whole curve can be divided into two stages. In stage 1 (strain range 0–0.2%), the increase of the resistance tends to be linear with a small slope. In stage II (strain range 0.2–1%), the resistance change exhibits a linear relationship with a large slope. The slope of the curve represents the gauge factor of the hybrid film which can be used to characterize the strain sensitivity of the hybrid film sensor. The gauge factors of the hybrid films with different GNP contents for stage I and II are listed in Table 3. It can be observed that the gauge factor is increasing with the increase of the GNP content as shown in Figure 8. Furthermore, gauge factor in stage I is larger than that of stage II. As the GNP content increases from 0 wt.% to 50 wt.%, the gauge factor increases from 1.16 to 2.34 in stage I, and increases from 1.54 to 3.56 in stage II. The mechanism corresponding to the increase of the resistance in the two stages can be explained as follows. The resistance of the hybrid film can be attributed to three main aspects, namely, contact resistance, tunneling resistance and intrinsic resistance. In stage I, the gauge factor is mainly affected by intrinsic resistance, the relative displacements of MWCNT and GNP are small, the wavy carbon nanotubes are straightened under strain due to its large flexibility, and a smaller gauge factor is acquired. However, in stage II, the normalized resistance change ($\Delta R/R_0$) of the hybrid film is mainly relied on the contact and tunneling resistances of adjacent nanomaterial sheets. The conductivity between neighboring flakes is determined by their overlap area and the contact resistance [37]. The assumption in the sensitivity change of MWCNT/GNP hybrid films can be further explained by the schematic diagram shown in Figure 9. Once a mechanical strain is applied to the hybrid film, the overlap area between neighboring flakes becomes smaller and the gap distance becomes larger, which results in an increase of the tunneling pathway between adjacent nanoplatelets so the tunneling resistance increases. In the process of mechanical loading, the tunneling resistance instead of the contact resistance becomes the dominant factor of the resistance. In addition, the more the GNP content, the more easily the conductive path gets disrupted by external strains, which results in higher strain sensitivity. Similar results were reported by Lu et al. [38]. They found that the sensitivity of the GNP/epoxy sensor was varied along with the applied strain and can be separated to three strain regions (0–0.2%), (0.2–0.6%) and (0.6–1.2%), respectively. The gauge factors of the GNP/epoxy sensor with 1.58 vol.% of GNP corresponding to these three strain regions were 2.53, 3.77 and 4.69, respectively.

Table 3. Gauge factor for different strain stage.

Gauge factor						
GNP wt%	0%	10%	20%	30%	40%	50%
0~0.2% strain	1.2	1.3	1.4	1.7	2.1	2.3
0.2~1% strain	1.5	2.1	2.3	2.9	3.0	3.6

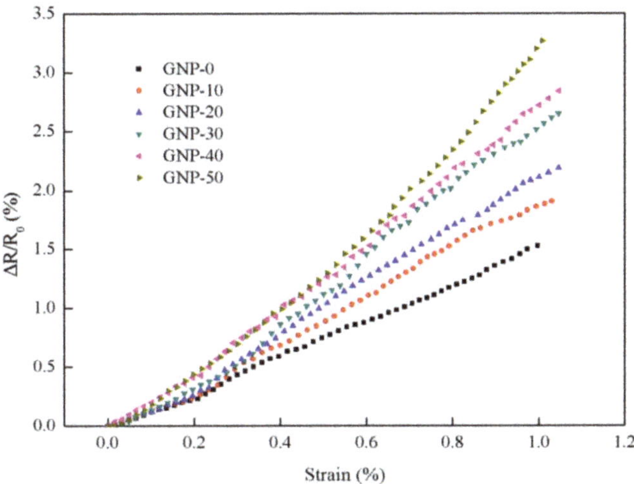

Figure 7. Normalized resistance change increases with the increase of the strain.

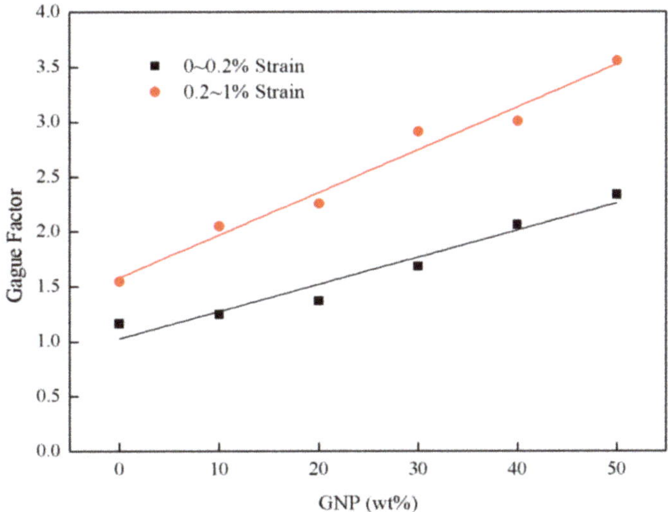

Figure 8. Gauge factor for different strain stages.

To investigate the stability, reversibility and reliability of the hybrid film sensor, the specimens were subjected to 200 cyclic loading–unloading tests. This test aimed to monitor the electric resistance response of the hybrid film under cyclic loading. The dynamic responses of the normalized resistance change and mechanical strain of hybrid films with 0 wt.% (GNP-0) and 50 wt.% (GNP-50) of GNP are plotted in Figure 10a,b, respectively. It can be observed that there is no obvious change during the 200 cycling tests for the hybrid film sensors. This demonstrates that the high durability and stability of the hybrid film sensor. Some researchers [39,40] also did the cycle loading–unloading tests for stability of the GNP/epoxy sensors, they are stable under certain cycles, but the cycles they tested were as low as 50 or even several cycles.

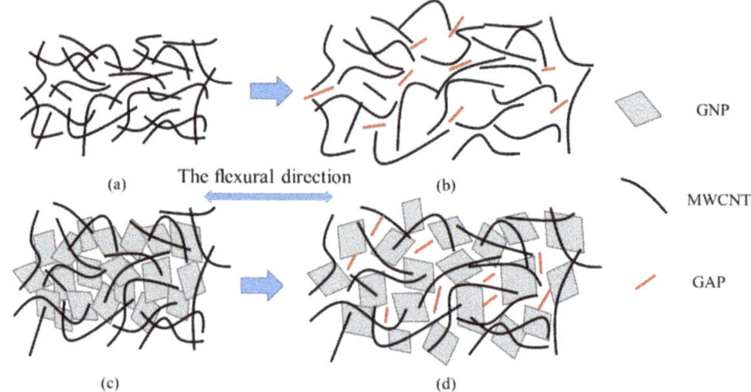

Figure 9. Schematic representation of microstructure changes in hybrid films subjected to mechanical strain. (**a**) MWCNT film; (**b**) Stretching of MWCNT film under flexural strain; (**c**) MWCNT/GNP hybrid film; (**d**) Stretching of MWCNT/GNP hybrid film under flexural strain.

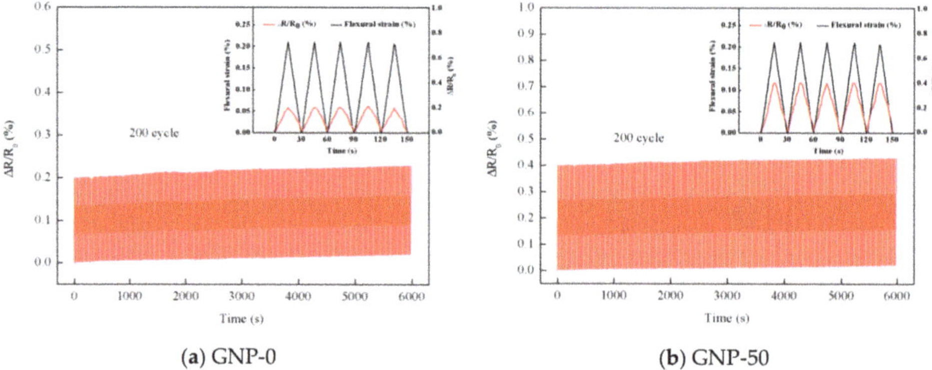

Figure 10. Normalized resistance change and mechanical strain of the hybrid film under cyclic loading-unloading test (**a**) 0 wt.% GNP-0 (**b**) 50 wt.% GNP-50.

4. Conclusions

MWCNT/GNP hybrid films were prepared with the aid of surfactant Triton X-100 and sonication through vacuum filtration process. SEM images show that MWCNTs and GNPs are successfully deposited to form densely packed film with layered structure. The effect of GNP content ranging from 0 to 50 wt.% on the mechanical and electrical properties of the hybrid films were characterized using the tensile test and Hall effect measurements, respectively. It can be observed that both the tensile strength and fracture strain are decreasing with the increase of GNP content. The electrical conductivity is increasing from 47.7 S/cm to 192.6 S/cm as the GNP loading increases from 0 to 50 wt.%. A series experimental tests were conducted to study the piezoresistive behavior and the strain-sensing capability of the hybrid film. The gauge factor defined as the ratio of relative change in resistance to applied strain was used to characterize the sensitivity of the strain sensor. There are two different linear strain-sensing stages (0–0.2% and 0.2%–1%) in the resistance of the hybrid film with applied strain. The gauge factor increases from 1.164 to 2.236 as the GNP loading increases from 0 to 50 wt.% in the strain-sensing range 0–0.2%. Moreover, the repeatability and stability of the strain sensitivity of the hybrid film were conformed through the cyclic loading and unloading tests. From the results obtained, it is demonstrated that the MWCNT/GNP hybrid film is very suitable for strain sensing.

Author Contributions: Conceptualization, J.R.H. and S.-C.H.; methodology, J.R.H. and X.X.Y.; validation, S.-C.H. and X.X.Y.; investigation, J.R.H. and M.N.Z.; writing: original draft preparation, J.R.H.; writing: review and editing S.-C.H.

Funding: This research was funded by Ministry of Science and Technology of the R.O.C, grant number MOST 104-2221-E155-057-MY3.

Acknowledgments: The authors would like to thank Mr. Yuan-Ming Liang and Wei-Chun Hsu at Yuan Ze University for their assistance of the experimental setup and tests.

Conflicts of Interest: The authors declare no conflict of interest.

References

1. Iijima, S. Helical microtubules of graphitic carbon. *Nature* **1991**, *354*, 56–58. [CrossRef]
2. Novoselov, K.S.; Geim, A.K.; Morozov, S.V.; Jiang, D.; Zhang, Y.; Dubonos, S.V.; Grigorieva, I.V.; Firsov, A.A. Electric field effect in atomically thin carbon films. *Science* **2004**, *306*, 666–669. [CrossRef] [PubMed]
3. Wang, X.; Lu, S.; Ma, K.; Xiong, X.; Zhang, H.; Xu, M. Tensile strain sensing of buckypaper and buckypaper composites. *Mater. Des.* **2015**, *88*, 414–419. [CrossRef]
4. Zhang, J.; Wang, X.; Ma, J.; Liu, S.; Yi, X. Preparation of cobalt hydroxide nanosheets on carbon nanotubes/carbon paper conductive substrate for supercapacitor application. *Electrochim. Acta* **2013**, *104*, 110–116. [CrossRef]
5. Liu, X.M.; Huang, Z.D.; Oh, S.W.; Zhang, B.; Ma, P.C.; Yuen, M.M.F.; Kim, J.Y. Carbon nanotube (CNT)-based composites as electrode material for rechargeable Li ion batteries: a review. *Compos. Sci. Technol.* **2012**, *72*, 121–144. [CrossRef]
6. Kang, S.J.; Kim, B.; Kim, K.S.; Zhao, Y.; Chen, Z.; Lee, G.H.; Hone, J.; Kim, P.; Nuckolls, C. Inking Elastomeric Stamps with Micro-Patterned; Single Layer Graphene to Create High-Performance OFETs. *Adv. Mater.* **2011**, *23*, 3531–3535. [CrossRef] [PubMed]
7. Che, J.F.; Chen, P.; Chan-Park, M.B. High-strength carbon nanotube buckypaper composites as applied to free-standing electrodes for supercapacitors. *J. Mater. Chem. A* **2013**, *1*, 4057–4066. [CrossRef]
8. Jia, X.; Chen, Z.; Suwarnasarn, A.; Rice, L.; Wang, X.; Sohn, H.; Zhang, Q.; Wu, B.M.; Wei, F.; Lu, Y. High-performance flexible lithium-ion electrodes based on robust network architecture. *Energy Environ. Sci.* **2012**, *5*, 6845–6849. [CrossRef]
9. Chatterjee, J.; Cardenal, J.; Shellikeri, A. Engineered carbon nanotube buckypaper: a platform for electrochemical biosensors. *J. Biomed. Nanotechnol.* **2015**, *11*, 150–156. [CrossRef] [PubMed]
10. Zhang, Z.; Wei, H.; Liu, Y.; Leng, J. Self-sensing properties of smart composite based on embedded buckypaper layer. *Struct. Health Monit.* **2015**, *14*, 127–136. [CrossRef]
11. Lu, X.; Dou, H.; Gao, B.; Yuan, C.; Yang, S.; Hao, L.; Shen, L.; Zhang, X. A flexible graphene/multiwalled carbon nanotube film as a high performance electrode material for supercapacitors. *Electrochim. Acta* **2011**, *56*, 5115–5121. [CrossRef]
12. Lou, C.; Wang, S.; Liang, T.; Pang, C.; Huang, L.; Run, M.; Liu, X. A Graphene-Based Flexible Pressure Sensor with Applications to Plantar Pressure Measurement and Gait Analysis. *Materials* **2017**, *10*, 1068. [CrossRef] [PubMed]
13. Lu, S.; Chen, D.; Wang, X.; Shao, J.; Ma, K.; Zhang, L.; Araby, S.; Meng, Q. Real-time cure behaviour monitoring of polymer composites using a highly flexible and sensitive CNT buckypaper sensor. *Compos. Sci. Technol.* **2017**, *152*, 181–189. [CrossRef]
14. Liu, G.; Tan, Q.; Kou, H.; Zhang, L.; Wang, J.; Lv, W.; Dong, H.; Xiong, J. A Flexible Temperature Sensor Based on Reduced Graphene Oxide for Robot Skin Used in Internet of Things. *Sensors* **2018**, *18*, 1400. [CrossRef] [PubMed]
15. Lu, H.; Zhang, J.; Luo, J.; Gong, W.; Li, C.; Li, Q.; Zhang, K.; Hu, M.; Yao, Y. Enhanced thermal conductivity of free-standing 3D hierarchical carbon nanotube-graphene hybrid paper. *Composites Part A* **2017**, *102*, 1–8. [CrossRef]
16. Meng, Q.; Wu, H.; Zhao, Z.; Araby, S.; Lu, S.; Ma, J. Free-standing, flexible, electrically conductive epoxy/grapheme composite films. *Composites: Part A* **2017**, *92*, 42–50. [CrossRef]
17. Wang, J.; Jin, X.; Wu, H.; Guo, S. Polyimide reinforced with hybrid graphene oxide @ carbon nanotube: Toward high strength, toughness, electrical conductivity. *Carbon* **2017**, *123*, 502–513. [CrossRef]

18. Arif, M.F.; Kumar, S.; Shah, T. Tunable morphology and its influence on electrical, thermal and mechanical properties of carbon nanostructure-buckypaper. *Mater. Des.* **2016**, *101*, 236–244. [CrossRef]

19. Lu, S.; Tian, C.; Wang, X.; Chen, D.; Ma, K.; Leng, J.; Zhang, L. Health monitoring for composite materials with high linear and sensitivity GnPs/epoxy flexible Strain sensors. *Sens. Actuators A* **2017**, *267*, 409–416. [CrossRef]

20. Liu, Y.; Zhang, D.; Wang, K.; Liu, Y.; Shang, Y. A novel strain sensor based on graphene composite films with layered structure. *Composites Part A* **2016**, *80*, 95–103. [CrossRef]

21. Moriche, R.; Sanchez, M.; Jimenez-Suarez, A.; Prolongo, S.G.; Urena, A. Strain monitoring mechanisms of sensors based on the addition of graphene nanoplatelets into an epoxy matrix. *Compos. Sci. Technol.* **2016**, *123*, 65–70. [CrossRef]

22. Sanli, A.; Benchirouf, A.; Müller, C.; Kanoun, O. Piezoresistive performance characterization of strain sensitive multi-walled carbon nanotube-epoxy nanocomposites. *Sens. Actuators A* **2017**, *254*, 61–68. [CrossRef]

23. Wang, Y.; Wang, S.; Li, M.; Gu, Y.; Zhang, Z. Piezoresistive response of carbon nanotube composite film under laterally compressive strain. *Sens. Actuators A* **2018**, *273*, 140–146. [CrossRef]

24. Wang, X.; Sparkman, J.; Gou, J. Strain sensing of printed carbon nanotube sensors on polyurethane substrate with spray deposition modeling. *Compos. Commun.* **2017**, *3*, 1–6. [CrossRef]

25. Natarajan, T.S.; Eshwaran, S.B.; Stöckelhuber, K.S.; Wießner, S.; Pötschke, P.; Heinrich, G.; Das, A. Strong Strain Sensing Performance of Natural Rubber Nanocomposites. *ACS Appl. Mater. Interfaces* **2017**, *9*, 4860–4872. [CrossRef] [PubMed]

26. Boland, C.S.; Khan, U.; Ryan, G.; Barwich, S.; Charifou, R.; Harvey, A.; Backes, C.; Li, Z.; Ferreira, M.S.; Möbius, M.E.; et al. Sensitive electromechanical sensors using viscoelastic graphene-polymer nanocomposites. *Science* **2016**, *354*, 1257–1260. [CrossRef] [PubMed]

27. Li, Y.; Samad, Y.A.; Liao, K. From cotton to wearable pressure sensor. *J. Mater. Chem. A* **2015**, *3*, 2181–2187. [CrossRef]

28. Samad, Y.A.; Li, Y.; Alhassan, S.M.; Liao, K. Novel Graphene Foam Composite with Adjustable Sensitivity for Sensor Applications. *ACS Appl. Mater. Interfaces* **2015**, *7*, 9195–9202. [CrossRef] [PubMed]

29. Samad, Y.A.; Li, Y.; Schiffer, A.; Alhassan, S.M.; Liao, K. Graphene Foam Developed with a Novel Two-Step Technique for Low and High Strains and Pressure-Sensing Applications. *Small* **2015**, *11*, 2380–2385. [CrossRef] [PubMed]

30. Girifalco, L.A.; Hodak, M.; Lee, R.S. Carbon nanotubes, buckyballs, ropes, and a universal graphitic potential. *Phys. Rev. B* **2000**, *62*, 13104–13110. [CrossRef]

31. Lu, H.; Liu, Y.; Gou, J.; Leng, J.; Du, S. Surface coating of multi-walled carbon nanotube nanopaper on shape-memory polymer for multifunctionalization. *Compos. Sci. Technol.* **2011**, *71*, 1427–1434. [CrossRef]

32. Lin, X.; Liu, X.; Jia, J.; Shen, X.; Kim, J.K. Electrical and mechanical properties of carbon nanofiber/graphene oxide hybrid papers. *Compos. Sci. Technol.* **2014**, *100*, 166–173. [CrossRef]

33. Hwang, S.H.; Park, H.W.; Park, Y.B. Piezoresistive behavior and multi-directional strain sensing ability of carbon nanotube-graphene nanoplatelet hybrid sheets. *Smart Mater. Struct.* **2013**, *22*, 05013. [CrossRef]

34. Shen, X.; Lin, X.; Yousefi, N.; Jia, J.; Kim, J.K. Wrinkling in graphene sheets and graphene oxide papers. *Carbon* **2014**, *66*, 84–92. [CrossRef]

35. Khan, U.; O'Connor, I.; Gun'ko, Y.K.; Coleman, J.N. The preparation of hybrid films of carbon nanotubes and nano-graphite/graphene with excellent mechanical and electrical properties. *Carbon* **2010**, *48*, 2825–2830. [CrossRef]

36. Fan, H.; Zhao, N.; Wang, H.; Xu, J.; Pan, F. 3D conductive network-based free-standing PANI-RGO-MWNTs hybrid film for high-performance flexible supercapacitor. *J. Mater. Chem. A* **2014**, *2*, 12340–12347. [CrossRef]

37. Kim, Y.J.; Cha, J.Y.; Ham, H.; Huh, H.; So, D.S.; Kang, I. Preparation of piezoresistive nano smart hybrid material based on graphene. *Curr. Appl. Phys.* **2011**, *11*, S350–S352. [CrossRef]

38. Lu, S.; Tian, C.; Wang, X.; Zhang, L.; Du, K.; Ma, K.; Xu, Y. Strain sensing behaviors of GnPs/epoxy sensor and health monitoring for composite materials under monotonic tensile and cyclic deformation. *Compos. Sci. Technol.* **2018**, *158*, 94–100. [CrossRef]

39. Zhao, J.; Zhang, G.Y.; Shi, D.X. Review of graphene-based strain sensors. *Chin. Phys. B* **2013**, *22*, 057701. [CrossRef]
40. Nakamura, A.; Hamanishi, T.; Kawakami, S.; Takeda, M. A piezo-resistive graphene strain sensor with a hollow cylindrical geometry. *Mater. Sci. Eng. B* **2017**, *219*, 20–27. [CrossRef]

 *nanomaterials*

Review

Cellulose Nanocrystals/Graphene Hybrids—A Promising New Class of Materials for Advanced Applications

Djalal Trache [1,*], Vijay Kumar Thakur [2,3] and Rabah Boukherroub [4]

[1] Energetic Materials Laboratory, Teaching and Research Unit of Energetic Processes, Ecole Militaire Polytechnique, BP 17, Bordj El-Bahri, 16046 Algiers, Algeria

[2] Biorefining and Advanced Materials Research Center, Scotland's Rural College (SRUC), Kings Buildings, Edinburgh EH9 3JG, UK; Vijay.kumar@sruc.ac.uk

[3] Department of Mechanical Engineering, School of Engineering, Shiv Nadar University, Greater Noida, Uttar Pradesh 201314, India

[4] Institut d'Electronique, de Microélectronique et de Nanotechnologie (IEMN-UMR CNRS 8520), University Lille, CNRS, Centrale Lille, University Polytechnique Hauts-de-France, UMR 8520—IEMN, F-59000 Lille, France; rabah.boukherroub@univ-lille.fr

* Correspondence: djalaltrache@gmail.com

Received: 20 July 2020; Accepted: 31 July 2020; Published: 4 August 2020

Abstract: With the growth of global fossil-based resource consumption and the environmental concern, there is an urgent need to develop sustainable and environmentally friendly materials, which exhibit promising properties and could maintain an acceptable level of performance to substitute the petroleum-based ones. As elite nanomaterials, cellulose nanocrystals (CNC) derived from natural renewable resources, exhibit excellent physicochemical properties, biodegradability and biocompatibility and have attracted tremendous interest nowadays. Their combination with other nanomaterials such as graphene-based materials (GNM) has been revealed to be useful and generated new hybrid materials with fascinating physicochemical characteristics and performances. In this context, the review presented herein describes the quickly growing field of a new emerging generation of CNC/GNM hybrids, with a focus on strategies for their preparation and most relevant achievements. These hybrids showed great promise in a wide range of applications such as separation, energy storage, electronic, optic, biomedical, catalysis and food packaging. Some basic concepts and general background on the preparation of CNC and GNM as well as their key features are provided ahead.

Keywords: cellulose nanocrystals; graphene; hybrids; applications

1. Introduction

The excessive consumption of fossil-based resources and resulting environmental problems issues coupled with the constancy growing global population requests the improvement of living standard and accelerating technology development. It has stimulated and attracted researchers worldwide to develop a sustainable bio-based alternative that can compete in performance with petroleum-based products expected to be employed in a wide range of applications [1–7]. Cellulose, as the most abundant bio-based material from the biosphere, has attracted more and more attention in different fields and could serve as a prominent alternative to the exhaustible fossil resources, owing to its renewability, biodegradability, biocompatibility, non-toxicity and environmental friendliness [8–12]. The advantages of cellulose can be also pushed forward through the exploration of its nonmetric size, which generates nanocellulose (NC), considered as a promising class for future materials due to its outstanding physicochemical properties [13–16]. NC displays low density, specific barrier properties and low thermal expansion coefficient, high strength, excellent stiffness, elongation morphology, inertness, large surface area and aspect ratio, abundance and ease of bio-conjugation [11,17–19]. The presence of several reactive chemical groups on its surface allows it to be modified by physical adsorption, covalent bonding or surface grafting to further extend its performance [20]. Research activities concerning NC had attracted growing interest over the past decade as reflected by the rapid increase of scientific publications and patents granted internationally [21]. According to Markets and Markets, the NC market is forecasted to achieve USD 783 Million by 2025 [13] and thus, NC production will have a high economic impact [22]. Moreover, an interesting review has been recently published by Charreau et al. dealing with the analysis of the evolution of patents involving nanocellulose since 2010 [23], demonstrating the increasing industrial interest in the this, which enabled the setting-up of the first facilities producing commercial quantities of NC.

Numerous nanocellulose types with outstanding features can be produced from different cellulosic sources employing various approaches [17,24–26]. NC can be divided into two main categories, that is, nanostructured materials (cellulose microfibrils and microcrystalline cellulose) and nanofibers (cellulose nanocrystals, cellulose nanofibrils and bacterial cellulose) [13]. Due to their excellent inherent characteristics, cellulose nanocrystals (CNC), as a subclass of NC, is commonly produced from cellulosic fibers and fibrils after the elimination of the amorphous regions by acid hydrolysis [23,27,28]. CNC have aroused wide scientific and technological interest from both academicians and industrials and can be utilized as an independent functional material, template support, stabilizer, filler or reinforcing agent [29–31]. CNC-based nanomaterials have been extensively investigated due to their unique physicochemical, mechanical, thermal, rheological and optical features. CNC could confer excellent properties to hybrids or nanocomposites (metallic, ceramics and polymeric) even at low concentration for different applications such as medical, pharmaceutics, catalysis, oil/water separation, decontamination, flame retardancy, electronic and optical devices, energy storage, sportswear, light weight armor systems, food packaging, to cite a few [10,11,13,14,20,32–39].

The combination of CNC and nanocarbons, such as fullerenes, nanotubes (single-walled, double-walled, few-walled or multi-walled), nanodiamonds and graphene-based materials (graphene, graphene oxide, reduced graphene oxide, graphene quantum dots), has recently emerged as a new class of hybrid materials for which a synergetic effect has been revealed in a wide range of applications, spanning from sensing and biosensing to catalysis, photonics and optics, energy and environment, water treatment, medical and optoelectronics. Other nanocarbons such as carbon black, activated carbon, carbon quantum dots and carbon nanofibers are less frequently used as CN-based hybrids [12,20,40–43].

Graphene-based nanomaterials (GNM), which have been considered as emerging and high efficient two-dimensional (2D) nanomaterials, play a crucial role in various research area since the discovery of graphene in 2004 [44,45]. They find applications in several fields such as thin-film transistors, ultra-sensitive chemical sensors and transparent conductive films, biomedical, microelectronics, composites, among others [46–49]. Recent investigations by Yang et al. provided general reviews of the whole graphene patenting activities and especially focused on the study of sustainable competitive advantages in the biomedical field [50,51]. A comprehensive review dealing with graphene, its related materials and properties have also been published [52]. Although the present development of industrial-scale graphene is still widely at the Research and Development (R&D) stage, the global graphene market reached ca. USD 78.7 million in 2019, with the request in nanocomposites, energy storage materials and semiconductor electronics, which are also underpinning future growth rate estimates of >30% per year and expected to reach >USD 221.4 million by 2025 [53,54]. Nowadays, graphene oxide (GO) materials account for >30% of the global graphene market share as progresses in GO, permitting for numerous of possibly scalable approaches to reach mass production of chemically modified graphene with a wide range of applications [54,55]. However, despite that many technically feasible approaches are currently being developed to produce efficient GNM, there are still numerous practical obstacles that need to be overcome. For instance, GNM are more frequently produced from aqueous dispersions but can easily aggregate. Such agglomeration behavior can reduce the surface area and negatively impact the mechanical, electrical and optical properties. Therefore, the incorporation of CNC not only surpasses such drawback through its excellent dispersive features but also confers further benefits to the produced GNM/CNC hybrids such as flexibility, stretchability, in addition to the improvement of the adsorption capacity, photothermal activity, stability, intrinsic luminescence and fluorescence, optical transparency and thermal conductivity [36,40,42,56].

Owing to the benefits of CNC and GNM materials as well as the numerous research works published during the last few years worldwide, a timely update on recent advancements in the field of CNC/GNM hybrid-based materials is an urgent need for both academic and industrial scientists. In this overview, we thoroughly review the recent progress made in the preparation, modification, properties and current applications of CNC/GNM hybrids in various fields. This work highlights a comprehensive overview with a forward-looking approach on CNC/GNM hybrids for numerous utilizations, which have emerged in the past five years. For the reader's comfort and to maintain lucidity, first, some of the basic concepts dealing with CNC and GNM, their preparation and features to further elucidate their unique attributes, are discussed in brief. We will then focus on state-of-the-art cellulose nanocrystals-graphene based materials, which have mainly emerged since 2015. Few articles before 2015 are succinctly summarized in some sections.

2. Cellulose Nanocrystals (CNC)

2.1. Fundamental of Nanocellulose

Cellulose, which was first extracted from wood by Enselme Payen in 1838, is a polysaccharide consisting of β-1,4-linked anhydroglucopyranoside units, in which every monomer unit is corkscrewed at 180° compared to its neighbors [5,57]. The annual production of this abundant biopolymer is estimated to be between 10^{10} and 10^{11} tons-per-year. It can be isolated from different sources such as wood, herbaceous plants, grass, crops and their byproducts, bacterial, algae and animal sources, among others [58–61]. The properties of this biomacromolecule are closely related to the natural source, its maturity and origin, pretreatment methods, processing approaches and reaction conditions. Typically, lignocellulosic biomass necessitates the removing of non-cellulosic constituents such as extractives, lignin and hemicellulose [62]. Trache et al. have recently reviewed the common pretreatment and processing methodologies, which allow the obtaining of pure cellulose from lignocellulosic [13]. As depicted in Figure 1, cellulose is semi-crystalline in nature, hence it contains both crystalline and amorphous regions. This latter is susceptible to hydrolysis; it can steadily be eliminated to generate crystalline parts upon phase segregation and is more prone to react with other molecular groups [57,63,64]. The intra- and intermolecular chemical groups (Figure 1) confer to this fascinating polymer its specific features such as infusibility, chirality, hydrophilicity, insolubility in several aqueous media and ease of chemical functionalization [65]. Cellulose can be classified into various polymorphs, that is, cellulose I, II, III$_I$, III$_{II}$, IV$_I$ and IV$_{II}$, which can be converted from one form to another through chemical or thermal treatments [42].

In their nano-size form, cellulose nanomaterials, also known as nanocellulose (NC), display outstanding physical, chemical, biological, magnetic, electrical and optical characteristics compared to the bulk materials [1,11,66]. Conceptually, NC can be produced by a top-down hydrolysis methodology through different steps, that is, (i) pretreatment processes of lignocellulosic biomass employing physical approaches concerning crushing, screening, washing and cooking to eliminate coarse particles, oily content and dust from the material surface, (ii) removing extractive/hemicellulose/lignin via chemical, physical, physicochemical, biological or the combination of two or more treatments, (iii) fragmentation and cleavage of cellulosic elementary fibrils or micro-fibrils to generate nanofibers through various approaches and (iv) post-treatments such as solvent removing, dialysis, sonication, centrifugation, surface modification, stabilization and drying. The three latter steps have received great interest from the scientific community for designing products with desired features [12,13,23,27,38,67–73]. NC possesses excellent useful properties such as renewability, eco-friendliness, biocompatibility, non-toxicity, hydrogen-bonding capacity, tunable crystallinity, high chemical resistance, tailored aspect ratios (100–150), low thermal expansion coefficient, reactive surface, low density (1.6 g/cm^3), high specific surface area (100–200 of m^2/g), high tensile strength (7.5–7.7 GPa) and elastic modulus (130–150 GPa) [10,30]. This promising polysaccharide has received tremendous attention during the last two decades in a wide range of applications such as sensors and biosensors, energy storage systems, oil and gas drilling and cementing, papermaking, filtration, decontamination, adsorption, separation, wood adhesives, Pickering emulsifiers, medical and nanocomposites, to cite a few [13,14,16,20,35,36,42,74–76]. Depending on the isolation method, morphology and size, NC is principally categorized into: (i) cellulose nanostructured materials such as cellulose microfibrils and microcrystalline cellulose and (ii) cellulose nano-objects, also known as nanofibers, such as cellulose nanocrystals (CNC), cellulose nanofibrils (CNF) and bacterial nanocellulose (BC).

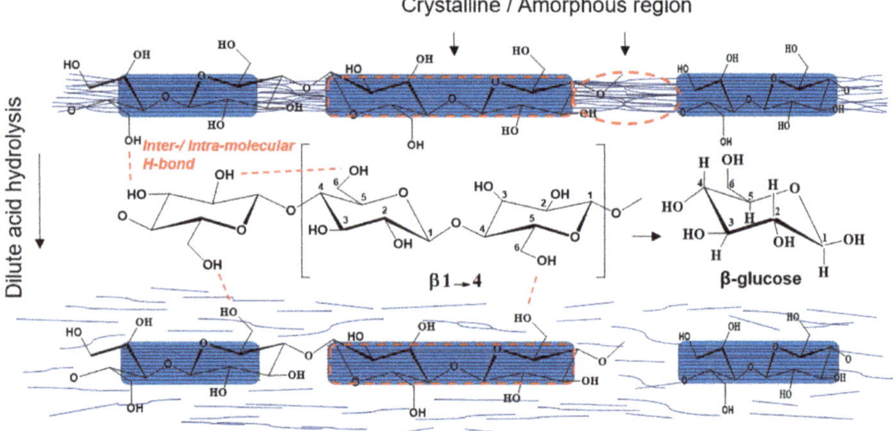

Figure 1. Hypothetical schematic of dilute acid pretreatment process to extract the crystalline regions of cellulose from amorphous domains. In the middle, the configuration of cellulose repeating unit with the β-(1,4) glycosidic linkage under the effect of intra/intermolecular hydrogen bonding is denoted. Reproduced with permission from Reference [36]. Copyright ©2019, Elsevier.

In contrast to nanostructured materials, nanofibers present more uniform particle size distribution, high specific surface area, amphiphilic nature, barrier properties, high crystallinity and tend to produce more stable self-assembled structures such as hydrogels and films [1,30,77]. In recent years, other types of nanofibers appeared such as amorphous nanocellulose, cellulose nanoyarn and cellulose nanoplatelets [17]. Several in-depth reviews with detailed discussions dealing with NC-based materials have been reported over the past few years, covering the nanocellulose sources, isolation methods, structure modification, potential uses, advantages and shortcomings [11,16,20,23,35,36,38,40,41,78–83]. In the following, we concisely go through current extraction methods of CNC and describe their outstanding features.

2.2. Extraction and Properties of CNC

CNC can be often obtained from different types of lignocellulose through a top-down hydrolysis approach by combining various procedures [9,17,23,84,85]. To extract pure cellulose (PC) through the elimination of extractives, lignins and hemicelluloses, some pretreatments (chemical, physical, physicochemical, biological or their combination) of the natural source are usually required [13,34,86]. Specific treatments can be then applied to PC to produce CNC through the removing of disordered regions from pristine cellulose. The crystalline domains remain intact because of their higher resistance to the hydrolytic action, whereas the amorphous parts dispersed as chain dislocations on segments along the cellulose fibrils are more susceptible to the hydrolysis process [19,78,82]. Afterwards, the elementary fibrils are transversely cleaved, producing short CNC with somewhat high crystallinity. Nonetheless, after this process, extra post-treatments such as solvent removing, sonication, fractionation, dialysis, centrifugation, filtration, washing, stabilization, surface modification, neutralization and drying are required to recover CNC product.

The most common hydrolysis method used to produce CNC relies on sulfuric acid, which can react with the surface hydroxyl groups of pristine cellulose through an esterification process, allowing the grafting of anionic ester groups [77,87]. This latter generates a negative electrostatic layer that covers nanocrystals, promoting the dispersion of CNC in water but reducing their thermal stability. Recently, as an alternative to sulfuric acid hydrolysis, other liquid inorganic acids such as nitric, hydrobromic, phosphoric and hydrochloric have been extensively reported [11,30]. The preparation of CNC from wood, for which the hydrolysis process causes preferential digestion of the amorphous part of cellulose while the ordered regions remain intact, is schematized in Figure 2. Both natural source and experimental conditions (acid concentration, reaction time, temperature, mass ratio, etc.) may influence the characteristic of the prepared CNC such as crystallinity, dimensional dispersity, thermal behavior, mechanical properties, density, aspect ratio and morphology. Although the hydrolysis process using mineral acids is simple and not time-consuming, certain drawbacks such as lower yield, high amount of water usage, severe environmental pollution and harsh corrosion of equipment should be overcome [30]. Therefore, to address the above issues, various recent procedures such as organic acid (oxalic, formic, etc.) hydrolysis [88], solid acid (phosphotungstic) [89], subcritical water hydrolysis [90], deep eutectic solvents [91], ionic liquids [92], oxidation [93], sonication [94], enzymatic [95] and combined approaches [5,17,31] have been applied and others continue to be developed worldwide to produce CNC with desired properties at lower costs and higher yield based on sustainability principle and environmentally friendly policy [5,13,17,31]. Nevertheless, scaling-up from laboratory to industrial scale remains one of the most important issues and considerable efforts should be made to prevail over the remaining constraints. Otherwise, some companies such as CelluForce and Alberta Innovates, among others, produce CNC at large scale [13,96].

CNC present unique features compared to the other classes of NC with the spotlight to characteristics such as physical, chemical, optical, thermal, mechanical, electrical properties [1]. CNC consist of an elongated, needle or rod-like nanoparticles. They are 4–70 nm in width and 100–6000 nm in length and aspect ratio of 5–70, as well as large surface area (150–500 $m^2 \cdot g^{-1}$), which allows it to be easily dispersed in water to generate a chiral nematic organization [13,16,97]. CNC also exhibit high crystallinity (50%–90%), a tensile strength of up to 7.5 GPa, a Young's modulus of ~170 GPa and a bending strength of about 10 GPa [9,13,68]. They also display good thermal stability up to 200 °C and can find applications in processes like thermoplastics [97]. Nevertheless, these features depend closely on the source of feedstock, extraction methods and experimental conditions, which will ultimately define their applicability [98].

It is worthy to note that the abundance of –OH or other reactive chemical groups and the high surface area to volume ratio render CNC highly reactive and easy to be functionalized [100]. Therefore, to improve their compatibility and ensure a good dispersion, CNC surface can be chemically, physically or enzymatically modified to impart stable negative or positive electrostatic charges on their surface [13,101]. Such modifications may allow tailoring the properties of the CNC-based materials depending on the intended application.

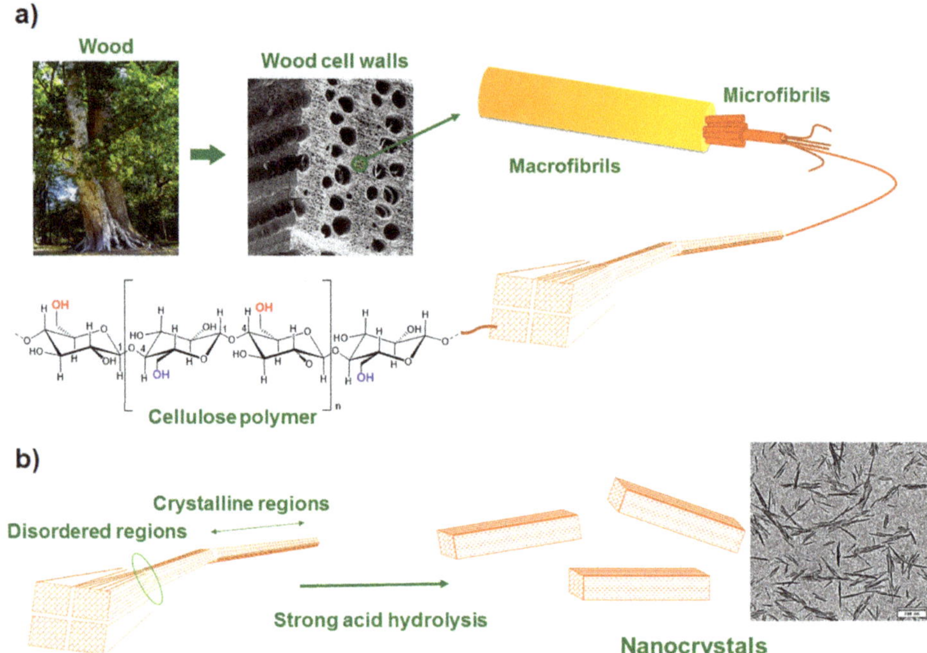

Figure 2. (**a**) Structural hierarchy of the cellulose fiber component from the tree to the anhydroglucose molecule. (**b**) Preparation of cellulose nanocrystals (CNC) by selective acid hydrolysis of cellulose microfibrils. Reproduced with permission from ref [99]. Licensed under a Creative Commons Attribution 3.0 International License (https://creativecommons.org/licenses/by/3.0/).

3. Graphene-Based Nanomaterials

3.1. Nomenclature and Fundamental Aspects

Graphene, discovered by Geim and Novoselov in 2004, is relatively a new two dimensional (2D) sheet-like material in which a honeycomb or hexagonal structure with a flat lattice configuration, completely composed of sp^2 hybridized carbon atoms that are covalently bound, is densely packed [102,103]. Graphene, an atomic layer of graphite, is the unique carbon's allotrope, where each atom is tightly linked to its neighbors by an only electronic cloud in which a C–C bond distance is 0.142 nm [104]. It is considered as a fundamental basis for all carbon allotropes and as the mother of a graphitic family for all the dimensions.

Graphene-based nanomaterials (GNMs), the first materials reported as examples of 2D nanocarbons, can be classified based on the number of sheet layers, surface modifications, total oxygen content or orientation [105]. Graphene is highly hydrophobic and is prone to agglomeration, owing to the strong van der Waals' interactions between the 2D graphene sheets, leading to low surface area and ineffective use of its outstanding features [106]. These latter are also closely dependent on the graphene availability as a single layer because if the layers are in close vicinity to each other, they are likely to restack or agglomerate due to π–π interactions. Hence, its functionalization is commonly required to surpass these issues. Typically, three types of functionalization approaches through covalent (nucleophilic substitution, electrophilic substitution, condensation and addition), noncovalent (π–π bonding, electrostatic attraction and hydrogen bonding, etc.) or a combination of both interactions can be used, where the aromaticity of graphene can be either lost or preserved [107]. As shown in Figure 3a–e, GNMs can be found in various forms for which the most important ones that will be the focus of the present review are graphene nanosheets (GNS), graphene nanoplatelets (GNP), graphene oxide (GO), reduced graphene oxide (RGO) and graphene and graphene oxide quantum dots (GQD). In the frame of the present review, the acronym GN will encompass GNS and GNP.

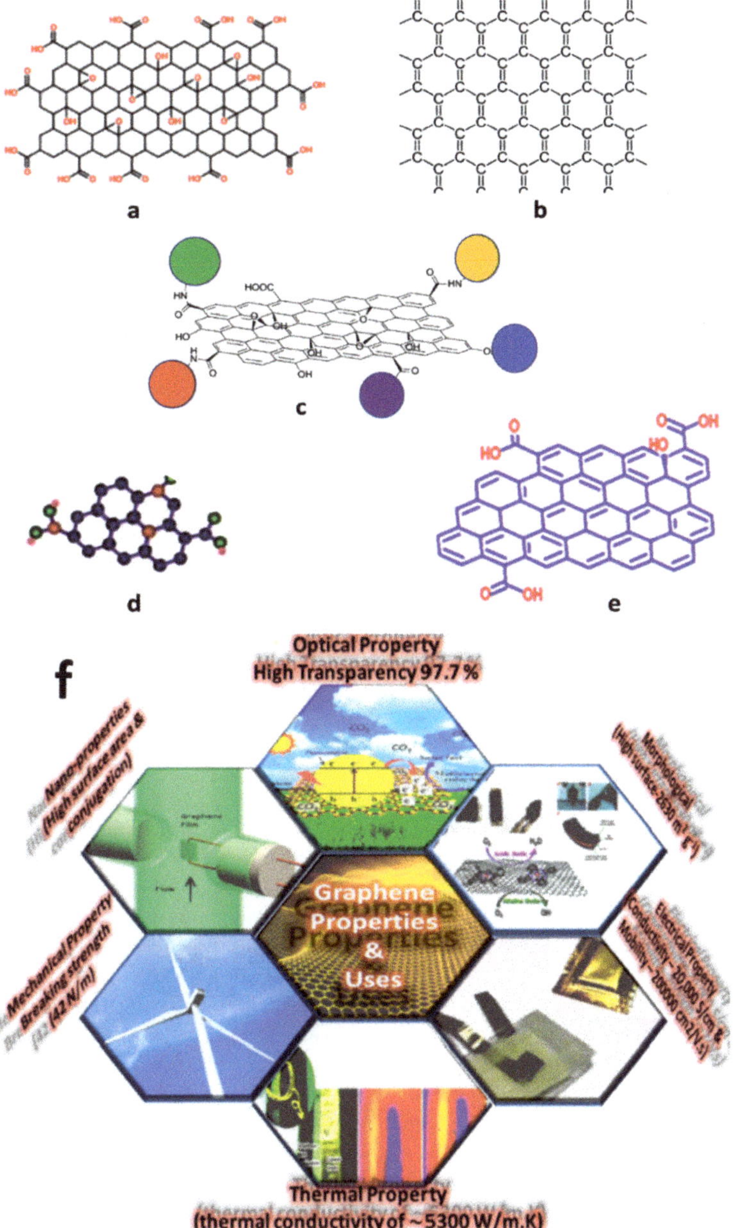

Figure 3. Some common forms of graphene: (**a**) graphene oxide, (**b**) pristine graphene, (**c**) functionalized graphene, (**d**) graphene quantum dot and (**e**) reduced graphene oxide. Reproduced with permission from Reference [46]. Licensed under a Creative Commons Attribution 3.0 International License (https://creativecommons.org/licenses/by/3.0/); (**f**) Different properties of graphene and its applications. Reproduced with permission from Reference [108]. Copyright ©2019, Elsevier.

Graphene oxide (GO), commonly prepared from the oxidation of graphite, consists of a few- or a single-layer sheet. GO sheets are rich in various oxygen-containing groups such as hydroxyl, epoxy, carboxyl, carbonyl, phenol, lactone and quinone, which can change the van der Waals interactions. The two former chemical groups are mostly present on the basal plane, whereas the others with small quantities are found at the sheet edges. These functional groups in GO can deeply influence its electrochemical, mechanical and electronic features. Despite the aromaticity of graphene is lost in GO, owing to exploitation of π electrons in the covalent bonding of these oxy groups on graphene backbone, the carbonyl, carboxyl and so forth groups at the edge render them more dispersible in both organic solvents and water [44,107]. The hydrophobic aromatic frameworks and the hydrophilic oxygen-containing groups make GO amphiphilic, allowing its interaction with inorganic and organic molecules.

Reduced graphene oxide (RGO), obtained by the reduction of GO [109], contains fewer oxygen atoms, hence, is less negatively charged [106]. During the reduction, RGO recovers the graphitic arrangements (partial recuperation of the sp^2 from sp^3 hybridization of GO) through the elimination of the oxygen-containing groups, which have been inserted in the oxidation step, thus, restoring the electronic properties of graphene [110]. This partial reduction and the exposure to some chemicals allow tailoring the conductivity, band-gap and optical features of the material [111,112].

Graphene quantum dots (GQDs), which can be found as single- or multiple layers, display interesting features such as good chemical stability, high surface area, tunable physical characteristics, stable photoluminescence and low toxicity [113,114]. They can be used in optoelectronic, electronic, biomedical, sensors and energy storage. They usually consist of up to 10 layers of 10–60 nm size RGO [46].

Graphene-based nanomaterials possess exceptional electrical, optical, mechanical, electrochemical and thermal features that make them versatile for a wide range of applications and have drawn worldwide attention in both academic and engineering fields [44,105]. They can be employed in industrial applications such as biomedical, solar cells, biosensors, supercapacitors, electromagnetic absorbers, optical devices, integrated circuit, protective coatings, organic light-emitting diodes, sound transducers, petroleum industry, automobile components, aerospace, energy storage, nanocomposites and contamination purification in wastewater management, to cite a few (Figure 3f) [107,115,116].

3.2. Synthesis Routes and Properties

Graphene can be produced from various sources such as graphitic, non-graphitic and waste materials using top-down or bottom-up approaches [117–120]. The common routes for its fabrication are summarized in Figure 4. The top-down synthesis routes encompass mechanical exfoliation, liquid phase-exfoliation (LPE), oxidative exfoliation-reduction, arc discharge, unzipping of carbon nanotubes, for which larger precursors such as carbon-based materials or graphite are destroyed to produce a single-, bi- and few-layer graphene. Broadly, some of these approaches can generate high-quality products and are likely scalable. Nevertheless, they provide limited yield and have complications in making nanomaterials with reliable characteristics, which are closely dependent on the carbon precursor. On the other hand, the bottom-up synthetic routes could produce graphene using atomic-sized precursors. These approaches comprise epitaxial growth, chemical vapor deposition (CVD), total organic synthesis, template route and substrate-free gas-phase synthesis. Despite the quality of the produced graphene is better than that generated using top-down methods, they often require advanced operational setup, high fabrication costs and are energy-consuming. Further advantages and the shortcomings of the most important methods have been reported elsewhere [117,118].

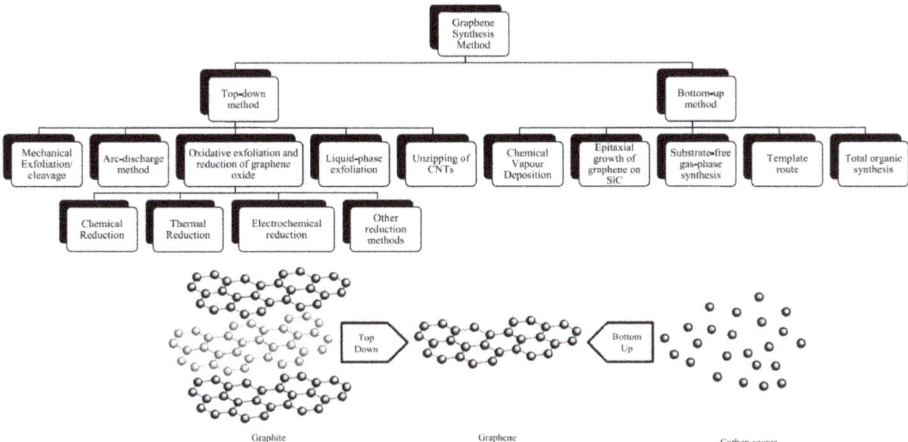

Figure 4. Production techniques of graphene materials. Reproduced with permission from Reference [119]. Licensed under a Creative Commons Attribution 3.0 International License (https://creativecommons.org/licenses/by/3.0/).

It is worthy to note that most of the studies have not usually utilized graphene in its pristine form, because of its lower yield from the production point of view. Therefore, its derivatives have received much attention. GO is commonly prepared using a chemical oxidation process of graphite with subsequent dispersion and exfoliation in a suitable solvent (e.g., water). Graphene oxide sheets can also be fabricated using a modified Hummers' method, which is described in several reports [121,122]. The oxidation processes can lead to fragmentation, crack, winkle, structure disorder, impurities and defects that may influence the adsorption, optical and electronic characteristics of GO. RGO, however, is usually produced by reducing graphene oxide employing different ways such as chemical, thermal, photocatalytic and electrochemical reductions [123]. Nonetheless, the obtained RGO may contain some impurities with the presence of structural defects. Besides that, the production strategies of GQDs comprise solvothermal, microwave, CVD and soft template processes, in-situ reduction of GO, electrochemical fabrication, chemical synthesis and electron beam lithography [113,114,124]. Among them, top-down approaches have been proved to be the most appropriate and cost-effective methods [46]. GQDs exhibit similar features compared to various types of quantum dots (QDs), particularly in the case of inorganic QDs [113].

It has been recently revealed that oxidative exfoliation-reduction, liquid-phase exfoliation and CVD are the most interesting production methods, which possess high potential for industrial implementation to produce graphene-based nanomaterials [45]. However, to develop effective synthesis processes of graphene and its derivatives, further research activities have to be conducted to improve the quality, yield of the products with tailorable properties using cost-effective, environmentally friendly, reliable and scalable approaches.

The properties of graphene-based materials are closely dependent on the number of layers as well as the extent of defects. Graphene, as the thinnest carbon material, presents outstanding features such as higher surface area of ~2630 m^2/g compared to GO and other derivatives. It has been reported that a single layer of graphene absorbs 2.3% of white light with a reflectance of less than 0.1%. At room temperature, the in-plane thermal conductivity of GN is about 2000–5000 W/m·K. Such dissimilarity is due to the dissemination of phonons pathway at the surface [46,108]. Some research works reported that the charge transporters and carriers mobility of 200,000 cm^2/V·s can be reached at electron densities of ~2 × 10^{11} cm^{-2} [108]. GN possesses good chemical stability and quantum Hall effect at ordinary temperature, intrinsic strength of 130 GPa, Young's modulus of 1.0 TPa, shear strength of 60 GPa and fracture stress of 97.54 GPa [123]. It is considered as one of the strongest materials ever tested

(200 times than steel) [125]. More details about the characterization methods and the properties of graphene and its derivatives have been extensively reviewed in recent years [105,123,124,126,127].

4. Preparation, Properties and Application of CNC/GNM Hybrids

CNC has aroused a tremendous amount of interest of the scientific community in recent years due to its outstanding features and can be employed as an independent functional material, template support, stabilizer, filler or reinforcing agent. Recently, it has been combined with numerous GNMs such as GN [128], GO [129], RGO [130], GQDs [131–133], free-standing graphene (FSG) [134] and graphene nanoscrolls [135] to produce hybrid materials with excellent thermal, mechanical, optical and electronic properties. However, several scientific and technical issues can be encountered during the production of such hybrid materials such as agglomeration, limited dispersion, process scalability and high costs, among others. Hence, many research works have been carried out and others continue to be conducted worldwide to overcome these problems and obtaining efficient CNC/GNM for a wide range of applications. The emphasis of the following subsections will be dedicated to the most important approaches used to produce CNC/GNM hybrids as well as their properties. Specific attention will be dedicated to the investigations performed during the last few years.

4.1. CNC/GN

GN possesses large surface area, exceptional electrocatalytic activity, high mechanical strength, good electronic transport characteristics, excellent optical properties and thermal performance, which has motivated its broad application prospect in several fields such as functional composites, electrochemical sensors and catalysis, among others [136]. To expand the number of applications of GN and enhance its inherent properties, numerous GN composites have been successfully produced and applied in several fields. Recently, nanocomposites of CNC/GN have attracted widespread attention, owing to their exceptional features and synergetic effects that develop new ways and opportunities for the production, characterization and application of new materials in nanotechnology. It has been recently demonstrated that the preparation of CNC/GN can be carried out with and without chemical functionalization for which the water-based dispersion is the common starting approach to produce composites with/without a combination with various types of materials such as metallic and ceramic nanoparticles or natural and synthetic polymers. Several processes can be further applied such as filtration, hot pressing, deposition and drying to generate a wide range of advanced materials. Thus, such CNC/GN-based materials hold a great promise for several applications ranging from packaging to biomedical fields. Nevertheless, the optimization of the composite compositions and tailoring of their properties can extend the number of applications and reduce the production cost for eventual scalability.

Carrasco at al. have employed CNC as an effective graphene stabilizer in aqueous dispersion at high concentration for which the exfoliation of graphite to generate graphene flakes has been carried out using a tip sonication [137]. Such an approach based on CNC-assisted liquid-phase exfoliation (LPE) produced graphene flakes decorated with CNC stabilizers with interesting properties. The authors proved that such hybrids could be employed in different applications such as composites and supercapacitors. In another study by Cui et al. an interesting efficient one-step mechanical-chemical method to in-situ produce CNC/GN hybrid, with rigid 2D structure and improved interfacial interactions, from micro fibrillated cellulose and graphite using ball milling has been developed [138]. A schematic illustration of the composite preparation is given in Figure 5A. This hybrid was successfully dispersed within poly(propylene carbonate) (PPC) with strong interfacial interactions which can increase its glass transition temperature (T_g) and enhance its mechanical and electrical features for practical uses. The obtained PPC/CNC/GN composite displayed a T_g of 51.3 °C, which is higher than that of pure PPC (34.0 °C). The percolation threshold considerably decreased from 15 to 5 wt.%, whereas the tensile strength and the Young's modulus reached 52.8 MPa and 731.2 MPa, respectively.

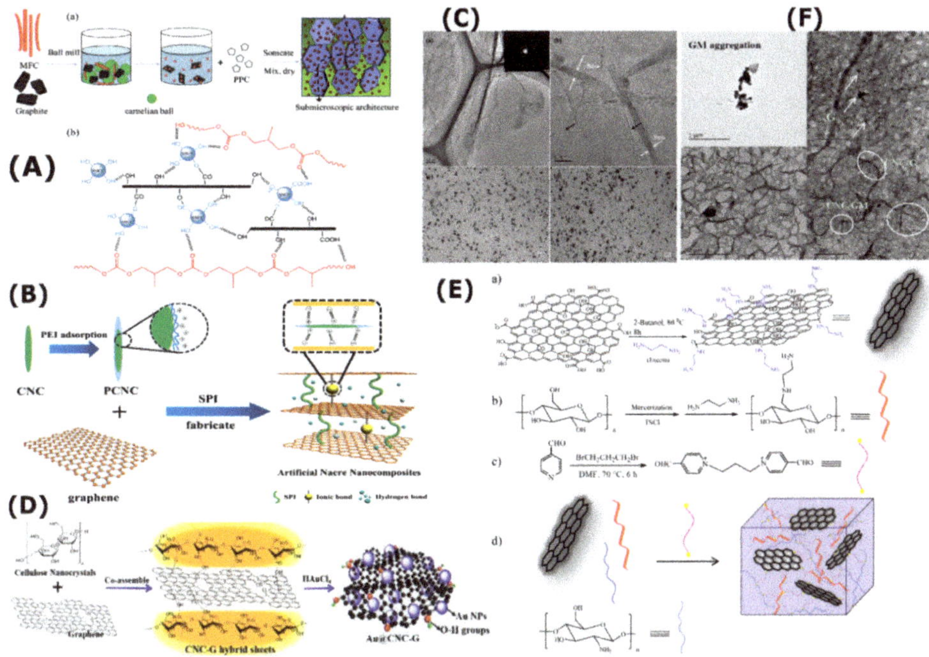

Figure 5. (**A**) Schematic illustration of the production of poly(propylene carbonate) (PPC)/cellulose nanocrystals (CNC)/aminated graphene (GN) composites and the available hydrogen bonding. Reproduced with permission from Reference [138]. Copyright ©2018, Elsevier; (**B**) Schematic presentation of the SPI-based nanocomposite film. Reproduced with permission from Reference [139]. Licensed under a Creative Commons Attribution 3.0 International License (https://creativecommons. org/licenses/by/3.0/); (**C**) Transmission electron microscopy (TEM) and optical micrographs of the CNC/GN solution showing good dispersion. Reproduced with permission from Reference [140]. Copyright ©2015, Elsevier; (**D**) Schematic synthesis of Au@CNC-GN catalyst. Reproduced with permission from Reference [141]. Copyright ©2018, The Royal Society of Chemistry (RSC) on behalf of the Centre National de la Recherche Scientifique (CNRS) and the RSC; (**E**) Preparation procedure of chitosan/WN/GN hydrogel. Reproduced with permission from Reference [142]. Copyright ©2020, Elsevier; (**F**) TEM images of GN and CNC/GN sol mixtures containing 2 wt.% of GN. Reproduced with permission from Reference [143]. Licensed under a Creative Commons Attribution 3.0 International License (https://creativecommons.org/licenses/by/3.0/).

Montes et al. have recently demonstrated the existence of a synergetic reinforcement of poly(vinyl alcohol) (PVA) nanocomposites with CNC-stabilized graphene [144]. They produced CNC/GN hybrid using a CNC-assisted LPE that allows the stabilization of the resulting GN in aqueous dispersion. Such hybrid was incorporated into a PVA aqueous solution by a direct blending to obtain a nanocomposite after casting evaporation. It was mentioned that the thermal stability of the composite is improved through the addition of 1 wt.% of CNC/GN hybrid nanofiller. Moreover, the mechanical features have been also enhanced compared to the neat PVA (20% improvement in tensile strength and 50% in Young's modulus). It was claimed that CNC played a dual role, where it acts as GN stabilizer and PVA reinforcement. Moreover, the synergetic effect of CNC/GN hybrid is notable, where interesting thermal, mechanical and electrical features can be attained through the tailoring of the nanofiller loading. Recently, this research group has studied the effect of CCN/GN hybrid on the properties of poly(lactic acid) (PLA) based film [145]. The composite was prepared using a melt blending method, a conventional technique for plastic compounding, at a total loading of 1wt.% and then processed by hot pressing to generate the film. Compared to the baseline, the film PLA/CNC/GN exhibited high thermal stability and better mechanical features with an increase of 11 and 8% in the tensile strength and Young's modulus, respectively. The investigation of the gas barrier properties as well as the antifungal activity of the prepared film revealed significant improvements, which make it a potential candidate in food packaging trays and agricultural film applications.

A few years ago, an interesting composite based on soy protein isolate (SBI) and CNC/GN has been developed by Li et al. as food packaging material [139]. A schematic presentation of the nanocomposite preparation is shown in Figure 5B. The authors exploited the high aspect ratio of 1D CNC with the flexible and strength 2D GN to manufacturing active interfacial adhesion laminate nanocomposites. To obtain a stable aqueous graphene dispersion via sonication, the negatively charged sulfate ester groups of CNC were firstly modified through the incorporation of positively charged surface functionalities using the cationic polyethyleneimine. Such modification enhanced the strong ionic interactions with negatively charged GN for efficient dispersion and later-by-layer assembly with SBI. The obtained composite film displayed interesting mechanical features and improved surface hydrophobicity for which the tensile strength increased from 3.75 to 7.49 MPa and the water contact angle augmented from 39° to 54° compared to the control film. Better thermal stability, water resistance and UV-visible light barrier ability were also exhibited by such composite film, making it a potential candidate as food packaging material.

Valentini et al. produced polymer solar cells using optically transparent conductive GN and CNC film [146]. They reported that the mixture containing 10 mL of CNC suspension (0.5 wt.%) and 10 mL of GN solution (1 wt.%), which was prepared in an ultrasonic bath at room temperature for 20 min and evaporated under a nitrogen stream, was the best composition. The obtained NG/CNC layer, which has a low surface roughness, was optically transparent and enabled light to go through. The measurement of the contact angle of GN/CNC demonstrated a lower contact angle value when compared to those of the neat glass or CNC film, which was assigned to the flatter surface morphology of the GN/CNC film. These authors produced a photovoltaic device by spin coating. The thickness of the spin-cast photosensitive layer was about 100 nm, as determined by atomic force microscopy. The manufactured polymer solar cell reached a higher short-circuit current density value, revealing its improved electron blocking action. The enhanced mechanical properties, the optical transparency as well as the electrical conductivity of the hybrid layer will certainly allow the development of the next generation of flexible and foldable printed optoelectronic devices. In another work, Wang et al. combined GN and CNC to produce flexible, electrically and thermally conductive hybrid thin film using a water-based approach and vacuum filtration [140]. Figure 5C shows the transmission electron microscopy (TEM) and optical micrographs of the obtained GN of about 15 layers as well as CNC/GN solution, revealing that GN was uniform without the appearance on any segregation after mixing with hydrophilic CNC. The better particle alignment with the removing of the internal pores was promoted by the use of the hot-press process. It was found that the hot-pressed 25 wt.% CNC hybrid paper exhibited interesting mechanical features for which the modulus was improved by 57% and tensile strength by 33% with respect to the neat GN paper. The electrical conductivity was negatively affected by the increase of the amount of CNC and the optimum CNC loading was 15%. Such hybrids can find application in heat and electrical-conducting fields.

The employment of CNC/GN hybrid as a supporting material to produce supported metal catalysts, which can be used as dispersing, capping or reducing agents, using a clean, simple and effective process has been reported. Wang et al. have deposited mono-dispersed gold nanoparticles (Au NPs) on multifunctional CNCN/GN hybrid sheets to generate catalysts with efficient catalytic activity, flexibility and stability [141]. The production procedure is briefly illustrated in Figure 5D. The hybrid structure allowed the reduction, growth and immobilization of Au NPs. The OH-groups of CNC coordinated with GN permitted creating narrow nanosized Au NPs anchored onto the surface of the hybrid through the in-situ reduction of Au^{3+}. Such a catalyst has been revealed to be effective for one-pot reaction of an alkyne, an amine and an aldehyde in water since it can minimize the environmental pollutions caused by heavy metallic ions and organic solvents. It can be reused for several times without significant deactivation. It is also expected to be employed in a wide range of applications such as energy storage and catalysis.

On the other hand, stimuli-responsive hydrogels, such as 3D polymeric networks, are considered prominent intelligent drug delivery systems to selectively release the drug at the desired sites. The emerging of CNC/GN hybrids has pushed the scientific community to develop a new generation of efficient hydrogels. Omidi et al. have successfully developed a pH-responsive hydrogel containing aminated CNC (WN), aminated graphene (GN) and chitosan via Schiff base reaction by a synthetic dialdehyde in a few minutes [142]. The preparation procedure is schematized in Figure 5E. The prepared hydrogel exhibited better sensitivity to different external stimuli encompassing pH and amino acids. More specifically, it displayed a pH-responsive release behavior for anticancer drugs. Also, the hydrogel presented strong antibacterial activity against gram-positive bacteria, revealing the efficiency of such hydrogel as a potential candidate for the localized drug delivery systems.

To extend the application of CNC/GN hybrids as anti-static or electromagnetic interference shielding materials, Liu et al. have prepared sandwiched films of epoxy resin and GNC/GN paper. Firstly, a hybrid paper of GN containing 10 wt.% of CNC was produced using ultrasonication process in aqueous suspension [147]. This hybrid displayed an electrical conductivity of 16,800 S/m and tensile strength of 31.3 MPa. To manufacture the sandwiched film of epoxy and CNC/GN, a dip coating method was applied through the introduction of the paper into epoxy resin solution followed by a curing process at ambient temperature. The authors revealed that the moduli of the films were about 300 folds and the tensile strength increased by two-folds concerning the pure resin. The glass transition of the composite increased as well when compared to that of the neat resin. Besides, the coated CMC/GN hybrid by epoxy resin displayed better dimensionality integrity after sonication in water for two hours. In another work, a composite containing water-born polyurethane/CNC/GN has been recently prepared through one-step sol process by Yang et al. as a thermosetting coating material for wood-based composites, which exhibited better energy-saving characteristics [143]. The better dispersion of CNC/GN has been optimized as shown in a TEM image in Figure 5F. The properties of the prepared composites have been improved through the incorporation of CNC/GN, where the thermal conductivity, abrasion resistance and hardness were enhanced and meanwhile, the coating adhesion was maintained at an acceptable level. The authors claimed that such findings can promote the development of wooden heating material with better-energy saving characteristics.

In another study, Nie et al. have introduced a small amount of CNC (also named CNWs) to GN (CNWs/GN = 1/20 *w/w*) to improve its uniform dispersion in a waterborne epoxy polymeric matrix (WEP), which is still challenging at a high GN loading, using a solution-casting approach [148]. A schematic illustration of the preparation of the composite is depicted in Figure 6. The obtained film at a GN loading of 1.0 wt.% achieved enhanced mechanical properties with a higher Young's modulus of 2820 MPa compared to the neat epoxy (2034 MPa). The glass transition of such composite increased by 4.3 °C when compared to the pure resin. The better dispersion of GN on the surface of epoxy owing to the effect of CNWs led to the increase of the water contact angle, confirming the improvement of the water-barrier behavior of the composite CNWs/GN/WEP. The authors have assessed the anticorrosion effectiveness of the prepared coating composite using potentiodynamic polarization and electrochemical impedance spectroscopy tools. The results demonstrated that CNWs played a double role through improving the dispersion of GN and the corrosion resistance for mild steel.

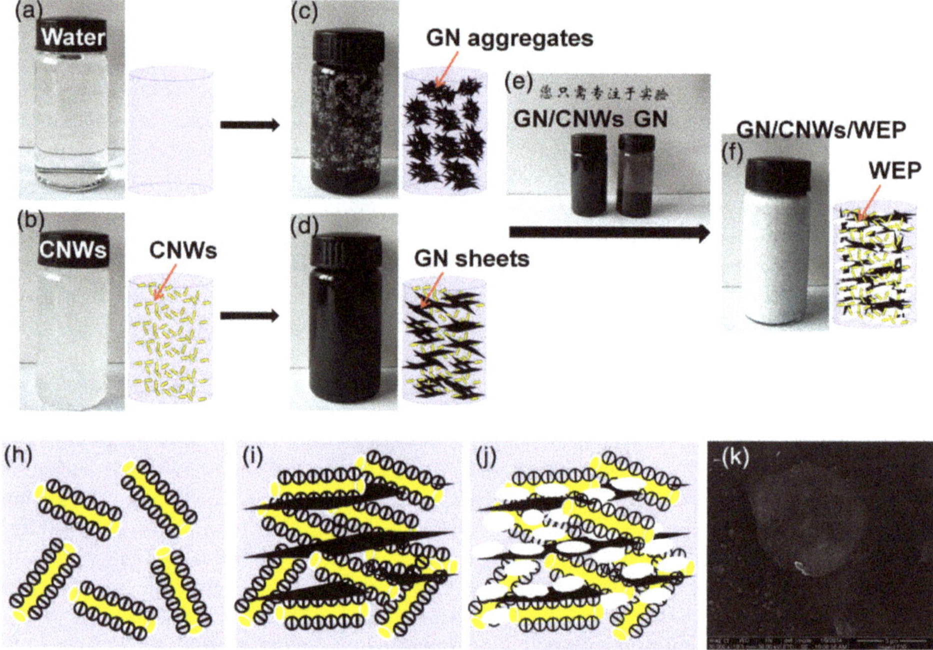

Figure 6. Neat water (**a**); aqueous dispersion of CNC (CNWs) (**b**); dispersion of GN in water (**c**); GN in CNWs aqueous dispersions (**d**); dispersion stability of GN in water and CNWs aqueous dispersions after the settlement of 30 min (**e**); dispersion of 1.0% GN/CNWs in waterborne epoxy polymeric matrix (WEP), (**f**); schematic of CNWs with negative charges (**h**); schematic of negatively charged CNWs adsorbed on graphene sheets (**i**); schematic of GN sheets stabilized in WEP assisted with CNWs (**j**); Field emission scanning electron microscopy (FE-SEM) micrograph of CNWs adsorbed on graphene sheet (**k**). Reproduced with permission from Reference [148]. Copyright ©2019, Wiley.

4.2. CNC/GO

GO, as one of the most important derivatives of graphene, contains a high density of oxygen-functional groups, which can allow covalent, ionic or hydrogen interactions with numerous polymeric matrices, paving the way to several technological applications. It displays interesting features such as high specific surface area, high binding potential, high hydrophilicity, high dispersibility, superior mechanical properties and surface functional groups that can be employed as attachment sites [149]. However, to fully exploit the potential of GO, the production of GO-based composites is significant. Various GO-based nanocomposites have become increasingly mature in several fields. Recently, studies on CNC/GO hybrids have been actively conducted. CNC with outstanding features such as high Young's modulus, high crystallinity, high surface chemical activity and tailorable surface characteristics can offer strong, non-toxic and flexible advanced GO-based hybrids, which further inherit the features of both CNC and GO for which desired properties can be obtained. CNC/GO-related material composites have been investigated both on their own and after incorporation of other components such as metals, ceramics or polymers to modulate their final properties for specific uses. These hybrids found applications in nano paper, food packaging, biomedical, energy storage, sensors, decontamination, catalysis, adsorption, shape memory devices, foams, fire retardants and insulating materials, to cite a few.

The aqueous GO solution with CNC is considered the common approach used to prepare CNC/GO composites. Kafy et al. have produced CNC/GO film as humidity sensor using this blending method, followed by the drying process [150]. They synthesized CNC using the conventional H_2SO_4 hydrolysis method, whereas they produced GO by way of the modified Hummer's method. Then, they mixed the two suspensions at a desired ratio followed by homogenization. After that, the solution was poured in a petri dish and dried. The obtained hybrid exhibited a good dispersion of GO in CNC matrix for which high dielectric constant and low dielectric loss have been revealed, owing to the special polarization dipoles in CNC. These authors prepared a renewable, flexible and cheap sensor using this hybrid and an interdigital transducer patterned electrode deposited on a polyethylene terephthalate (PET) substrate. This sensor displayed a good sensitivity to humidity even under different temperatures. Similarly, Chen et al. produced CNC/GO using an aqueous suspension of CNC with either GO suspension or GO powder (Figure 7) [151]. The mixing process generated a stable solution for the first, whereas a metastable solution was obtained when GO powder was used. The drying process carried out via vacuum-assisted self-assembly technique (VASA), engendered non-iridescence and iridescence films, respectively, for the hybrids containing GO suspension and GO powder. It is worthy to note that CNC-based iridescent films found applications in optical functional materials. It was demonstrated that self-organized film was obtained from stable solution, while the separated structure was generated from the metastable solution. Interestingly, the later film, which displayed iridescent optical properties, consisted of self-assembled liquid crystals phase of CNC with embedded GO sheets. The authors claimed that such iridescent hybrid can be applied in security materials, reflective filters, sensors and other photonic materials.

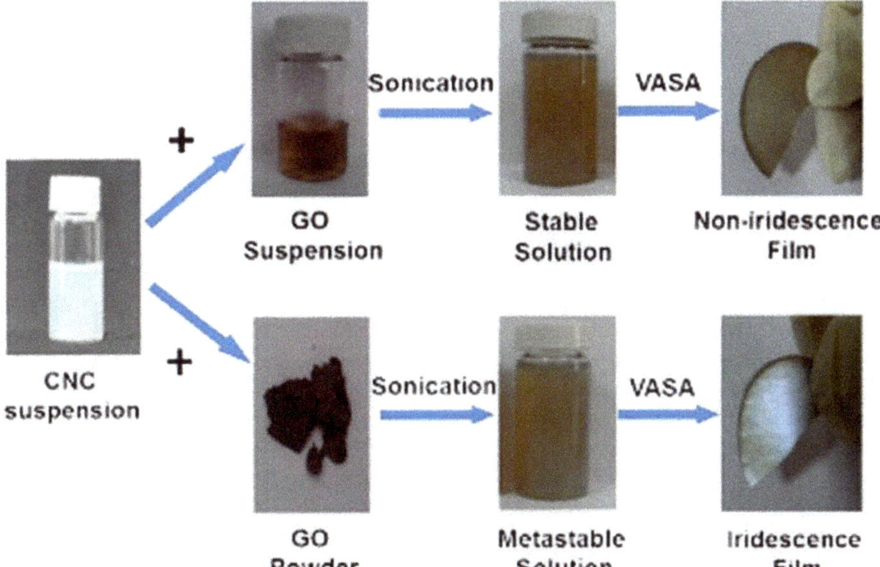

Figure 7. Schematic illustration of the preparation process of GO/CNC hybrid films with or without iridescence. Reproduced with permission from Reference [151]. Copyright ©2014, The Royal Society of Chemistry.

Although the water-based dispersion is the widely adopted method to produce CNC/GO, it represents an unavoidable issue of the higher resistance. Valentini et al. have developed a method to produce CNC/GO with reduced electrical resistivity [129]. They employed the same approach of the mixing of the CNC suspension with the GO solution but assisted by an external electric field. This latter induced de-oxygenation of GO and hence its conductivity can be recovered to some extent. Such electrical conductivity was rather moderate, because of the presence of CNC as an insulating matrix. In a separate work of the same authors, the above approach, which is based on the drop-casting of an aqueous solution of CNC/GO between two metal electrodes, was found to be efficient to produce a resistive memory device based on CNC/GO thin hybrid [152]. Such thin film-based device exhibited a transition between low and high conductivity states upon changing the polarity of the applied external electric field. The authors claimed that such an achievement could promote the development of post-silicon electronic devices based on the integration of CNC/GN thin hybrids. Recently, Pan et al. developed a new method to produce chiral smectic structures through self-assembling 2D GO and 1D CNC nanorods [153]. Such a structure is closely dependent on the ratio of nanorods and nanosheets as well as the concentration of the composite colloid. The authors initially mixed CNC and GO suspensions at low concentration (<1%) and incorporated cross-linked polyacrylate hydrogel to concentrate the blend suspension. The CNC/GO was recovered by spin-coating of the colloid on PES substrate and dried at 60 °C. This method was considered timesaving compared to traditional approaches. It was demonstrated that such advancement can pave the way to develop optical metamaterials for optical modulation and mechanochromic sensors.

It was reported that poor dispersion of reinforcements at the nanoscale in addition to the weak interfacial interactions can negatively affect the material strength, toughness and other properties. Thus, several physical or chemical modifications can be employed to overcome such issues. In the case of CNC/GO composites, several approaches have been proposed to improve their efficiency for numerous applications. The common ones used to enhance the interfacial of such hybrids were based on the modification of CNC surface features, whereas few modifications have been simultaneously applied to CNC and GO.

One of the interesting production methods was that developed by Xiong et al. to manufacture ultra-robust transparent CNC/GO membrane with high electrical conductivity [154]. These authors improved the interfacial interactions of anionic CNC, prepared by H_2SO_4 hydrolysis and anionic GO sheets obtained through the modification of CNC with 10 wt.% cationic polyethyleneimine to introduce positive surface charge functionalities. This modification enhanced the ionic interactions between the strongly positively charged polymer and negatively charged flexible GO, which consequently improved the layer-by-layer assembly, carried out on a sacrificial layer of cellulose acetate on a silicon wafer, to design laminated nanohybrids with high flexibility, outstanding mechanical strength, high optical transparency along with excellent toughness. The authors claimed that such CNC/GO hybrids could be used for a wide range of technological applications, encompassing wearable electronic devices, biofluid separation, electromagnetic interference shielding and ballistic protection. The authors also employed the same approach to produce CNC/RGO hybrids after the electrochemical reduction of the former membrane [155]. In another work, Kabiri et al. produced acetylated CNC (CNCA), which was further used to prepare well dispersed CNCA/GO hybrid by a solvent casting method (Figure 8A) [156]. It was stipulated that the modification of CNC will promote its interfacial adhesion and miscibility with GO via hydrogen bonding. It was proved that composite supplemented with 0.8M of GO offered better thermal stability, interesting mechanical properties with an increase in the tensile strength of 61.92% with respect to CNCA. Moreover, the barrier characteristics against water were improved. The authors claimed that such a composite could find potential application in electrical and electrochemical fields.

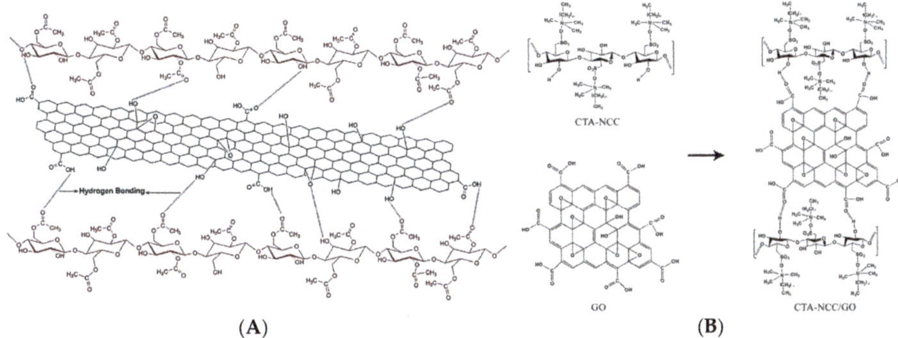

(A)   **(B)**

Figure 8. (**A**) Schematic interaction of CNCA/GO. Reproduced with permission from Reference [156]. Copyright ©2014, Springer; (**B**) Schematic diagram of CTA-NCC/GO nanocomposite. Reproduced with permission from Reference [157]. Copyright ©2019, Elsevier.

Recently, Daniyal et al. have prepared hexadecyltrimethylammonium bromide (CTA) modified CNC/GO thin film and assessed its potential in sensing copper and nickel ions based on surface plasmon resonance (SPR) technique [157,158]. The authors initially prepared CTA-CNC solution and then 0.1 wt.% of GO was dispersed within the solution and sonicated at 70 °C for 1 h (Figure 8B). The obtained solution of CTA-CNC/GO was spin-coated and deposited as a thin layer on the glass substrate modified with a thin gold film. The authors demonstrated that the presence of CTA improved the sensitivity of the SPR. They revealed that the combination of SPR and CTA-CNC/GO has the potential to be employed as effective sensors, which can detect copper and nickel ions. In another research work, Beyranvand et al. produced hydrogel based on CNC/GO hybrid as a new adsorbent for methylene blue. During the preparation, azide-functionalized CNC was synthesized after CNC tosylation [159]. Then CNC-N_3/GO was obtained via nitrene chemistry [160]. The production process of CNC-N_3/GO, as well as its mechanism of action, is illustrated in Figure 9. It was demonstrated that the prepared hybrid was an excellent adsorbent of methylene blue owing to the higher adsorption capacity, reasonable contact time and recyclability. More recently, Zheng et al. have synthesized a modified CNC/GO hybrid as an efficient adsorbent of Dy (III). The authors used the evaporation-induced self-assembly (EISA) method to spontaneously form an imprinted film. Beforehand, they carried out an in situ selective oxidation of CNC using 2,2,3,3-tetramethylpiperidine-1-oxyl (TEMPO) for which C6 hydroxyl group was primarily oxidized to the carboxyl group (-COOH) [161]. It was found that such modification improved the stability of the TEMPO-modified CNC through strong electrostatic repulsion on one hand and on the other hand, it offered more surface active sites for the adsorption of Dy (III). The latter was further improved by the introduction of GO, which created extra bonding sites to Dy (III) and enhanced the adsorption capacity of TEMPO-CNC/GO hybrid. These authors reported that the developed green hybrid was efficient and had a strong regeneration performance.

The preparation and design of molecularly imprinted polymers (MIPs) is a multidisciplinary field, which encompasses various aspects of molecular recognition, biomimetic biology and polymer chemistry. MIPs preparation involves arranging functional monomers around a template, followed by polymerization with the presence of cross-linkers and a suitable initiator through covalent, semi-covalent or non-covalent intermolecular interactions and finally template removal. Such an approach has been recently explored to produce CNC/GO-based composites as molecular imprinted electrochemical sensors, which exhibited outstanding features. For instance, Anirudhan et al. have prepared MIP of silylated GO and chemically modified CNC using a drop cast method for the selective sensing of cholesterol [162]. The authors incorporated ZnO to CNC to enhance their conductivity. The electrochemical studies were carried out using cyclic voltammetry and differential pulse voltammetry. This sensor achieved good stability and reproducibility, low detection limit and wide linear range. The optimum pH, equivalent to the blood pH, was 7.4 and the optimum response time was only 10 min. In another work,

Wang et al. manufactured a CNC/GO-based MIP for the selective extraction and fat adsorption of synthetic antibiotics (fluoroquinolones, FQs), which can accumulate as residues in river water, causing a hazard for living organisms [163]. The preparation process of the magnetic@GO-grafted-CNC@MIP is depicted in Figure 10. It was found that the utilization of CNC and GO as substrates can improve the properties such as the stability, selectivity and affinity of MIPs compared to the conventional ones. The authors demonstrated that the prepared hybrid displayed an ultra-fast adsorption profile for FQs with high recognition and large detection limit range. They claimed that this method is accurate, effective, sensitive and simple, thereby appropriate for the detection of residual FQs in water sample.

Figure 9. Schematic representation of the production procedure, as well as the exhibition of the carboxylate and phenoxide adsorption, GO sites of the hybrids towards methylene blue. Reproduced with permission from Reference [159]. Copyright ©2019, American Chemical Society.

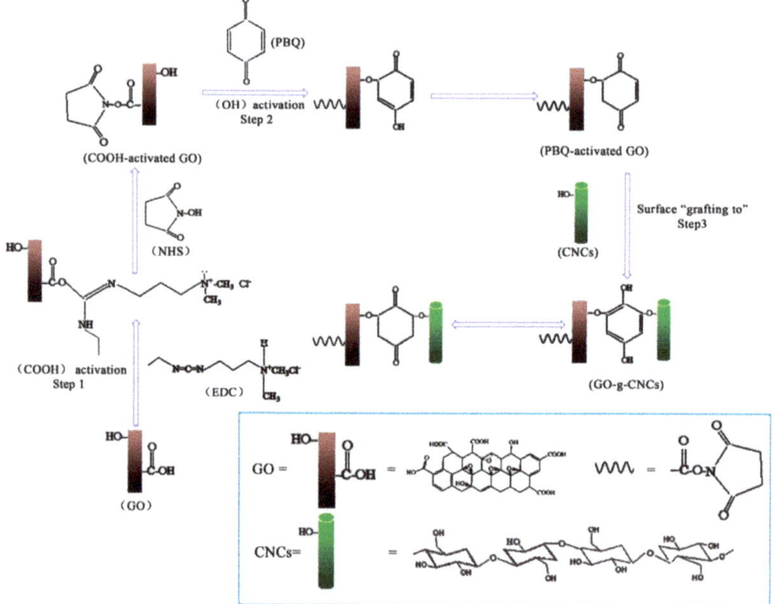

Figure 10. Schematic procedure of the preparation of the magnetic@GO-grafted-CNC@MIP. Reproduced with permission from Reference [163]. Copyright ©2017, Springer.

The simultaneous incorporation of CNC and GO to numerous polymers such poly(3-hydroxybutyrate-co-3-hydroxy valerate) [164], poly-N-isopropyl acrylamide [165], poly(3,4-ethylenedioxythiophene) [166], poly(vinylidene fluoride) [167], polyacrylamide [168], poly(ε-caprolactone) [169], polylactic acid [170], poly(vinyl alcohol) (PVA) [171] and chitosan [172] was adopted as an efficient method to produce composites with excellent features since these nanofillers offer outstanding synergetic effects. Some surface modifications can be applied to CNC or GO to improve their dispersion and compatibility within the polymeric matrices. This type of nanocomposite found a wide range of applications in biosensing [165], plastic masks [172], tissue engineering [168,169], wastewater treatment [167], food packaging [173], supercapacitors [166], to cite a few. For instance, El Miri et al. evaluated the synergetic effect of CNC/GO as a functional hybrid to enhance the properties of PVA nanocomposites (Figure 11I). The nanocomposites were prepared via solvent casting method. The authors demonstrated that the tensile strength, toughness and Young's modulus were respectively enhanced by 124%, 159% and 320% compared to the neat PVA. The strong interfacial interactions and the synergetic effect of 1D elongated CNC and 2D exfoliated GO, which improved the dispersion and avoided the agglomeration of the nanofillers, were also highlighted compared to the incorporation of pure CNC or GO. Such nanocomposite may find application in food packaging materials. In another recent study, Kumar et al. produced hybrid hydrogels containing polyacrylamide-sodium carboxymethylcellulose (PMC), GO and CNC via in situ free-radical polymerization (Figure 11II) [168]. The obtained composite displayed outstanding mechanical performance, self-healing behavior and shape-recovery feature. The authors claimed that such highly hydrated hybrid hydrogel with tailorable properties might provide a 3D microenvironment for tissue engineering applications.

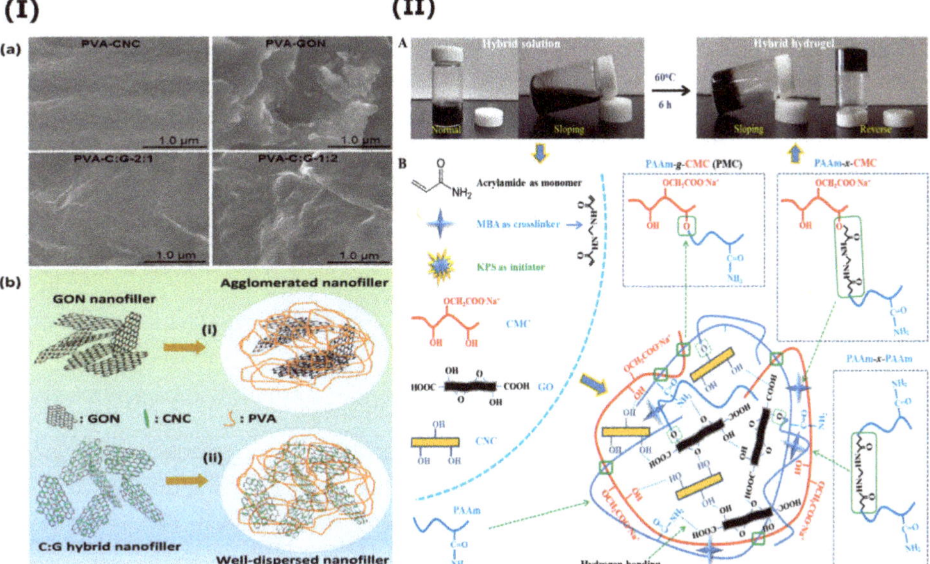

Figure 11. (**Ia**) Scanning electron microscopy (SEM) micrographs of PVA nanocomposites with CNC, GO and their hybrid (C: G-2:1 and C:G-1:2) and (**Ib**) schematic representations of the dispersion state of (i) GO and (ii) C: G hybrid within the PVA polymeric matrix. Reproduced with permission from Reference [171]. Copyright ©2016, Elsevier; (**II**) Schematic of the formation of the hydrogel: (**A**) Before and after heat treatment of PMC-GO1/CNC10.0 hybrid solution and (**B**) A suggested mechanism of physical and chemical interactions in the hybrid hydrogel system. Reproduced with permission from Reference [168]. Copyright ©2018, Elsevier.

4.3. CNC/RGO

The production of RGO, which is commonly performed by the exfoliation of pristine graphene followed by oxidation and reduction, may generate numerous defects such as grain boundaries, vacancies, Stone-Wales defects and macroscopic defects. These defects not only restrict its production at the industrial scale but also limit the full exploitation of its outstanding properties. Another obstacle in the practical use of RGO is the formation of irreversible agglomeration caused by the strong van der Waals interactions between graphene planes. Therefore, various attempts were made to reduce these drawbacks through RGO functionalization or incorporation of other additives such as vitamin C, green tea, protein bovine serum albumin and CNC, among others, to improve the properties and performance of the final derived nanocomposites and increase the number of its applications in several fields [28,173].

Nowadays, various unmodified CNC/RGO hybrids were actively explored for different applications such as sensors, flexible electronics, supercapacitors and photonic devices [174]. Several approaches used to produce some unmodified CNC/RGO hybrids have been reported for which the common method used to produce CNC/GO can be applied. Wan Khalid prepared COC/RGO nanocomposite by dispersion/ultrasonication of 1 mg RGO in ethanol and 1 mg of CNC in deionized water [175]. The supernatant was eliminated by centrifugation to recover the final hybrid. This latter, re-dispersed in ethanol, was drop-coated onto an electrode surface, which was intended to be used for electrochemical sensing of methyl paraben. The authors revealed that the obtained sensor exhibited good stability, reproducibility, selectivity toward methyl paraben and reusability compared to RGO-based sensor. Similarly, Nan et al. produced iridescent RGO/CNC film with advanced optical properties [176]. They prepared a suspension containing 1 wt.% of CNC and RGO floccules, which underwent an ultrasound treatment and dried using vacuum-assisted self-assembly (VASA) technique. The obtained films displayed regularly metallic iridescence owing to the homogeneous dispersion of RGO within the chiral nematic liquid crystals of CNC. The key factors to tune such behavior were the duration of ultrasonic treatment and the drying process. This iridescent hybrid exhibited better electrical properties in addition to the reversible change in color during the adsorption/desorption of water. The authors claimed that such a hybrid might find applications in photonic devices and biosensors. Recently, Wang et al. adopted a facile one-pot technique to prepare CNC/GO nanocomposite that was followed by a reduction using L-ascorbic acid to form CNC/RGO conductive paper [177]. The process, compared to the well-known ones, is schematically represented in Figure 12. Briefly, the exfoliation of graphite and the hydrolysis of cellulose occurred simultaneously in the reaction system, followed by subsequent reduction using green L-ascorbic acid. A conductive paper (CP) with high conductivity, excellent mechanical properties and thermal stability was then formed using ultrafiltration. It was stated that such CP can be used in implantable biosensors, smart textiles and portable micropower devices. In another research activity, Chen et al. proposed a new method to produce CNC/RGO hybrid, which was based on non-liquid-crystal spinning followed by a reduction using hydrogen iodide (HI), as schematized in Figure 13 [130]. The authors revealed that the incorporation of an alkaline media during the dispersion of CNC/RGO caused the electrostatic repulsion between CNC and GO sheets, leading to weaker hydrogen-bonding interaction and rendering the flowing process during spinning more homogeneous and easier. The authors demonstrated that the strength of RGO/CNC hybrid (230.6 MPa) was improved compared to pure RGO (157.5 Mpa). The hydrophilicity of the hybrid was also improved in addition to the high capacitive performance and conductivity. After that, such a hybrid was immersed in a polyvinyl alcohol acidic solution to fabricate flexible all-solid-state supercapacitor. The assembled supercapacitor achieved excellent bending stability, better flexibility, high energy density (5.1 mW h cm^{-3}) and power density (496.4 mW cm^{-3}). The authors claimed that the prepared hybrid easily meets the requirements of flexible or even wearable supercapacitor.

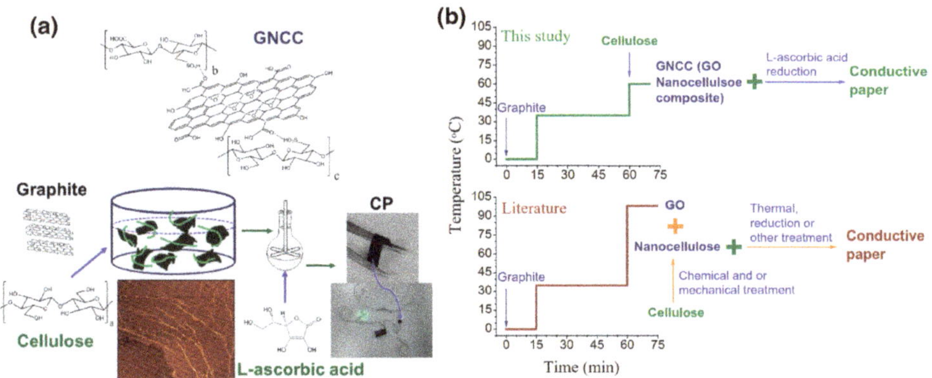

Figure 12. (**a**) A schematic flow diagram illustrates CNC/GO (GNCC) production using the one-pot method and further reduced to form conductive paper (CP, CNC/RGO). (**b**) A comparison of the present one-pot process with the conventional approach in literature. Reproduced with permission from Reference [177]. Copyright ©2019, Springer.

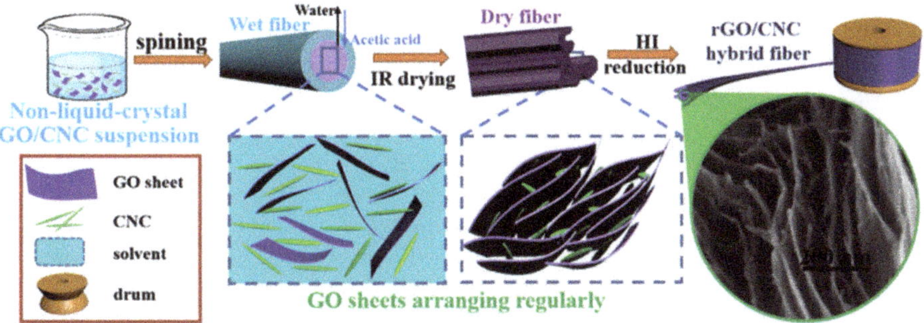

Figure 13. Schematic illustration of the preparation of CNC/RGO hybrid fiber. Reproduced with permission from Reference [130]. Copyright ©2018, Elsevier.

To further extend the number of applications, improve the different properties of CNC/RGO hybrids as well as their efficiency, numerous modifications of either CNC, RGO or both of them have been recently assessed. For instance, Zhao et al. produced electro-conductive nanocomposite based on CNC and TiO_2-RGO. Firstly, GO prepared by the modified Hummers method was subjected to the photocatalytic reduction via TiO_2. The obtained TiO_2-RGO suspension was mixed with CNC suspension under ultrasonication. CNC/TiO_2-RGO was then vacuum-filtered and dried. The obtained flexible transparent hybrid displayed improved electro-conductivity (9.3 S/m) with enhanced elastic modulus (3998 MPa) and tensile strength (18.1 MPa), stipulating that it can be used as a transparent flexible substrate for future electronic devices. In another work, Zhang et al. demonstrated the feasibility of the spinning of conductive filaments from oppositely charged nano-species, that is, cationic CNC and anionic RGO using interfacial nanoparticle complexation [178]. Initially, 2,3 dialdehyde cellulose was prepared by periodate oxidation, subjected to cationization with Girard's reagent or aminoguanidine hydrochloride and passed through the double-chamber system of a microfluidizer to form cationic CNC. Droplets of aqueous suspensions (cationic CNC and anionic GN), placed adjacent to each other, generated continuous CNC/GO filaments, as shown in Figure 14. These latter were immersed into hydrogen iodide solution, washed and dried, to afford CNC/RGO hybrid filaments. These hybrids displayed an electrical conductivity of 3298 ± 167 S/m and tensile strength of 190.3 ± 8 MPa.

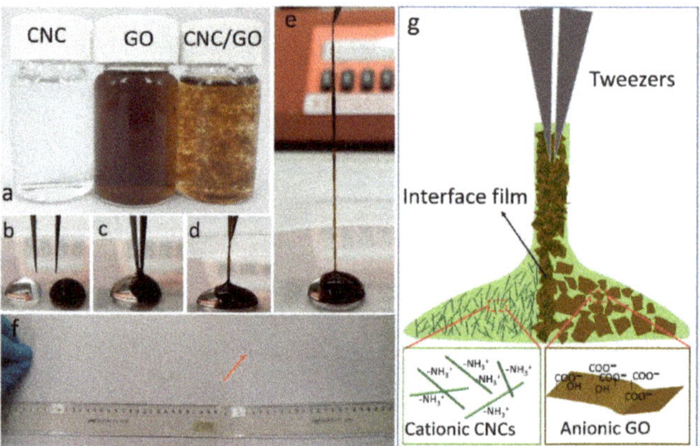

Figure 14. (**a**) suspensions of cationic CNC and GO and a dual precipitated complex after simple mixing; (**b–e**) CNC/GO hybrid filament drawing process; (**f**) a single dried CNC/GO hybrid filament with a diameter of ~33 µm and a length of 53 cm; (**g**) a scheme illustrating the CNC/GO hybrid filament drawing process. Reproduced with permission from Reference [178]. Licensed under a Creative Commons Attribution 3.0 International License (https://creativecommons.org/licenses/by/3.0/).

Kabiri and Namari described another interesting process for the preparation of CNC/RGO hybrid. They functionalized RGO with CNC via "click" coupling between terminated propargyl-functionalized CNC (PG-CNC) and azide-functionalized GO (GO-N_3) [179]. After the surface azidation of GO, the "click" reaction between GO-N_3 and PG-CNC, already synthesized by the Peng method [180], was performed using copper-catalyzed azide-alkyne cycloaddition. The reduction of the final nanocomposite dispersed in deionized water under sonication was carried out using hydrazine at 70 °C. The obtained free-dried CNC/RGO hybrid exhibited interesting physicochemical properties and thermal stability. In another study done by Sadasivyni et al., a transparent and eco-friendly CNC/RGO film for proximity sensing was developed [181]. The authors employed layer-by-layer spraying of modified CNC/GO nanocomposite, which was obtained as detailed in Figure 15, on lithographic patterns of interdigitated electrodes on polymer substrates. The modified nanocomposite was reduced using anhydrous hydrazine at 80 °C to generate a hydrophobic CNC/RGO hybrid. The obtained sensitive sensor allowed detecting a human finger interface within a distance of 6 mm with interesting response and recovery time interval. This sensor has potential to be used in various applications such as robotics, punching machines, smart phones, electronics and optoelectronics.

On the other hand, several CNC/RGO-based polymer composites have been produced and assessed in several applications such as sensors, scaffolds in tissue engineering, food and drug packaging. The polymeric matrices tested include polyvinylidene chloride (PVDC) [173], natural rubber (NR) [182], polyethylene oxide [183], poly-lactic acid (PLA) [184,185] and polyamide 6 [186]. It was demonstrated that the incorporation of CNC/RGO conferred to the polymeric nanocomposites outstanding mechanical and thermal properties, interesting barrier features, low toxicity, high conductivity and so forth [183,184,186]. For instance, Cao et al. prepared a 3D interconnected CNC/RGO/NR network using a latex assembly method for which NR latex was incorporated into a CNC/RGO suspension [182]. The solid formed through the co-coagulation induced by an acidic solution was vacuum filtered. A schematic illustration of the process is provided in Figure 16. The obtained conductive structure displayed higher electric conductivity and better mechanical features with superior resistivity responses for organic liquids. Such nanocomposite can find application in sensing to discriminate various solvents leakage in chemical industries and environmental monitoring.

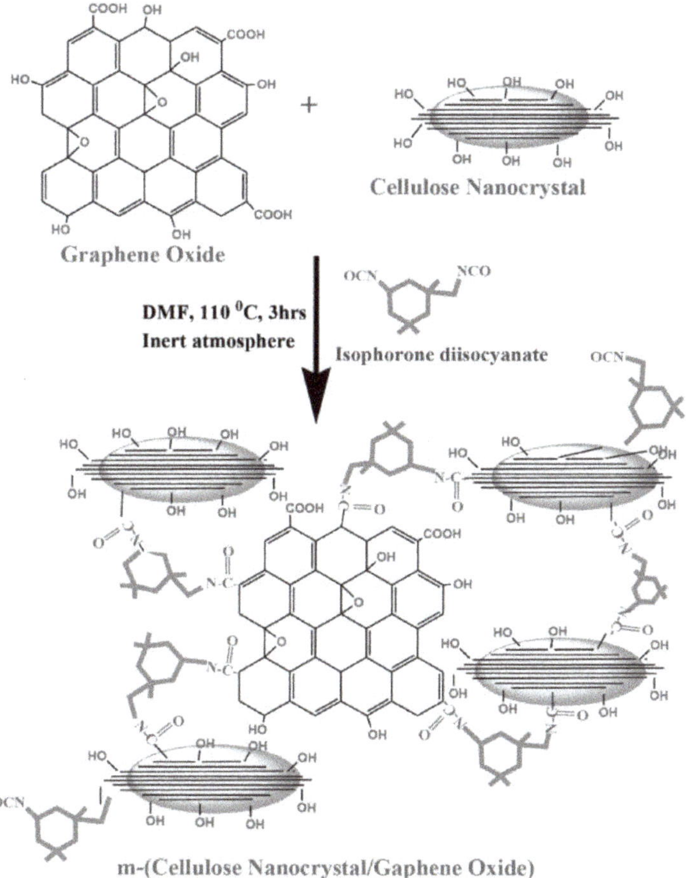

Figure 15. Reaction mechanism involved in modified CNC/GO synthesis. Reproduced with permission from Reference [181]. Copyright ©2015, Elsevier.

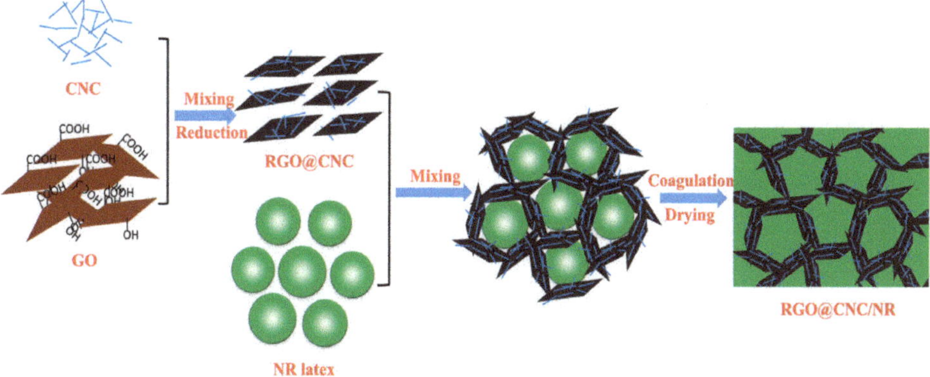

Figure 16. Schematic presentation of the synthesis of CNC/RGO/NR nanocomposite. Reproduced with permission from Reference [182]. Copyright ©2016, Elsevier.

In other research, Pal et al. assessed the combined effect of CNC and RGO in PLA nanocomposite as a scaffold in tissue engineering [185]. They initially prepared CNC and RGO via acid hydrolysis and modified Hummer's method, respectively and then employed a solution casting approach to produce CNC/RGO/PLA hybrid. The detailed preparation procedure is schematized in Figure 17. Compared to pristine PLA, the developed hydrophilic hybrid film revealed higher thermal stability, significantly increased tensile strength up to 23% with and enhancement in elongation at break, showing its ductile behavior. Moreover, the antibacterial activity against both Gram-negative *Escherichia coli* (*E. coli*) and Gram-positive *Staphylococcus aureus* (*S. aureus*) bacterial strains was highlighted. The *in-vitro* cytotoxicity assay indicated the non-toxicity of the nanocomposite film toward fibroblast cell line (NIH-3T3) as well. More recently, interesting research has been conducted by You et al. [173], who introduced the hybrid CNC/RGO to the solution of PVDC (Figure 18). The precipitated sample was vacuum dried to produce the CNCN/RGO/PVDC nanocomposite. The transparency of the CNC/RGO/PVDC coated on PET substrate was determined as 84% at 550 nm wavelength by UV–visible spectrometer in the regular transmission mode (Figure 18). It was proved that the utilization of stable dispersion of CNC/RGO enabled the fabrication of optically clear and thermostable nanohybrid film with improved barrier characteristics against water and oxygen. The authors claimed that the developed approach to produce CNCN/RGO/PVDC nanocomposite was effective and the obtained hybrid film is considered a potential candidate for food and drug packaging.

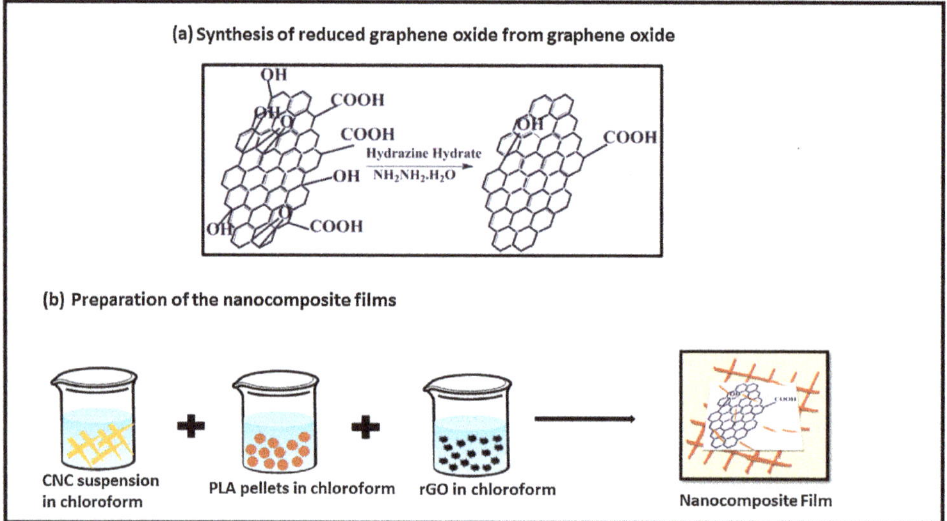

Figure 17. Schematic illustration of (**a**) the synthesis of reduced graphene oxide (RGO) from GO, (**b**) the method employed to prepare nanocomposite films. Reproduced with permission from Reference [185]. Copyright ©2017, Elsevier.

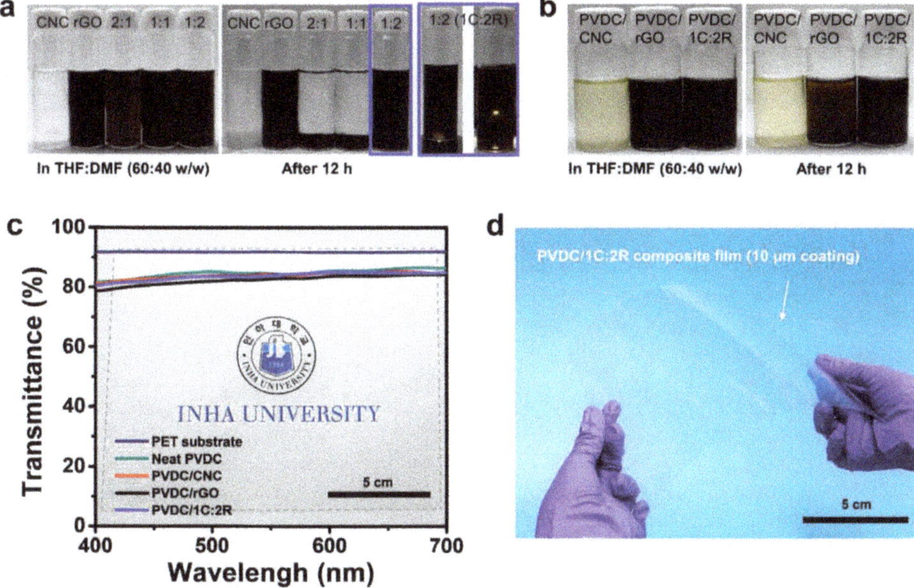

Figure 18. Dispersion stability of nanofillers. CNC, RGO and 1CNC:2RGO (1C:2R) hybrid in (**a**) THF:DMF co-solvent and (**b**) PVDC nanocomposite solutions of 0.1 wt.% fillers loading to PVDC. (**c**) Transmittance results of 0.1 wt.% PVDC/CNC, RGO and 1C:2R nanocomposite films. The 10 mm thick nanocomposite films were deposited on 125 mm thick PET substrates. (**d**) Large area (17 cm × 21 cm) PVDC/1C:2R–0.1 wt% nanocomposite film was obtained. Reproduced with permission from Reference [173]. Copyright ©2020, Elsevier.

5. Summary and Outlook

During the last decade, significant advances have been made in the preparation, characterization and application of CNC/GNM hybrids. This article is a brief review of this fast-growing research area and intended to highlight the up-to-date studies and utilization of CNC/GNM hybrids. Firstly, we have introduced some basic concepts of nanocellulose and GNM, then summarized their preparation methods and properties, with a particular focus on their outstanding features to elucidate their unique attributes. The different preparation processes of CNC/GNM have been discussed as well as their properties. Furthermore, to well understand the characteristics of theses hybrids, their different applications have been provided.

CNC/GNM hybrid-based materials displayed interesting innovative features due to synergetic effects, which are unachievable by taking CNC and GNM materials separately. It is shown that the combination of the diversity and specificity of both CNC and GNM not only expands the number of applications but also has indisputable advantages to benefit their unique attributes. These hybrids hold a cornucopia of favorable properties that warrant their employment in the fields of sensing, catalysis, separation, electronics, optics, biomedical, energy storage, to name a few. Nonetheless, the development of CNC/GNM hybrid-based materials is relatively a new concept, which is mostly limited to academic discipline but is expected that CNC/GNM hybrids will certainly be commercially available in the future, which will attract more research attention not only in various applications but also to achieve multifunctional multi-systems and open new perspectives. Moreover, the practical application of such hybrids as next-generation materials requires further improvements in functionality and performance in addition to the reduction of the production costs and the environmental impacts.

Author Contributions: Conceptualization, D.T.; Methodology, D.T.; Investigation, D.T.; Resources, D.T.; Writing—original draft preparation, D.T., V.K.T. and R.B.; Writing-review and editing, D.T., V.K.T. and R.B.; Supervision, D.T., V.K.T. and R.B. All authors have read and agreed to the published version of the manuscript.

Funding: This work was financially supported by the Ecole Militaire Polytechnique.

Conflicts of Interest: The authors declare no conflict of interest.

References

1. Klemm, D.; Cranston, E.D.; Fischer, D.; Gama, M.; Kedzior, S.A.; Kralisch, D.; Kramer, F.; Kondo, T.; Lindström, T.; Nietzsche, S. Nanocellulose as a natural source for groundbreaking applications in materials science: Today's state. *J. Mater. Today* **2018**. [CrossRef]

2. Thakur, V.K.; Voicu, S.I. Recent advances in cellulose and chitosan based membranes for water purification: A concise review. *Carbohydr. Polym.* **2016**, *146*, 148–165. [CrossRef]

3. Thakur, V.K. *Lignocellulosic Polymer Composites: Processing, Characterization, and Properties*; John Wiley & Sons: Hoboken, NJ, USA, 2015.

4. Lin, N.; Tang, J.; Dufresne, A.; Tam, M.K. *Advanced Functional Materials from Nanopolysaccharides*; Springer: Berlin, Germany, 2019.

5. Trache, D.; Tarchoun, A.F.; Derradji, M.; Mehelli, O.; Hussin, M.H.; Bessa, W. Cellulose fibers and nanocrystals: Preparation, characterization and surface modification. In *Functionalized Nanomaterials i: Fabrication*; Kumar, V., Guleria, P., Dasgupta, N., Ranjan, S., Eds.; Taylor & Francis: Oxfordshire, UK, 2020.

6. Mohammadi, F.; Moeeni, M.; Li, C.; Boukherroub, R.; Szunerits, S. Interaction of cellulose and nitrodopamine coated superparamagnetic iron oxide nanoparticles with alpha-lactalbumin. *RSC Adv.* **2020**, *10*, 9704–9716. [CrossRef]

7. Manoj, D.; Saravanan, R.; Santhanalakshmi, J.; Agarwal, S.; Gupta, V.K.; Boukherroub, R. Towards green synthesis of monodisperse cu nanoparticles: An efficient and high sensitive electrochemical nitrite sensor. *Sens. Actuators B* **2018**, *266*, 873–882. [CrossRef]

8. Oun, A.A.; Shankar, S.; Rhim, J.-W. Multifunctional nanocellulose/metal and metal oxide nanoparticle hybrid nanomaterials. *Crit. Rev. Food Sci. Nutr.* **2020**, *60*, 435–460. [CrossRef] [PubMed]

9. Mokhena, T.; John, M. Cellulose nanomaterials: New generation materials for solving global issues. *Cellulose* **2020**, 1–46. [CrossRef]

10. Fang, Z.; Hou, G.; Chen, C.; Hu, L. Nanocellulose-based films and their emerging applications. *Curr. Opin. Solid State Mater. Sci.* **2019**, *23*, 100764. [CrossRef]

11. Dufresne, A. Nanocellulose processing properties and potential applications. *Curr. For. Rep.* **2019**, *5*, 76–89. [CrossRef]

12. Thomas, B.; Raj, M.C.; Joy, J.; Moores, A.; Drisko, G.L.; Sanchez, C. Nanocellulose, a versatile green platform: From biosources to materials and their applications. *Chem. Rev.* **2018**, *118*, 11575–11625. [CrossRef]

13. Trache, D.; Tarchoun, A.F.; Derradji, M.; Hamidon, T.S.; Masruchin, N.; Brosse, N.; Hussin, M.H. Nanocellulose: From fundamentals to advanced applications. *Front. Chem.* **2020**, *8*, 392. [CrossRef]

14. Lasrado, D.; Ahankari, S.; Kar, K. Nanocellulose-based polymer composites for energy applications—A review. *J. Appl. Polym. Sci.* **2020**, 48959. [CrossRef]

15. Lengowski, E.C.; Júnior, E.A.B.; Kumode, M.M.N.; Carneiro, M.E.; Satyanarayana, K.G. Nanocellulose-reinforced adhesives for wood-based panels. In *Sustainable Polymer Composites and Nanocomposites*; Springer: Berlin, Germany, 2019; pp. 1001–1025.

16. Kim, J.H.; Lee, D.; Lee, Y.H.; Chen, W.; Lee, S.Y. Nanocellulose for energy storage systems: Beyond the limits of synthetic materials. *Adv. Mater.* **2019**, *31*, 1804826. [CrossRef] [PubMed]

17. Trache, D.; Hussin, M.H.; Haafiz, M.M.; Thakur, V.K. Recent progress in cellulose nanocrystals: Sources and production. *Nanoscale* **2017**, *9*, 1763–1786. [CrossRef] [PubMed]

18. Chen, Y.; Gan, L.; Huang, J.; Dufresne, A. Reinforcing mechanism of cellulose nanocrystals in nanocomposites. *Nanocellulose Fundam. Adv. Mater.* **2019**, 201–249.

19. Kargarzadeh, H.; Ahmad, I.; Thomas, S.; Dufresne, A. *Handbook of Nanocellulose and Cellulose Nanocomposites*; John Wiley & Sons: Hoboken, NJ, USA, 2017.

20. Zhang, Q.; Zhang, L.; Wu, W.; Xiao, H. Methods and applications of nanocellulose loaded with inorganic nanomaterials: A review. *Carbohydr. Polym.* **2020**, *229*, 115454. [CrossRef]

21. Tan, K.; Heo, S.; Foo, M.; Chew, I.M.; Yoo, C. An insight into nanocellulose as soft condensed matter: Challenge and future prospective toward environmental sustainability. *Sci. Total Environ.* **2019**, *650*, 1309–1326. [CrossRef]

22. ResearchAndMarkets. The Global Market for Nanocellulose to 2030: Cellulose Nanofibers(cnf), Cellulose Nanocrystals (cnc) and Bacterial Cellulose Particles (bc). Report id: 4701515. Available online: https://www.researchandmarkets.com/reports/4701515/the-global-market-for-nanocellulose-to-2030 (accessed on 26 July 2020).

23. Charreau, H.; Cavallo, E.; Foresti, M.L. Patents involving nanocellulose: Analysis of their evolution since 2010. *Carbohydr. Polym.* **2020**, 116039. [CrossRef]

24. Nechyporchuk, O.; Belgacem, M.N.; Bras, J. Production of cellulose nanofibrils: A review of recent advances. *Ind. Crop. Prod.* **2016**, *93*, 2–25. [CrossRef]

25. Trache, D.; Hussin, M.H.; Chuin, C.T.H.; Sabar, S.; Fazita, M.N.; Taiwo, O.F.; Hassan, T.; Haafiz, M.M. Microcrystalline cellulose: Isolation, characterization and bio-composites application—A review. *Int. J. Biol. Macromol.* **2016**, *93*, 789–804. [CrossRef]

26. Platnieks, O.; Gaidukovs, S.; Barkane, A.; Sereda, A.; Gaidukova, G.; Grase, L.; Thakur, V.K.; Filipova, I.; Fridrihsone, V.; Skute, M. Bio-based poly (butylene succinate)/microcrystalline cellulose/nanofibrillated cellulose-based sustainable polymer composites: Thermo-mechanical and biodegradation studies. *Polymers* **2020**, *12*, 1472. [CrossRef]

27. Salimi, S.; Sotudeh-Gharebagh, R.; Zarghami, R.; Chan, S.Y.; Yuen, K.H. Production of nanocellulose and its applications in drug delivery: A critical review. *ACS Sustain. Chem. Eng.* **2019**, *7*, 15800–15827. [CrossRef]

28. Khili, F.; Borges, J.; Almeida, P.L.; Boukherroub, R.; Omrani, A.D. Extraction of cellulose nanocrystals with structure i and ii and their applications for reduction of graphene oxide and nanocomposite elaboration. *Wast. Biomass Valori.* **2019**, *10*, 1913–1927. [CrossRef]

29. Shojaeiarani, J.; Bajwa, D.; Shirzadifar, A. A review on cellulose nanocrystals as promising biocompounds for the synthesis of nanocomposite hydrogels. *Carbohydr. Polym.* **2019**, *216*, 247–259. [CrossRef] [PubMed]

30. Du, H.; Liu, W.; Zhang, M.; Si, C.; Zhang, X.; Li, B. Cellulose nanocrystals and cellulose nanofibrils based hydrogels for biomedical applications. *Carbohydr. Polym.* **2019**, *209*, 130–144. [CrossRef] [PubMed]

31. Xie, H.; Du, H.; Yang, X.; Si, C. Recent strategies in preparation of cellulose nanocrystals and cellulose nanofibrils derived from raw cellulose materials. *Int. J. Polym. Sci.* **2018**, *2018*, 1–25. [CrossRef]

32. Dufresne, A. *Nanocellulose: From Nature to High Performance Tailored Materials*; Walter de Gruyter: Berlin, Germany, 2013.

33. Rojas, O.J. *Cellulose Chemistry and Properties: Fibers, Nanocelluloses and Advanced Materials*; Springer: Berlin, Germany, 2016; Volume 271.

34. Thakur, V.K. *Nanocellulose Polymer Nanocomposites: Fundamentals and Applications*; John Wiley & Sons: Hoboken, NJ, USA, 2015.

35. Ramasamy, J.; Amanullah, M. Nanocellulose for oil and gas field drilling and cementing applications. *J. Pet. Sci. Eng.* **2020**, *184*, 106292. [CrossRef]

36. Tayeb, P.; Tayeb, A.H. Application of nanocellulose in sustainable electrochemical and piezoelectric systems: A review. *Carbohydr. Polym.* **2019**, 115149. [CrossRef]

37. Sharma, A.; Thakur, M.; Bhattacharya, M.; Mandal, T.; Goswami, S. Commercial application of cellulose nano-composites-a review. *Biotechnol. Rep.* **2019**, e00316. [CrossRef]

38. Naz, S.; Ali, J.S.; Zia, M. Nanocellulose isolation characterization and applications: A journey from non-remedial to biomedical claims. *Bio Des. Manuf.* **2019**, 1–16. [CrossRef]

39. Kim, D.; Islam, M.S.; Tam, M.K. The use of nano-polysaccharides in biomedical applications. In *Advanced Functional Materials from Nanopolysaccharides*; Springer: Berlin/Heidelberg, Germany, 2019; pp. 171–219.

40. Bacakova, L.; Pajorova, J.; Tomkova, M.; Matejka, R.; Broz, A.; Stepanovska, J.; Prazak, S.; Skogberg, A.; Siljander, S.; Kallio, P. Applications of nanocellulose/nanocarbon composites: Focus on biotechnology and medicine. *Nanomaterials* **2020**, *10*, 196. [CrossRef]

41. Miyashiro, D.; Hamano, R.; Umemura, K. A review of applications using mixed materials of cellulose, nanocellulose and carbon nanotubes. *Nanomaterials* **2020**, *10*, 186. [CrossRef] [PubMed]

42. Tshikovhi, A.; Mishra, S.B.; Mishra, A.K. Nanocellulose-based composites for the removal of contaminants from wastewater. *Int. J. Biol. Macromol.* **2020**, *152*, 616–632. [CrossRef] [PubMed]

43. Neibolts, N.; Platnieks, O.; Gaidukovs, S.; Barkane, A.; Thakur, V.; Filipova, I.; Mihai, G.; Zelca, Z.; Yamaguchi, K.; Enachescu, M. Needle-free electrospinning of nanofibrillated cellulose and graphene nanoplatelets based sustainable poly (butylene succinate) nanofibers. *Mater. Today Chem.* **2020**, *17*, 100301. [CrossRef]

44. Lawal, A.T. Graphene-based nano composites and their applications. A review. *Biosens. Bioelectron.* **2019**, 111384. [CrossRef]

45. Lee, X.J.; Hiew, B.Y.Z.; Lai, K.C.; Lee, L.Y.; Gan, S.; Thangalazhy-Gopakumar, S.; Rigby, S. Review on graphene and its derivatives: Synthesis methods and potential industrial implementation. *J. Taiwan Inst. Chem. Eng.* **2019**, *98*, 163–180. [CrossRef]

46. Tiwari, S.K.; Sahoo, S.; Wang, N.; Huczko, A. Graphene research and their outputs: Status and prospect. *J. Sci. Adv. Mater. Devices* **2020**, *5*, 10–29. [CrossRef]

47. Halouane, F.; Oz, Y.; Meziane, D.; Barras, A.; Juraszek, J.; Singh, S.K.; Kurungot, S.; Shaw, P.K.; Sanyal, R.; Boukherroub, R. Magnetic reduced graphene oxide loaded hydrogels: Highly versatile and efficient adsorbents for dyes and selective Cr (VI) ions removal. *J. Colloid Interface Sci.* **2017**, *507*, 360–369. [CrossRef]

48. Chnadel, N.; Dutta, V.; Sharma, S.; Raizada, P.; Hosseini-Bandegharaei, A.; Kumar, R.; Singh, P.; Thakur, V.K. Z-scheme photocatalytic dye degradation on AgBr/Zn (Co) Fe_2O_4 photocatalysts supported on nitrogen-doped graphene. *Mater. Today Sustain.* **2020**, 100043. [CrossRef]

49. Chandel, N.; Sharma, K.; Sudhaik, A.; Raizada, P.; Hosseini-Bandegharaei, A.; Thakur, V.K.; Singh, P. Magnetically separable zno/znfe2o4 and zno/cofe2o4 photocatalysts supported onto nitrogen doped graphene for photocatalytic degradation of toxic dyes. *Arab. J. Chem.* **2020**, *13*, 4324–4340. [CrossRef]

50. Yang, X.; Liu, X.; Song, J. A study on technology competition of graphene biomedical technology based on patent analysis. *Appl. Sci.* **2019**, *9*, 2613. [CrossRef]

51. Yang, X.; Yu, X.; Liu, X. Obtaining a sustainable competitive advantage from patent information: A patent analysis of the graphene industry. *Sustainability* **2018**, *10*, 4800. [CrossRef]

52. Ferrari, A.C.; Bonaccorso, F.; Fal'Ko, V.; Novoselov, K.S.; Roche, S.; Bøggild, P.; Borini, S.; Koppens, F.H.; Palermo, V.; Pugno, N. Science and technology roadmap for graphene, related two-dimensional crystals, and hybrid systems. *Nanoscale* **2015**, *7*, 4598–4810. [CrossRef]

53. MarketDataForecast. Global Graphene Market Size, Share, Trends&Forecast (2020–2025). Report id: 8845. Available online: https://www.marketdataforecast.com/market-reports/graphene-market (accessed on 26 July 2020).

54. GrandViewResearch. Graphene Market Size, Share & Trends Analysis Report by Application (Electronics, Composites, Energy), by Product (Graphene Nanoplatelets, Graphene Oxide), by Region, and Segment Forecasts, 2020–2027. Report id: 978-1-68038-788-9. Available online: https://www.grandviewresearch.com/industry-analysis/graphene-industry (accessed on 6 July 2020).

55. An, S.; Wu, J.; Nie, Y.; Li, W.; Fortner, J.D. Free chlorine induced phototransformation of graphene oxide in water: Reaction kinetics and product characterization. *Chem. Eng. J.* **2020**, *381*, 122609. [CrossRef]

56. Xing, J.; Tao, P.; Wu, Z.; Xing, C.; Liao, X.; Nie, S. Nanocellulose-graphene composites: A promising nanomaterial for flexible supercapacitors. *Carbohydr. Polym.* **2019**, *207*, 447–459. [CrossRef] [PubMed]

57. Tarchoun, A.F.; Trache, D.; Klapötke, T.M. Microcrystalline cellulose from posidonia oceanica brown algae: Extraction and characterization. *Int. J. Biol. Macromol.* **2019**, *138*, 837–845. [CrossRef]

58. Tarchoun, A.F.; Trache, D.; Klapötke, T.M.; Chelouche, S.; Derradji, M.; Bessa, W.; Mezroua, A. A promising energetic polymer from posidonia oceanica brown algae: Synthesis, characterization, and kinetic modeling. *Macromol. Chem. Phys.* **2019**, *220*, 1900358. [CrossRef]

59. Trache, D.; Khimeche, K.; Mezroua, A.; Benziane, M. Physicochemical properties of microcrystalline nitrocellulose from alfa grass fibres and its thermal stability. *J. Therm. Anal. Calorim.* **2016**, *124*, 1485–1496. [CrossRef]

60. Trache, D.; Donnot, A.; Khimeche, K.; Benelmir, R.; Brosse, N. Physico-chemical properties and thermal stability of microcrystalline cellulose isolated from alfa fibres. *Carbohydr. Polym.* **2014**, *104*, 223–230. [CrossRef]

61. Trache, D.; Khimeche, K.; Donnot, A.; Benelmir, R. *Thermal Analysis of Microcrystalline Cellulose Prepared from Esparto Grass*; MATEC Web of Conferences; EDP Sciences: Les Ulis, France, 2013; p. 01067.

62. Liao, J.J.; Abd Latif, N.H.; Trache, D.; Brosse, N.; Hussin, M.H. Current advancement on the isolation, characterization and application of lignin. *Int. J. Biol. Macromol.* **2020**, *162*, 985–1024. [CrossRef]

63. Tarchoun, A.F.; Trache, D.; Klapötke, T.M.; Derradji, M.; Bessa, W. Ecofriendly isolation and characterization of microcrystalline cellulose from giant reed using various acidic media. *Cellulose* **2019**, *26*, 7635–7651. [CrossRef]

64. Trache, D.; Khimeche, K.; Donnot, A.; Benelmir, R. Ftir spectroscopy and x-ray powder diffraction characterization of microcrystalline cellulose obtained from alfa fibers. *MATEC Web Conf.* **2013**, *3*, 01023. [CrossRef]

65. Habibi, Y.; Lucia, L.A.; Rojas, O.J. Cellulose nanocrystals: Chemistry, self-assembly, and applications. *Chem. Rev.* **2010**, *110*, 3479–3500. [CrossRef] [PubMed]

66. Trache, D. Nanocellulose as a promising sustainable material for biomedical applications. *AIMS Mater. Sci.* **2018**, *5*, 201–205. [CrossRef]

67. Vineeth, S.; Gadhave, R.V.; Gadekar, P.T. Chemical modification of nanocellulose in wood adhesive. *Open J. Polym. Chem.* **2019**, *9*, 86. [CrossRef]

68. Pires, J.R.; Souza, V.G.; Fernando, A.L. Valorization of energy crops as a source for nanocellulose production–current knowledge and future prospects. *Ind. Crop. Prod.* **2019**, *140*, 111642. [CrossRef]

69. Phanthong, P.; Reubroycharoen, P.; Hao, X.; Xu, G.; Abudula, A.; Guan, G. Nanocellulose: Extraction and application. *Carbon Resour. Convers.* **2018**, *1*, 32–43. [CrossRef]

70. Chen, W.; Yu, H.; Lee, S.-Y.; Wei, T.; Li, J.; Fan, Z. Nanocellulose: A promising nanomaterial for advanced electrochemical energy storage. *Chem. Soc. Rev.* **2018**, *47*, 2837–2872. [CrossRef]

71. Zhang, S.; Sun, G.; He, Y.; Fu, R.; Gu, Y.; Chen, S. Preparation, characterization, and electrochromic properties of nanocellulose-based polyaniline nanocomposite films. *ACS Appl. Mater. Interfaces* **2017**, *9*, 16426–16434. [CrossRef]

72. Ng, H.-M.; Sin, L.T.; Bee, S.-T.; Tee, T.-T.; Rahmat, A. Review of nanocellulose polymer composite characteristics and challenges. *Polym. Plast. Technol. Eng.* **2017**, *56*, 687–731. [CrossRef]

73. Mahfoudhi, N.; Boufi, S. Nanocellulose as a novel nanostructured adsorbent for environmental remediation: A review. *Cellulose* **2017**, *24*, 1171–1197. [CrossRef]

74. Dai, L.; Wang, Y.; Zou, X.; Chen, Z.; Liu, H.; Ni, Y. Ultrasensitive physical, bio, and chemical sensors derived from 1-, 2-, and 3-d nanocellulosic materials. *Small* **2020**, 1906567. [CrossRef]

75. Balea, A.; Monte, M.C.; Merayo, N.; Campano, C.; Negro, C.; Blanco, A. Industrial application of nanocelluloses in papermaking: A review of challenges, technical solutions, and market perspectives. *Molecules* **2020**, *25*, 526. [CrossRef] [PubMed]

76. Cheng, H.; Kilgore, K.; Ford, C.; Fortier, C.; Dowd, M.K.; He, Z. Cottonseed protein-based wood adhesive reinforced with nanocellulose. *J. Adhes. Sci. Technol.* **2019**, *33*, 1357–1368. [CrossRef]

77. Nascimento, D.M.; Nunes, Y.L.; Figueirêdo, M.C.; de Azeredo, H.M.; Aouada, F.A.; Feitosa, J.P.; Rosa, M.F.; Dufresne, A. Nanocellulose nanocomposite hydrogels: Technological and environmental issues. *Green Chem.* **2018**, *20*, 2428–2448. [CrossRef]

78. Kargarzadeh, H.; Mariano, M.; Gopakumar, D.; Ahmad, I.; Thomas, S.; Dufresne, A.; Huang, J.; Lin, N. Advances in cellulose nanomaterials. *Cellulose* **2018**, *25*, 2151–2189. [CrossRef]

79. Dufresne, A. Cellulose nanomaterials as green nanoreinforcements for polymer nanocomposites. *Philos. Trans. R. Soc.* **2018**, *376*, 20170040. [CrossRef]

80. Abouzeid, R.E.; Khiari, R.; El-Wakil, N.; Dufresne, A. Current state and new trends in the use of cellulose nanomaterials for wastewater treatment. *Biomacromolecules* **2018**, *20*, 573–597. [CrossRef]

81. Dufresne, A. Cellulose nanomaterial reinforced polymer nanocomposites. *Curr. Opin. Colloid Interface Sci.* **2017**, *29*, 1–8. [CrossRef]

82. Dufresne, A. Nanocellulose: A new ageless bionanomaterial. *Mater. Today* **2013**, *16*, 220–227. [CrossRef]

83. Köse, K.; Mavlan, M.; Youngblood, J.P. Applications and impact of nanocellulose based adsorbents. *Cellulose* **2020**, 1–24. [CrossRef]

84. Yahya, M.; Chen, Y.W.; Lee, H.V.; Hassan, W.H.W. Reuse of selected lignocellulosic and processed biomasses as sustainable sources for the fabrication of nanocellulose via ni (ii)-catalyzed hydrolysis approach: A comparative study. *J. Polym. Environ.* **2018**, *26*, 2825–2844. [CrossRef]

85. Yin, F.; Lin, L.; Zhan, S. Preparation and properties of cellulose nanocrystals, gelatin, hyaluronic acid composite hydrogel as wound dressing. *J. Biomater. Sci. Polym. Ed.* **2019**, *30*, 190–201. [CrossRef] [PubMed]

86. Rezania, S.; Oryani, B.; Cho, J.; Talaiekhozani, A.; Sabbagh, F.; Hashemi, B.; Rupani, P.F.; Mohammadi, A.A. Different pretreatment technologies of lignocellulosic biomass for bioethanol production: An overview. *Energy* **2020**, 117457. [CrossRef]

87. Mao, J.; Abushammala, H.; Brown, N.; Laborie, M. In Comparative assessment of methods for producing cellulose i nanocrystals from cellulosic Sources, nanocelluloses: Their preparation, properties, and applications. *ACS Symp. Ser.* **2017**, *2*, 19–53.

88. Chen, L.; Zhu, J.; Baez, C.; Kitin, P.; Elder, T. Highly thermal-stable and functional cellulose nanocrystals and nanofibrils produced using fully recyclable organic acids. *Green Chem.* **2016**, *18*, 3835–3843. [CrossRef]

89. Liu, Y.; Wang, H.; Yu, G.; Yu, Q.; Li, B.; Mu, X. A novel approach for the preparation of nanocrystalline cellulose by using phosphotungstic acid. *Carbohydr. Polym.* **2014**, *110*, 415–422. [CrossRef]

90. Novo, L.P.; Bras, J.; García, A.; Belgacem, N.; da Silva Curvelo, A.A. A study of the production of cellulose nanocrystals through subcritical water hydrolysis. *Ind. Crop. Prod.* **2016**, *93*, 88–95. [CrossRef]

91. Sirviö, J.A.; Visanko, M.; Liimatainen, H. Acidic deep eutectic solvents as hydrolytic media for cellulose nanocrystal production. *Biomacromolecules* **2016**, *17*, 3025–3032. [CrossRef]

92. Pang, Z.; Wang, P.; Dong, C. Ultrasonic pretreatment of cellulose in ionic liquid for efficient preparation of cellulose nanocrystals. *Cellulose* **2018**, *25*, 7053–7064. [CrossRef]

93. Wang, H.; Pudukudy, M.; Ni, Y.; Zhi, Y.; Zhang, H.; Wang, Z.; Jia, Q.; Shan, S. Synthesis of nanocrystalline cellulose via ammonium persulfate-assisted swelling followed by oxidation and their chiral self-assembly. *Cellulose* **2019**, *27*, 657–676. [CrossRef]

94. Chowdhury, Z.Z.; Hamid, S.B.A. Preparation and characterization of nanocrystalline cellulose using ultrasonication combined with a microwave-assisted pretreatment process. *BioResources* **2016**, *11*, 3397–3415. [CrossRef]

95. Tong, X.; Shen, W.; Chen, X.; Jia, M.; Roux, J.C. Preparation and mechanism analysis of morphology-controlled cellulose nanocrystals via compound enzymatic hydrolysis of eucalyptus pulp. *J. Appl. Polym. Sci.* **2020**, *137*, 48407. [CrossRef]

96. Mishra, S.; Kharkar, P.S.; Pethe, A.M. Biomass and waste materials as potential sources of nanocrystalline cellulose: Comparative review of preparation methods (2016–till date). *Carbohydr. Polym.* **2019**, *207*, 418–427. [CrossRef]

97. Karimian, A.; Parsian, H.; Majidinia, M.; Rahimi, M.; Mir, M.; Smadi-Kafil, H.; Shafiei-Irannejad, V.; Kheyrollah, M.; Ostadi, H.; Yousefi, B. Nanocrystalline cellulose: Preparation, physicochemical properties, and applications in drug delivery systems. *Int. J. Biol. Macromol.* **2019**, *133*, 850–859. [CrossRef] [PubMed]

98. Lavoine, N.; Bergström, L. Nanocellulose-based foams and aerogels: Processing, properties, and applications. *J. Mater. Chem. A* **2017**, *5*, 16105–16117. [CrossRef]

99. Kaushik, M.; Fraschini, C.; Chauve, G.; Putaux, J.-L.; Moores, A. Transmission electron microscopy for the characterization of cellulose nanocrystals. In *The Transmission Electron Microscope: Theory and Applications*; Maaz, K., Ed.; Intech Open: London, UK, 2015; pp. 129–163.

100. Tang, J.; Sisler, J.; Grishkewich, N.; Tam, K.C. Functionalization of cellulose nanocrystals for advanced applications. *J. Colloid Interface Sci.* **2017**, *494*, 397–409. [CrossRef]

101. Eyley, S.; Thielemans, W. Surface modification of cellulose nanocrystals. *Nanoscale* **2014**, *6*, 7764–7779. [CrossRef]

102. Novoselov, K.S.; Geim, A.K.; Morozov, S.V.; Jiang, D.; Zhang, Y.; Dubonos, S.V.; Grigorieva, I.V.; Firsov, A.A. Electric field effect in atomically thin carbon films. *Science* **2004**, *306*, 666–669. [CrossRef]

103. Geim, A.K. Graphene: Status and prospects. *Science* **2009**, *324*, 1530–1534. [CrossRef]

104. Tiwari, S.K.; Mishra, R.K.; Ha, S.K.; Huczko, A. Evolution of graphene oxide and graphene: From imagination to industrialization. *Chem. Nano. Mat.* **2018**, *4*, 598–620. [CrossRef]

105. Kumar, A.; Sharma, K.; Dixit, A.R. A review of the mechanical and thermal properties of graphene and its hybrid polymer nanocomposites for structural applications. *J. Mater. Sci.* **2019**, *54*, 5992–6026. [CrossRef]

106. Li, M.-F.; Liu, Y.-G.; Zeng, G.-M.; Liu, N.; Liu, S.-B. Graphene and graphene-based nanocomposites used for antibiotics removal in water treatment: A review. *Chemosphere* **2019**, *226*, 60–380. [CrossRef] [PubMed]

107. Thakur, K.; Kandasubramanian, B. Graphene and graphene oxide-based composites for removal of organic pollutants: A review. *J. Chem. Eng. Data* **2019**, *64*, 833–867. [CrossRef]

108. Karthick, R.; Chen, F. Free-standing graphene paper for energy application: Progress and future scenarios. *Carbon* **2019**, *150*, 292–310. [CrossRef]

109. Cao, X.; Qi, D.; Yin, S.; Bu, J.; Li, F.; Goh, C.F.; Zhang, S.; Chen, X. Ambient fabrication of large-area graphene films via a synchronous reduction and assembly strategy. *Adv. Mater.* **2013**, *25*, 2957–2962. [CrossRef]

110. Siwal, S.S.; Zhang, Q.; Devi, N.; Thakur, V.K. Carbon-based polymer nanocomposite for high-performance energy storage applications. *Polymers* **2020**, *12*, 505. [CrossRef]

111. Schöche, S.; Hong, N.; Khorasaninejad, M.; Ambrosio, A.; Orabona, E.; Maddalena, P.; Capasso, F. Optical properties of graphene oxide and reduced graphene oxide determined by spectroscopic ellipsometry. *Appl. Surf. Sci.* **2017**, *421*, 778–782. [CrossRef]

112. Rydzkowski, T.; Reszka, K.; Szczypiński, M.; Szczypiński, M.M.; Kopczyńska, E.; Thakur, V.K. Manufacturing and evaluation of mechanical, morphological, and thermal properties of reduced graphene oxide-reinforced expanded polystyrene (eps) nanocomposites. *Adv. Polym. Technol.* **2020**, *2020*, 1–9. [CrossRef]

113. Bacon, M.; Bradley, S.J.; Nann, T. Graphene quantum dots. *Part. Part. Syst. Charact.* **2014**, *31*, 415–428. [CrossRef]

114. Zheng, X.T.; Ananthanarayanan, A.; Luo, K.Q.; Chen, P. Glowing graphene quantum dots and carbon dots: Properties, syntheses, and biological applications. *Small* **2015**, *11*, 1620–1636. [CrossRef]

115. Neuberger, N.; Adidharma, H.; Fan, M. Graphene: A review of applications in the petroleum industry. *J. Pet. Sci. Eng.* **2018**, *167*, 152–159. [CrossRef]

116. Hanafi, S.; Trache, D.; He, W.; Xie, W.-X.; Mezroua, A.; Yan, Q.-L. Thermostable energetic coordination polymers based on functionalized go and their catalytic effects on the decomposition of ap and rdx. *J. Phys. Chem. C* **2020**, *124*, 5182–5195. [CrossRef]

117. Whitener, K.E., Jr.; Sheehan, P.E. Graphene synthesis. *Diam. Relat. Mater.* **2014**, *46*, 25–34. [CrossRef]

118. Deng, J.; You, Y.; Sahajwalla, V.; Joshi, R.K. Transforming waste into carbon-based nanomaterials. *Carbon* **2016**, *96*, 105–115. [CrossRef]

119. Edwards, R.S.; Coleman, K.S. Graphene synthesis: Relationship to applications. *Nanoscale* **2013**, *5*, 38–51. [CrossRef] [PubMed]

120. Rosli, N.N.; Ibrahim, M.A.; Ludin, N.A.; Teridi, M.A.M.; Sopian, K. A review of graphene based transparent conducting films for use in solar photovoltaic applications. *Renew. Sust. Energ. Rev.* **2019**, *99*, 83–99. [CrossRef]

121. Hummers, W.S., Jr.; Offeman, R.E. Preparation of graphitic oxide. *J. Am. Chem. Soc.* **1958**, *80*, 1339. [CrossRef]

122. Park, S.; Lee, K.-S.; Bozoklu, G.; Cai, W.; Nguyen, S.T.; Ruoff, R.S. Graphene oxide papers modified by divalent ions—enhancing mechanical properties via chemical cross-linking. *ACS Nano* **2008**, *2*, 572–578. [CrossRef]

123. Mohan, V.B.; Lau, K.-T.; Hui, D.; Bhattacharyya, D. Graphene-based materials and their composites: A review on production, applications and product limitations. *Compos. Part B* **2018**, *142*, 200–220. [CrossRef]

124. Kim, Y.J.; Jeong, B. Graphene-based nanomaterials and their applications in biosensors. In *Biomimetic Medical Materials: From Nanotechnology to 3d Bioprinting*; Noh, I., Ed.; Springer: Berlin, Germany, 2018; pp. 61–71.

125. Shekhawat, A.; Ritchie, R.O. Toughness and strength of nanocrystalline graphene. *Nat. Commun.* **2016**, *7*, 1–8. [CrossRef]

126. Zhu, Y.; Murali, S.; Cai, W.; Li, X.; Suk, J.W.; Potts, J.R.; Ruoff, R.S. Graphene and graphene oxide: Synthesis, properties, and applications. *Adv. Mater.* **2010**, *22*, 3906–3924. [CrossRef]

127. Bottari, G.; Herranz, M.Á.; Wibmer, L.; Volland, M.; Rodríguez-Pérez, L.; Guldi, D.M.; Hirsch, A.; Martín, N.; D'Souza, F.; Torres, T. Chemical functionalization and characterization of graphene-based materials. *Chem. Soc. Rev.* **2017**, *46*, 4464–4500. [CrossRef] [PubMed]

128. Liu, D.; Dong, Y.; Liu, Y.; Ma, N.; Sui, G. Cellulose nanowhisker (cnw)/graphene nanoplatelet (gn) composite films with simultaneously enhanced thermal, electrical and mechanical properties. *Front. Mater.* **2019**, *6*, 235. [CrossRef]

129. Valentini, L.; Cardinali, M.; Fortunati, E.; Torre, L.; Kenny, J.M. A novel method to prepare conductive nanocrystalline cellulose/graphene oxide composite films. *Mater. Lett.* **2013**, *105*, 4–7. [CrossRef]

130. Chen, G.; Chen, T.; Hou, K.; Ma, W.; Tebyetekerwa, M.; Cheng, Y.; Weng, W.; Zhu, M. Robust, hydrophilic graphene/cellulose nanocrystal fiber-based electrode with high capacitive performance and conductivity. *Carbon* **2018**, *127*, 218–227. [CrossRef]

131. Khabibullin, A.; Alizadehgiashi, M.; Khuu, N.; Prince, E.; Tebbe, M.; Kumacheva, E. Injectable shear-thinning fluorescent hydrogel formed by cellulose nanocrystals and graphene quantum dots. *Langmuir* **2017**, *33*, 12344–12350. [CrossRef]

132. Ruiz-Palomero, C.; Soriano, M.L.; Benítez-Martínez, S.; Valcarcel, M. Photoluminescent sensing hydrogel platform based on the combination of nanocellulose and s, n-codoped graphene quantum dots. *Sens. Actuators B* **2017**, *245*, 946–953. [CrossRef]

133. Alizadehgiashi, M.; Khuu, N.; Khabibullin, A.; Henry, A.; Tebbe, M.; Suzuki, T.; Kumacheva, E. Nanocolloidal hydrogel for heavy metal scavenging. *ACS Nano.* **2018**, *12*, 8160–8168. [CrossRef]

134. Zhou, X.; Liu, Y.; Du, C.; Ren, Y.; Li, X.; Zuo, P.; Yin, G.; Ma, Y.; Cheng, X.; Gao, Y. Free-standing sandwich-type graphene/nanocellulose/silicon laminar anode for flexible rechargeable lithium ion batteries. *ACS Appl. Mater. Interfaces* **2018**, *10*, 29638–29646. [CrossRef]

135. Dhar, P.; Gaur, S.S.; Kumar, A.; Katiyar, V. Cellulose nanocrystal templated graphene nanoscrolls for high performance supercapacitors and hydrogen storage: An experimental and molecular simulation study. *Sci. Rep.* **2018**, *8*, 1–15. [CrossRef]

136. Xie, F.; Yang, M.; Jiang, M.; Huang, X.-J.; Liu, W.-Q.; Xie, P.-H. Carbon-based nanomaterials—A promising electrochemical sensor toward persistent toxic substance. *TrAC Trends Anal. Chem.* **2019**, *119*, 115624. [CrossRef]

137. Carrasco, P.M.; Montes, S.; García, I.; Borghei, M.; Jiang, H.; Odriozola, I.; Cabañero, G.; Ruiz, V. High-concentration aqueous dispersions of graphene produced by exfoliation of graphite using cellulose nanocrystals. *Carbon* **2014**, *70*, 157–163. [CrossRef]

138. Cui, S.; Wei, P.; Li, L. Preparation of poly (propylene carbonate)/graphite nanoplates-spherical nanocrystal cellulose composite with improved glass transition temperature and electrical conductivity. *Compos. Sci. Technol.* **2018**, *168*, 63–73. [CrossRef]

139. Li, K.; Jin, S.; Han, Y.; Li, J.; Chen, H. Improvement in functional properties of soy protein isolate-based film by cellulose nanocrystal–graphene artificial nacre nanocomposite. *Polymers* **2017**, *9*, 321. [CrossRef] [PubMed]

140. Wang, F.; Drzal, L.T.; Qin, Y.; Huang, Z. Multifunctional graphene nanoplatelets/cellulose nanocrystals composite paper. *Compos. Part B* **2015**, *79*, 521–529. [CrossRef]

141. Wang, Y.; Zhang, H.; Lin, X.; Chen, S.; Jiang, Z.; Wang, J.; Huang, J.; Zhang, F.; Li, H. Naked au nanoparticles monodispersed onto multifunctional cellulose nanocrystal–graphene hybrid sheets: Towards efficient and sustainable heterogeneous catalysts. *New J. Chem.* **2018**, *42*, 2197–2203. [CrossRef]

142. Omidi, S.; Pirhayati, M.; Kakanejadifard, A. Co-delivery of doxorubicin and curcumin by a ph-sensitive, injectable, and in situ hydrogel composed of chitosan, graphene, and cellulose nanowhisker. *Carbohydr. Polym.* **2020**, *231*, 115745. [CrossRef]

143. Montes, S.; Carrasco, P.M.; Ruiz, V.; Cabañero, G.; Grande, H.J.; Labidi, J.; Odriozola, I. Synergistic reinforcement of poly (vinyl alcohol) nanocomposites with cellulose nanocrystal-stabilized graphene. *Compos. Sci. Technol.* **2015**, *117*, 26–31. [CrossRef]

144. Montes, S.; Etxeberria, A.; Mocholi, V.; Rekondo, A.; Grande, H.; Labidi, J. Effect of combining cellulose nanocrystals and graphene nanoplatelets on the properties of poly (lactic acid) based films. *Express Polym. Lett.* **2018**, *12*, 543–555. [CrossRef]

145. Valentini, L.; Bon, S.B.; Fortunati, E.; Kenny, J.M. Preparation of transparent and conductive cellulose nanocrystals/graphene nanoplatelets films. *J. Mater. Sci.* **2014**, *49*, 1009–1013. [CrossRef]

146. Liu, D.; Liu, Y.; Sui, G. Synthesis and properties of sandwiched films of epoxy resin and graphene/cellulose nanowhiskers paper. *Compos. Part A* **2016**, *84*, 87–95. [CrossRef]

147. Yang, F.; Wu, Y.; Zhang, S.; Zhang, H.; Zhao, S.; Zhang, J.; Fei, B. Mechanical and thermal properties of waterborne polyurethane coating modified through one-step cellulose nanocrystals/graphene materials sols method. *Coatings* **2020**, *10*, 40. [CrossRef]

148. Nie, J.; Liu, D.; Li, S.; Qiu, Z.; Ma, N.; Sui, G. Improved dispersion of the graphene and corrosion resistance of waterborne epoxy–graphene composites by minor cellulose nanowhiskers. *J. Appl. Polym. Sci.* **2019**, *136*, 47631. [CrossRef]

149. Shandilya, P.; Sudhaik, A.; Raizada, P.; Hosseini-Bandegharaei, A.; Singh, P.; Rahmani-Sani, A.; Thakur, V.; Saini, A.K. Synthesis of eu3+− doped zno/bi2o3 heterojunction photocatalyst on graphene oxide sheets for visible light-assisted degradation of 2, 4-dimethyl phenol and bacteria killing. *Solid State Sci.* **2020**, *102*, 106164. [CrossRef]

150. Kafy, A.; Akther, A.; Shishir, M.I.; Kim, H.C.; Yun, Y.; Kim, J. Cellulose nanocrystal/graphene oxide composite film as humidity sensor. *Sens. Actuators A* **2016**, *247*, 221–226. [CrossRef]

151. Chen, Q.; Liu, P.; Sheng, C.; Zhou, L.; Duan, Y.; Zhang, J. Tunable self-assembly structure of graphene oxide/cellulose nanocrystal hybrid films fabricated by vacuum filtration technique. *RSC Adv.* **2014**, *4*, 39301–39304. [CrossRef]

152. Valentini, L.; Cardinali, M.; Fortunati, E.; Kenny, J.M. Nonvolatile memory behavior of nanocrystalline cellulose/graphene oxide composite films. *Appl. Phys. Lett.* **2014**, *105*, 153111. [CrossRef]

153. Pan, H.; Zhu, C.; Lu, T.; Lin, J.; Ma, J.; Zhang, D.; Zhu, S. A chiral smectic structure assembled from nanosheets and nanorods. *Chem. Commun.* **2017**, *53*, 1868–1871. [CrossRef]

154. Xiong, R.; Hu, K.; Grant, A.M.; Ma, R.; Xu, W.; Lu, C.; Zhang, X.; Tsukruk, V.V. Ultrarobust transparent cellulose nanocrystal-graphene membranes with high electrical conductivity. *Adv. Mater.* **2016**, *28*, 1501–1509. [CrossRef]

155. Hu, K.; Tolentino, L.S.; Kulkarni, D.D.; Ye, C.; Kumar, S.; Tsukruk, V.V. Written-in conductive patterns on robust graphene oxide biopaper by electrochemical microstamping. *Angew. Chem. Int. Ed.* **2013**, *52*, 13784–13788. [CrossRef]

156. Kabiri, R.; Namazi, H. Nanocrystalline cellulose acetate (ncca)/graphene oxide (go) nanocomposites with enhanced mechanical properties and barrier against water vapor. *Cellulose* **2014**, *21*, 3527–3539. [CrossRef]

157. Daniyal, W.M.E.M.M.; Fen, Y.W.; Abdullah, J.; Sadrolhosseini, A.R.; Saleviter, S.; Omar, N.A.S. Label-free optical spectroscopy for characterizing binding properties of highly sensitive nanocrystalline cellulose-graphene oxide based nanocomposite towards nickel ion. *Spectrochim. Acta A* **2019**, *212*, 25–31. [CrossRef] [PubMed]

158. Daniyal, W.M.E.M.M.; Fen, Y.W.; Abdullah, J.; Saleviter, S.; Omar, N.A.S. Preparation and characterization of hexadecyltrimethylammonium bromide modified nanocrystalline cellulose/graphene oxide composite thin film and its potential in sensing copper ion using surface plasmon resonance technique. *Optik* **2018**, *173*, 71–77. [CrossRef]

159. Beyranvand, N.S.; Samiey, B.; Tehrani, A.D.; Soleimani, K. Graphene oxide–cellulose nanowhisker hydrogel nanocomposite as a novel adsorbent for methylene blue. *J. Chem. Eng. Data* **2019**, *64*, 5558–5570. [CrossRef]

160. Soleimani, K.; Tehrani, A.D.; Adeli, M. Bioconjugated graphene oxide hydrogel as an effective adsorbent for cationic dyes removal. *Ecotoxicol. Environ. Saf.* **2018**, *147*, 34–42. [CrossRef] [PubMed]

161. Zheng, X.; Zhang, Y.; Bian, T.; Zhang, Y.; Li, Z.; Pan, J. Oxidized carbon materials cooperative construct ionic imprinted cellulose nanocrystals films for efficient adsorption of dy (iii). *Chem. Eng. J.* **2020**, *381*, 122669. [CrossRef]

162. Anirudhan, T.; Deepa, J. Electrochemical sensing of cholesterol by molecularly imprinted polymer of silylated graphene oxide and chemically modified nanocellulose polymer. *Mater. Sci. Eng. C* **2018**, *92*, 942–956. [CrossRef]

163. Wang, N.; Wang, Y.-F.; Omer, A.M.; Ouyang, X.-k. Fabrication of novel surface-imprinted magnetic graphene oxide-grafted cellulose nanocrystals for selective extraction and fast adsorption of fluoroquinolones from water. *Anal. Bioanal. Chem.* **2017**, *409*, 6643–6653. [CrossRef]

164. Li, F.; Yu, H.-Y.; Wang, Y.-Y.; Zhou, Y.; Zhang, H.; Yao, J.-M.; Abdalkarim, S.Y.H.; Tam, K.C. Natural biodegradable poly (3-hydroxybutyrate-co-3-hydroxyvalerate) nanocomposites with multifunctional cellulose nanocrystals/graphene oxide hybrids for high-performance food packaging. *J. Agric. Food Chem.* **2019**, *67*, 10954–10967. [CrossRef]

165. Burrs, S.; Vanegas, D.; Bhargava, M.; Mechulan, N.; Hendershot, P.; Yamaguchi, H.; Gomes, C.; McLamore, E. A comparative study of graphene–hydrogel hybrid bionanocomposites for biosensing. *Analyst* **2015**, *140*, 1466–1476. [CrossRef]

166. Kulandaivalu, S.; Shukur, R.A.; Sulaiman, Y. Improved electrochemical performance of electrochemically designed layered poly (3, 4-ethylenedioxythiophene)/graphene oxide with poly (3,4-ethylenedioxythiophene)/ nanocrystalline cellulose nanocomposite. *Synth. Met.* **2018**, *245*, 24–31. [CrossRef]

167. Lv, J.; Zhang, G.; Zhang, H.; Yang, F. Graphene oxide-cellulose nanocrystal (go-cnc) composite functionalized pvdf membrane with improved antifouling performance in mbr: Behavior and mechanism. *Chem. Eng. J.* **2018**, *352*, 765–773. [CrossRef]

168. Kumar, A.; Rao, K.M.; Han, S.S. Mechanically viscoelastic nanoreinforced hybrid hydrogels composed of polyacrylamide, sodium carboxymethylcellulose, graphene oxide, and cellulose nanocrystals. *Carbohydr. Polym.* **2018**, *193*, 228–238. [CrossRef]

169. Patel, D.K.; Seo, Y.-R.; Dutta, S.D.; Lim, K.-T. Enhanced osteogenesis of mesenchymal stem cells on electrospun cellulose nanocrystals/poly (ε-caprolactone) nanofibers on graphene oxide substrates. *RSC Adv.* **2019**, *9*, 36040–36049. [CrossRef]

170. Chan, C.; Chia, C.; Zakaria, S.; Ahmad, I.; Dufresne, A.; Tshai, K. Low filler content cellulose nanocrystal and graphene oxide reinforced polylactic acid film composites. *Polym. Res. J.* **2015**, *9*, 165–177.

171. El Miri, N.; El Achaby, M.; Fihri, A.; Larzek, M.; Zahouily, M.; Abdelouahdi, K.; Barakat, A.; Solhy, A. Synergistic effect of cellulose nanocrystals/graphene oxide nanosheets as functional hybrid nanofiller for enhancing properties of pva nanocomposites. *Carbohydr. Polym.* **2016**, *137*, 239–248. [CrossRef]

172. Yang, M.-C.; Tseng, Y.-Q.; Liu, K.-H.; Cheng, Y.-W.; Chen, W.-T.; Chen, W.-T.; Hsiao, C.-W.; Yung, M.-C.; Hsu, C.-C.; Liu, T.-Y. Preparation of amphiphilic chitosan–graphene oxide–cellulose nanocrystalline composite hydrogels and their biocompatibility and antibacterial properties. *Appl. Sci.* **2019**, *9*, 3051. [CrossRef]

173. You, J.; Won, S.; Jin, H.-J.; Yun, Y.S.; Wie, J.J. Nano-patching defects of reduced graphene oxide by cellulose nanocrystals in scalable polymer nanocomposites. *Carbon* **2020**, *165*, 18–25. [CrossRef]

174. Zaid, M.H.M.; Abdullah, J.; Yusof, N.A.; Wasoh, H.; Sulaiman, Y.; Noh, M.F.M.; Issa, R. Reduced graphene oxide/tempo-nanocellulose nanohybrid-based electrochemical biosensor for the determination of mycobacterium tuberculosis. *J. Sens.* **2020**, *4051474*, 1–11. [CrossRef]

175. Wan Khalid, W.E.F.; Mat Arip, M.N.; Jasmani, L.; Lee, Y.H. A new sensor for methyl paraben using an electrode made of a cellulose nanocrystal–reduced graphene oxide nanocomposite. *Sensors* **2019**, *19*, 2726. [CrossRef]

176. Nan, F.; Chen, Q.; Liu, P.; Nagarajan, S.; Duan, Y.; Zhang, J. Iridescent graphene/cellulose nanocrystal film with water response and highly electrical conductivity. *RSC Adv.* **2016**, *6*, 93673–93679. [CrossRef]

177. Wang, R.; Ma, Q.; Zhang, H.; Ma, Z.; Yang, R.; Zhu, J. Producing conductive graphene–nanocellulose paper in one-pot. *J. Polym. Environ.* **2019**, *27*, 148–157. [CrossRef]

178. Zhang, K.; Ketterle, L.; Järvinen, T.; Lorite, G.S.; Hong, S.; Liimatainen, H. Self-assembly of graphene oxide and cellulose nanocrystals into continuous filament via interfacial nanoparticle complexation. *Mater. Des.* **2020**, *193*, 108791. [CrossRef]

179. Kabiri, R.; Namazi, H. Surface grafting of reduced graphene oxide using nanocrystalline cellulose via click reaction. *J. Nanopart. Res.* **2014**, *16*, 2474. [CrossRef]

180. Peng, P.; Cao, X.; Peng, F.; Bian, J.; Xu, F.; Sun, R. Binding cellulose and chitosan via click chemistry: Synthesis, characterization, and formation of some hollow tubes. *J. Polym. Sci. A* **2012**, *50*, 5201–5210. [CrossRef]

181. Sadasivuni, K.K.; Kafy, A.; Zhai, L.; Ko, H.U.; Mun, S.; Kim, J. Transparent and flexible cellulose nanocrystal/reduced graphene oxide film for proximity sensing. *Small* **2015**, *11*, 994–1002. [CrossRef]

182. Cao, J.; Zhang, X.; Wu, X.; Wang, S.; Lu, C. Cellulose nanocrystals mediated assembly of graphene in rubber composites for chemical sensing applications. *Carbohydr. Polym.* **2016**, *140*, 88–95. [CrossRef]

183. Ye, Y.-S.; Zeng, H.-X.; Wu, J.; Dong, L.-Y.; Zhu, J.-T.; Xue, Z.-G.; Zhou, X.-P.; Xie, X.-L.; Mai, Y.-W. Biocompatible reduced graphene oxide sheets with superior water dispersibility stabilized by cellulose nanocrystals and their polyethylene oxide composites. *Green Chem.* **2016**, *18*, 1674–1683. [CrossRef]

184. Pal, N.; Banerjee, S.; Roy, P.; Pal, K. Reduced graphene oxide and peg-grafted tempo-oxidized cellulose nanocrystal reinforced poly-lactic acid nanocomposite film for biomedical application. *Mater. Sci. Eng. C* **2019**, *104*, 109956. [CrossRef]

185. Pal, N.; Dubey, P.; Gopinath, P.; Pal, K. Combined effect of cellulose nanocrystal and reduced graphene oxide into poly-lactic acid matrix nanocomposite as a scaffold and its anti-bacterial activity. *Int. J. Biol. Macromol.* **2017**, *95*, 94–105. [CrossRef]

186. Teodoro, K.B.; Migliorini, F.L.; Facure, M.H.; Correa, D.S. Conductive electrospun nanofibers containing cellulose nanowhiskers and reduced graphene oxide for the electrochemical detection of mercury (ii). *Carbohydr. Polym.* **2019**, *207*, 747–754. [CrossRef] [PubMed]

Review

Applications of Nanocellulose/Nanocarbon Composites: Focus on Biotechnology and Medicine

Lucie Bacakova [1,*], Julia Pajorova [1], Maria Tomkova [2], Roman Matejka [1], Antonin Broz [1], Jana Stepanovska [1], Simon Prazak [1], Anne Skogberg [3], Sanna Siljander [4] and Pasi Kallio [3]

1. Department of Biomaterials and Tissue Engineering, Institute of Physiology of the Czech Academy of Sciences, Videnska 1083, 14220 Prague, Czech Republic; Julia.Pajorova@fgu.cas.cz (J.P.); Roman.Matejka@fgu.cas.cz (R.M.); Antonin.Broz@fgu.cas.cz (A.B.); Jana.Stepanovska@fgu.cas.cz (J.S.); Simon.Prazak@fgu.cas.cz (S.P.)
2. Faculty of Biotechnology and Food Sciences, Slovak University of Agriculture in Nitra, Tr. A. Hlinku 2, 94976 Nitra, Slovakia; xtomkovam2@uniag.sk
3. BioMediTech Institute and Faculty of Medicine and Health Technology, Tampere University, Korkeakoulunkatu 3, 33014 Tampere, Finland; anne.skogberg@tuni.fi (A.S.); pasi.kallio@tuni.fi (P.K.)
4. Automation Technology and Mechanical Engineering, Faculty of Engineering and Natural Sciences, Tampere University, Korkeakoulunkatu 6, 33720 Tampere, Finland; sanna.siljander@tuni.fi
*   Correspondence: Lucie.Bacakova@fgu.cas.cz; Tel.: +420-2-9644-3743

Received: 27 December 2019; Accepted: 21 January 2020; Published: 23 January 2020

Abstract: Nanocellulose/nanocarbon composites are newly emerging smart hybrid materials containing cellulose nanoparticles, such as nanofibrils and nanocrystals, and carbon nanoparticles, such as "classical" carbon allotropes (fullerenes, graphene, nanotubes and nanodiamonds), or other carbon nanostructures (carbon nanofibers, carbon quantum dots, activated carbon and carbon black). The nanocellulose component acts as a dispersing agent and homogeneously distributes the carbon nanoparticles in an aqueous environment. Nanocellulose/nanocarbon composites can be prepared with many advantageous properties, such as high mechanical strength, flexibility, stretchability, tunable thermal and electrical conductivity, tunable optical transparency, photodynamic and photothermal activity, nanoporous character and high adsorption capacity. They are therefore promising for a wide range of industrial applications, such as energy generation, storage and conversion, water purification, food packaging, construction of fire retardants and shape memory devices. They also hold great promise for biomedical applications, such as radical scavenging, photodynamic and photothermal therapy of tumors and microbial infections, drug delivery, biosensorics, isolation of various biomolecules, electrical stimulation of damaged tissues (e.g., cardiac, neural), neural and bone tissue engineering, engineering of blood vessels and advanced wound dressing, e.g., with antimicrobial and antitumor activity. However, the potential cytotoxicity and immunogenicity of the composites and their components must also be taken into account.

Keywords: nanofibrillated cellulose; cellulose nanocrystals; fullerenes; graphene; carbon nanotubes; diamond nanoparticles; sensors; drug delivery; tissue engineering; wound dressing

---

1. Introduction

Nanocellulose/nanocarbon composites are hybrid materials containing cellulose and carbon nanoparticles. Integration of nanocarbon materials with nanocellulose provides functionality of nanocarbons, using an eco-friendly, low-cost, strong, dimension-stable, nonmelting, nontoxic and nonmetal matrix or carrier, which alone has versatile applications in industry, biotechnology and biomedicine (for a review, see [1,2]). In addition to its advantageous combination with nanocarbon materials, nanocellulose is an appealing material for biomedical applications due to its tunable chemical properties, nonanimal origin, and resemblance to biological molecules in dimension, chemistry and viscoelastic properties, etc. [3–6].

Cellulose nanomaterials include cellulose nanofibrils (CNFs) and cellulose nanocrystals (CNCs) [3]. CNFs are manufactured using either a bottom-up or a top-down approach. The bottom-up approach involves bacterial (*Gluconacetobacter*) biosynthesis to obtain bacterial cellulose (BC), while, in the top-down method, cellulosic biomass from plant fibers is disintegrated into smaller CNFs [7] that contain amorphous and crystalline regions [3]. The fibrillation of cellulose is achieved using mechanical forces, chemical treatments, enzymes or combinations of these. After fibrillation, the width of CNFs is typically between 3 and 100 nm, and the length can be several micrometers [8]. Separation of the crystalline parts from the amorphous regions of the fibers or fibrils to obtain CNCs typically requires acid hydrolysis, which destroys the amorphous regions [9]. Entangled CNFs are longer, while CNCs possess shorter needle- or rod-like morphology with a similar diameter and a more rigid molecule due to their higher crystallinity [3,9]. In general, the properties of nanocelluloses are variable and depend on their origin, type, processing, pretreatments and functionalization. Integration with other materials, as well as fabrication of the final product, further affects the properties of the resulting composite or hybrid structure.

Carbon nanoparticles include fullerenes (usually C_{60}), graphene-based particles (graphene, graphene oxide, reduced graphene oxide, graphene quantum dots), nanotubes (single-walled, double-walled, few-walled or multi-walled) and nanodiamonds (for a review, see [10–20]). The most frequently used nanocellulose/nanocarbon composites contain graphene or carbon nanotubes, while composites of nanocellulose with nanodiamond, and particularly with fullerenes, are less frequently used. Other carbon nanostructures, which are less frequently used in nanocellulose/nanocarbon composites, at least for biomedical applications, include carbon nanofibers [21–25], carbon quantum dots [26–28] activated carbon [29,30] and carbon black [31–33].

Nanocellulose/nanocarbon composites can be prepared in one-dimensional (1D), two-dimensional (2D) or three-dimensional (3D) forms. 1D composites are represented, for example, by C_{60} fullerenes grafted onto cellulose nanocrystals that have undergone amination or oxidation [34,35]. 2D composites are represented by films, which can be self-standing or supported, i.e., in the form of free-standing membranes [29,36–41] or in the form of coatings deposited on bulk materials [33,42]. The films can be formed by depositing carbon nanoparticles on a nanocellulose layer [43,44]. More frequently, however, they are fabricated from aqueous dispersions of nanocellulose and carbon nanoparticles [39,42]. It should be pointed out that cellulose nanoparticles are excellent dispersive agents for carbon nanoparticles, as they prevent the aggregation of these nanoparticles and maintain them in long-term stable homogeneous suspensions without the need to subject them to chemical functionalization [45,46]. Suspensions of cellulose and carbon nanoparticles are also starting materials for the creation of 3D nanocellulose/nanocarbon composites in the form of aerogels, foams or sponges [45,47–50]. In addition, composite 3D scaffolds, especially for tissue engineering and for regenerative medicine, can be fabricated by 3D printing using bioinks based on cellulose and carbon nanoparticles [51,52]. Both 2D composites and 3D composites can also be created by adding carbon nanoparticles to cultures of cellulose-producing bacteria, such as *Gluconacetobacter xylinus*. These nanoparticles are then incorporated into bacterial nanocellulose in situ during its growth [53–57]. Another approach is via the electrospinning or wet spinning of solutions containing cellulose and carbon nanoparticles [58–60].

Nanocellulose/nanocarbon composites exhibit several more advantageous properties than materials containing only cellulose nanoparticles or only carbon nanoparticles. Adding carbon nanoparticles to nanocellulose materials can further increase their mechanical strength [59,61]. At the same time, the presence of nanocellulose promotes the flexibility and stretchability of the materials [52,62,63]; for a review, see [64]. Adding graphene, carbon nanotubes or boron-doped diamond nanoparticles endows nanocellulose materials with electrical conductivity [39,50,57,65,66]. Other advantageous properties of nanocellulose/nanocarbon composites include their thermal stability [67–69], tunable thermal conductivity and optical transparency [48,57,70], intrinsic fluorescence and luminescence [26,71,72] photothermal activity [56], hydrolytic stability [61], nanoporous character and high adsorption capacity [49,61]. Nanocellulose/nanocarbon composites can therefore be used in a wide range of industrial and technological applications, such as water purification [22,29,43,49,54,56, 61,73–76], the isolation and separation of various molecules [22,74,77–79], energy generation, storage and conversion [21,23,44,47,64,80–85], biocatalysis [86], food packaging [67–69,87], construction of fire retardants [48], heat spreaders [70] and shape memory devices [38,88–90]. These composites are also used as fillers for various materials, usually polymers, in order to improve their mechanical, electrical and other physical and chemical properties [67–69,87,91].

In addition, nanocellulose/nanocarbon composites are promising for biomedical applications, though these applications are less frequent than industrial applications. Biomedical applications include radical scavenging [34,92], photothermal ablation of pathogenic bacteria [93], photodynamic and combined chemophotothermal therapy against cancer [35,94], drug delivery [16,28,65,72,95–97], biosensorics [31–33,63,66,71,91,98–104], and particularly tissue engineering and wound dressings. Hybrid materials containing nanocellulose and nanocarbons stimulated the growth and osteogenic differentiation of human bone marrow mesenchymal stem cells [37,59]. They provided good substrates for the attachment, growth and differentiation of SH-SHY5Y human neuroblastoma cells [51] and PC12 neural cells, particularly under electrical stimulation [105]. They enhanced the outgrowth of neurites from rat dorsal root ganglions in vitro and stimulated nerve regeneration in rats in vivo [106]. They also promoted the growth of vascular endothelial cells, enhanced angiogenesis and arteriogenesis in a chick chorioallantoic membrane model [107], and improved cardiac conduction when applied to surgically disrupted myocardium in dogs [52]. In addition, these materials supported the growth of human dermal fibroblasts [108] and mouse subcutaneous L929 fibroblasts [58,62], promoted wound healing in vivo in mice [109] and showed an antibacterial effect [30]. These materials are therefore promising for bone, neural and vascular tissue engineering, for creating cardiac patches and for advanced wound dressings. The biomedical applications of nanocellulose/nanocarbon composites are summarized in Table 1.

Table 1. Biomedical applications of nanocellulose/nanocarbon composites.

Application	Nanocellulose/Nanocarbon Composites Containing:				
	Fullerenes	Graphene	CNTs	Nanodiamonds	Others
Radical scavenging	NH_2-CNC/C_{60} [34]; CNC/C_{60}(OH)$_{30}$ [92]				
Photodynamic cancer therapy	TEMPO-oxidized CNC/C_{60}-NH_2 [35]				
Photothermal, chemo-photothermal therapy		Bacteria: [93]			
		Cancer: [94]			
Drug delivery		Anticancer drugs (doxorubicin) [72,95,96]	Anticancer and other drugs [16]	Anticancer drugs (doxorubicin) [97]	Carbon quantum dots: Anticancer drugs (temozolomide) [28]
(Bio)sensors		Electrochemical: cholesterol [98]; glucose and bacteria [110]; avian leucosis virus [111]; organic liquids [112]	Electrochemical: ATP metabolites [102]; oxygen [84]	Electrochemical: Biotin [66]	Carbon black: Electrochemical aptasensor for *S. aureus* [32]; electrochemical sensor for H_2O_2 [33]
		Piezoelectric: strain, human motion [63,99,113]	Piezoresistance and thermoelectric-based: pressure and temperature [103]; pressure [17]; strain, human motion [90,91,114]; humidity, human breath [104]		Carbon black: Strain, human motion [31,33]
	Optical: oxygen and temperature [115]; oxygen [116]	Optical: *SERS*: bilirubin [100]; *Fluorescence*: laccase [71]			Carbon quantum dots: optical sensor for biothiols [26]
		Acoustic: ammonia [101]			
Isolation of biomolecules		Histidine-rich proteins, hemoglobin [77]; bovine serum albumin [79]			
Electrical stimulation of tissues			Cardiac tissue [52]; neural tissue [106]		
Tissue engineering (TE)		General cell biocompatibility [68,69,87]; bone TE [37,59]; neural TE [105]; vascular TE [107]	Neural tissue engineering [51]; TE in general [117]		
Wound dressing/healing	Polysaccharides/fullerene C_{60} derivatives [118]	Human dermal fibroblasts in vitro [108]; mouse model in vivo [109]		L929 fibroblasts in vitro [58,62]; HeLa cells in vitro, wound dressings delivering doxorubicin [97]	Activated carbon: antibacterial wound dressing [30]

This review summarizes recent knowledge on the types, properties and applications of nanocellulose/nanocarbon-based hybrid materials, particularly in biotechnology, biomedicine and tissue engineering, and reports on the experience acquired by our group.

2. Nanocellulose/Fullerene Composites

2.1. Characterization of Fullerenes

Fullerenes are spheroidal cage-like molecules composed entirely of carbon atoms (Figure 1a). Fullerenes with 60 and 70 carbon atoms (C_{60} and C_{70}) are the most stable molecules, and they are therefore most frequently used in industrial and biomedical applications. Fullerenes were discovered in 1985 by Sir Harold Walter Kroto (1939–2016) and his co-workers Richard Smalley, Robert Curl, James Heath and Sean O'Brien. Kroto, Smalley and Curl were awarded the Nobel Prize in 1996. Fullerenes were named after Richard Buckminster "Bucky" Fuller (1895–1983), an American architect, designer, futurist, inventor, poet and visionary, who designed his geodesic dome on similar structural principles (for a review, see [12,13,18,119,120]). Fullerenes are carbon nanoparticles with diverse biological activities. This is due to the fact that they can act as either acceptors or donors of electrons (for a review, see [121]). The acceptor activity can lead to oxidative damage to cell components, such as DNA, cell membrane, mitochondria and various enzymes, to the activation of inflammatory reactions, and to cell apoptosis. These harmful effects of fullerenes can, however, be utilized for photodynamic therapy against tumors and pathogenic microorganisms (for a review, see [122]). The electron donor activity is associated with quenching oxygen radicals, which can be used in protecting skin against UV irradiation, in anti-inflammatory therapy against osteoarthritis, in cardioprotection during ischemia, in neuroprotection during amyloid-related diseases, damage by alcohol or heavy metals, in obesity treatment and in the treatment of diabetes-related disorders. Due to their structural analogy with clathrin-coated vesicles, fullerenes are also promising candidates for drug and gene delivery (for a review, see [12,13,18,119]).

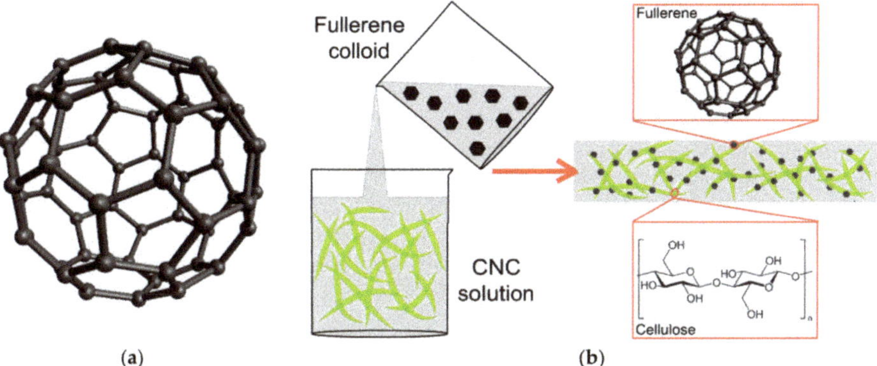

Figure 1. Scheme of fullerene C_{60} (**a**) and of the preparation and structure of nanocellulose/fullerene composites (**b**).

However, fullerenes have low solubility in many solvents, especially in water. This is a major drawback for their wider application in biomedical applications. The water solubility of fullerenes can be achieved by functionalizing them with hydrophilic groups, but this approach does not solve problems arising from the aggregation and clustering of fullerenes. In addition, the formation of singlet oxygen, which is needed for photodynamic therapy, decreases after functionalization due to the perturbation of the fullerene π system. These problems can be mitigated by the complexation of fullerenes with water-soluble agents, including nanocellulose [35].

2.2. Preparation and (Bio)Application of Nanocellulose/Fullerene Composites

Cellulose nanocrystals (CNCs) were used to create nanocellulose/fullerene composites. These nanocrystals are typically produced by acid hydrolysis of cellulose fibers, employing either sulfuric acid or hydrochloric acid in order to destroy the amorphous regions of the cellulose, while the crystalline segments remain intact. CNCs can have a needle-like or rod-like morphology, and are also referred to as nanowhiskers or nanorods. This morphology is characterized by a high aspect ratio (i.e., high length to diameter ratio), and thus by a relatively large surface area. In addition, CNCs have a wide range of other advantageous properties, such as high mechanical resistance, broad chemical-modifying capacity, renewability, biodegradability and low cytotoxicity [34,35]; for a review, see [2]. From these points of view, CNCs were considered ideal for immobilization of fullerene nanoparticles [92]. A scheme of preparation of nanocellulose/fullerene composites is depicted in Figure 1b. Composites of CNCs with fullerenes C_{60} were prepared by amine functionalization of CNCs and by subsequently grafting C_{60} onto the surface of amine-terminated CNCs [34]. Conversely, functionalized fullerenes, e.g., polyhydroxylated fullerenes $C_{60}(OH)_{30}$, were conjugated with the surface of CNCs [92]. Both of these composites showed a higher radical scavenging capacity in vitro than fullerenes alone, and therefore are promising for biomedical application in antioxidant therapies, e.g., as components of skin care products. In the third type of composites, both cellulose nanocrystals and fullerenes were functionalized, i.e., amino-fullerene C_{60} derivatives were covalently grafted onto the surface of 2, 2, 6, 6-tetramethylpiperidine-1-oxylradical (TEMPO)-oxidized nanocrystalline cellulose [35]. These composites hold promise for photodynamic cancer therapy (Table 1). When these composites were added to the culture medium of human breast cancer MCF-7 cells in the dark, they were taken up by these cells without changes in the cell viability, as revealed by a resazurin assay. However, when irradiated with light, these composites showed dose-dependent toxicity for MCF-7 cells [35].

However, fullerenes are less widely used in nanocellulose/nanocarbon composites than other carbon allotropes, particularly graphene and carbon nanotubes. More frequently, fullerenes are incorporated into a non-nanostructured cellulose matrix. For example, fullerene C_{70}, characterized by a strong thermally activated delayed fluorescence at elevated temperatures, which is extremely oxygen sensitive, was incorporated into ethyl cellulose, i.e., a highly oxygen-permeable polymer. This composite was used for construction of an optical dual sensor for oxygen and temperature [115]. An oxygen sensor was constructed using isotopically enriched carbon-13 fullerene C_{70}, dissolved in an ethyl cellulose matrix [116]. Mixed-matrix membranes, consisting of ethyl cellulose as a continuous matrix and fullerenes C_{60} as a dispersed phase, were prepared for propylene/propane separation [78]. Electrospun cellulose acetate nanofibers reinforced with fullerenes were used in the construction of dry-type actuators [123]. Cellulose impregnated with fullerenes C_{60} dissolved in o-xylene showed greater extraction efficiency for Cu^{2+}, Ni^{2+} and Cd^{2+} ions from an aqueous environment than the pure polymer [124]. Biocompatible composites containing polysaccharides (cellulose, chitosan and gamma-cyclodextrin) and fullerene derivatives (amino-C_{60} and hydroxy-C_{60}) were developed for various applications ranging from dressing and treating chronically infected wounds to nonlinear optics, biosensors, and therapeutic agents [118].

3. Nanocellulose/Graphene Composites

3.1. Characterization of Graphene

Graphene is a single layer of sp^2-hybridized carbon atoms arranged into a two-dimensional honeycomb-like lattice (Figure 2a). In other words, graphene is a one-atom-thick layer of graphite. It is a basic building block for other carbon allotropes, such as fullerenes, carbon nanotubes and graphite. Graphene is a very thin, nearly transparent sheet, but it is remarkably strong (about 100 times stronger than steel), and highly electrically and thermally conductive (for a review, see [19,20,125,126]). Graphene can be prepared by various methods, which can be divided into two main categories, namely the top-down approach and the bottom-up approach. The top-down approaches include treatment of graphite by mechanical or electrochemical exfoliation, intercalation or sonication, and also nanotube slicing. The bottom-up approaches include growth of graphene from carbon-metal melts, epitaxial growth of graphene on silicon carbide, the dry ice method, and deposition methods such as chemical vapor deposition or dip coating a substrate with graphene oxide (GO), followed by GO reduction (for a review, see [19,20,125,126]). Graphene can be prepared in the form of monolayer or bilayer sheets, nanoplatelets, nanoflakes, nanoribbons and nanoscrolls. Chemically, graphene-based materials include pure graphene sheets, GO or reduced graphene oxide (rGO). Pure graphene sheets can be produced by mechanical exfoliation of graphite or by chemical vapor deposition. GO, a highly oxidative and water-soluble form of graphene, can be obtained by the exfoliation of graphite oxide. Reduced GO can be prepared by chemical, thermal or pressure reduction, and even by bacteria-mediated reduction of GO, which improves its electrical properties (for a review, see [13,18,19,37,125,126]). Graphene and graphene-based materials hold a great promise not only for a wide range of industrial and technology applications, but also for biomedical applications, such as drug, gene and protein delivery, photothermal therapy, construction of biosensors, bioimaging, antimicrobial treatment, and also as scaffolds for tissue engineering (for a review, see [20]).

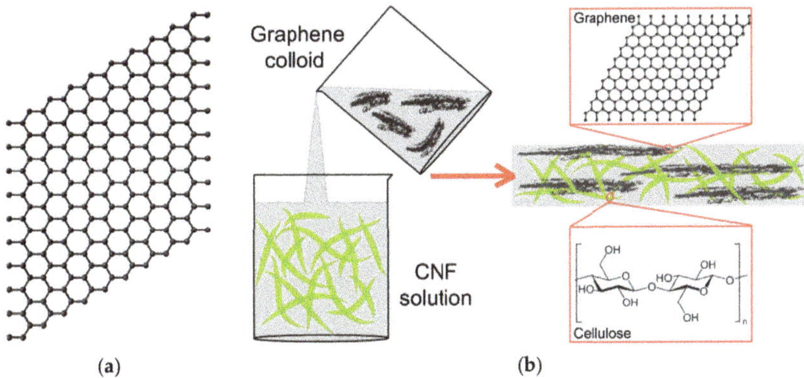

Figure 2. Scheme of graphene (**a**) and of the preparation and structure of nanocellulose/graphene composites (**b**).

3.2. Preparation and Industrial Application of Nanocellulose/Graphene Composites

Similarly as in fullerenes, cellulose nanoparticles in the form of nanofibrils and nanocrystals increase the dispersion of graphene nanoparticles in water-based environments and prevent their aggregation without the need to subject them to chemical functionalization [46]. A water-based dispersion is the starting material for fabricating nanocellulose/graphene composites (Figure 2b). These composites can be created by filtration [127], filtration combined with hot pressing for fabricating films [128], or by freeze-drying [75] and freeze-casting [48] for fabricating 3D materials, such as aerogels and foams. Other methods are deposition of graphene on a nanocellulose layer [43] and incorporation of graphene into nanocellulose during its biological synthesis by bacteria [54–56].

All forms of nanocellulose and graphene have been used for constructing nanocellulose/graphene composites, i.e., CNFs, CNCs, unmodified graphene, GO and rGO. In order to modulate the properties of nanocellulose/graphene composites for specific applications, these materials can be further enriched by various substances, such as metallic or ceramic nanoparticles, oxides, carbides, sulfides, vitamin C, synthetic and natural polymers, enzymes and antibodies. For example, nanocellulose/graphene composites have high adsorption, filtration and photocatalytic ability, and they are therefore widely used for water purification, e.g., for removing antibiotics [75], dyes [43], heavy metals, such as Cu^{2+}, Hg^{2+}, Ni^{2+} and Ag^+ [61,76], or for their bactericidal effect [56]. The water-cleaning capacity of these composites can be further enhanced by introducing additional photocatalytic agents, i.e., palladium nanoparticles [54] or zinc oxide (ZnO) nanoparticles [73]. An optimized ultrafiltration membrane for water purification was constructed from polyvinylidene fluoride (PVDF), modified by cellulose nanocrystals functionalized with common bactericides, such as dodecyl dimethyl benzyl ammonium chloride, ZnO and GO nanosheets [129]. Another important additive is vitamin C, which reduces the GO in nanocellulose/GO composites, increases the surface area of the material and increases pore formation, and thus enhances the capacity of the composites for water purification [49]. A combination of rGO-coated cellulose nanofibers with hydrophobic and oleophilic trimethyl chlorosilane enhanced the adsorption capacity of this composite, which is necessary for effective removal of oil-based pollutants from water [74].

Another important industrial application of nanocellulose/graphene composites is in energy storage, generation and conversion. Devices for these purposes include supercapacitors [64,80, 130], hydrogen storage devices [131], electrodes for hydrogen evolution reaction [132], lithium ion batteries [133], actuators [81], solar steam generators [82] and electric heating membranes [83]. These devices can be based on pure nanocellulose/nanocarbon composites without additives [80–83]. However, they often contain additives such as manganese oxide (MnO), which contributes to faradaic pseudocapacitance in supercapacitors [130] or polypyrrole, which acts as an insulator, but its oxidized derivatives are good electrical conductors [134]. Other additives are palladium or platinum nanoparticles for enhanced hydrogen storage [131], nitrogen-doped molybdenum carbide nanobelts in electrocatalysts for hydrogen evolution reaction [132], and silicon oxide nanoparticles in lithium ion batteries [133].

Other important industrial applications of nanocellulose/graphene composites are in the construction of fire retardants, shape memory devices, biocatalysts and materials for food packaging. Super-insulating, fire-retardant, mechanically strong anisotropic foams were produced by freeze-casting suspensions of cellulose nanofibers, GO and sepiolite nanorods, and they performed better than traditional polymer-based insulating materials [48]. Shape memory devices are based on GO/CNC thin films and nanomembranes [38,88] or on GO introduced into a nanocellulose paper made of nanofibers extracted from sisal fibers [89]. An example of a biocatalyst is a nanocellulose/polypyrrole/GO nanocomposite for immobilization of lipase, a versatile hydrolytic enzyme. This biocatalyst was employed for synthesizing ethyl acetoacetate, a fruit flavor compound [86]. Food packaging materials were constructed by filling CNCs and rGO, either separately or in the form of CNC/rGO nanohybrids, into poly (lactic acid) (PLA) matrix or in a poly (3-hydroxybutyrate-*co*-3-hydroxyvalerate) (PHBV) matrix. These composite materials exhibited better mechanical properties than the pristine polymers, and possessed antibacterial activity. In addition, the composites with CNC/rGO nanohybrids performed better than those with a single component nanofiller, i.e., either CNCs or rGO. Due to their antibacterial activity, antioxidant properties and good in vitro cytocompatibility, these composites are also promising for biomedical applications, e.g., as scaffolds for tissue engineering [67–69,87].

3.3. Biomedical Application of Nanocellulose/Graphene Composites

The biomedical applications of nanocellulose/nanographene composites include photothermal ablation of pathogenic bacteria, combined chemo-photothermal therapy against cancer, drug delivery, biosensorics, isolation and separation of various biomolecules, wound dressing and particularly tissue engineering (Table 1).

For photothermal ablation of pathogenic bacteria, a composite paper containing nanocellulose with Au linked to GO using quaternized carboxymethyl chitosan was developed. When excited by near-infrared laser irradiation, the paper generated a rise in temperature of more than 80 °C, sufficient for photothermal ablation, both on Gram-positive bacteria (*Bacillus subtilis* and *Staphylococcus aureus*) and on Gram-negative bacteria (*Escherichia coli* and *Pseudomonas aeruginosa*). Additionally, the composite paper showed a remarkable enhancement in tensile strength, bursting index and tear index in comparison with the properties of pure nanocellulose paper [93].

For chemophotothermal synergistic therapy of colon cancer cells, dual stimuli-responsive polyelectrolyte nanoparticles were developed by layer-by-layer (LbL) assembly of aminated nanodextran and carboxylated nanocellulose on the surface of chemically modified GO. Tests on the HCT116 human colon cancer cell line revealed that these nanoparticles allowed for the intracellular delivery of curcumin, which was released in response either to acidic environments or to near-infrared excitation [94]. In this context, nanocellulose/graphene composites are good candidates as carriers for controlled drug delivery, particularly of anticancer drugs. Systems releasing doxorubicin, a model drug with broad-spectrum anticancer properties, were developed. These systems included nanocomposite carboxymethyl cellulose/GO hydrogel beads [95], nanocomposite films made of graphene quantum dots incorporated into a carboxymethyl cellulose hydrogel [72] or macroporous polyacrylamide hydrogels. These hydrogels were prepared using an oil-in-water Pickering emulsion, containing GO and hydroxyethyl cellulose with a quaternary ammonium group [96].

Sensing and biosensing is another important application of nanocellulose/graphene composites. These sensors can be divided into electrochemical, piezoelectric, optical and acoustic wave-based sensors. Electrochemical sensors were constructed for detecting cholesterol [98], glucose and pathogenic bacteria [110], avian leucosis virus [111] and organic liquids [112]. The biosensor for detecting cholesterol was based on chemically-modified nanocellulose, grafted with silylated GO and enriched with ZnO nanoparticles in order to enhance its electrical conductivity [98]. The biosensor for detecting glucose and pathogenic bacteria was based on nanocellulose paper coated with GO, reduced by vitamin C and functionalized with platinum nanoparticles with a cauliflower-like morphology in order to enhance the electrical conductivity of the composite. The platinum surface was functionalized either with glucose oxidase (via chitosan encapsulation) or with an RNA aptamer [110]. The biosensor for the avian leucosis virus was an immunosensor, based on graphene-perylene-3, 4, 9, 10-tetracarboxylic acid nanocomposites as carriers for primary antibodies, on composites of nanocellulose and Au nanoparticles as carriers for secondary antibodies, and on the alkaline phosphatase catalytic reaction [111]. The sensor for organic liquids, mainly organic solvents, was based on cellulose nanocrystal-graphene nanohybrids, selectively located in the interstitial space between the natural rubber latex microspheres [112].

Piezoelectric nanocellulose/graphene-based sensors have usually been designed for strain sensing, i.e., as wearable electronics for monitoring the motion of various parts of the human body, e.g., fingers [63,99,113]. For these purposes, the flexibility and stretchability of nanocellulose was further enhanced by adding other polymers, such as elastomers, represented e.g., by polydimethylsiloxane (PDMS) [113], or hydrogels, represented e.g., by poly(vinyl alcohol) (PVA), crosslinked (together with cellulose nanofibers and graphene) with borax [63].

Optical sensors can be based on surface-enhanced Raman spectroscopy (SERS) or fluorescence. Cellulose SERS strips decorated with plasmonic nanoparticles, termed graphene-isolated-Au-nanocrystals (GIANs), were developed for constructing portable sensors for detecting complex biological samples, e.g., for detecting free bilirubin in the blood of newborns [100]. A fluorescence sensor, based on sulfur and nitrogen-co-doped graphene quantum dots, immersed into nanocellulosic hydrogels, was developed for detecting laccase. This enzyme is widely used in industrial and technological applications, such as bleaching of fabrics, tooth whitening, decoloration of hair, water purification and in oxidizing dyes in beer, must and wines [71].

An example of an acoustic wave-based sensor is an ammonia sensor, based on a quartz crystal microbalance (QCM) with a sensing coating. This coating is composed of negatively-charged electrospun cellulose acetate nanofibers, positively-charged polyethylenimine and negatively-charged GO, and it was created by the electrostatic LbL self-assembly technique [101].

For protein isolation, a metal affinity carboxymethyl cellulose-functionalized magnetic graphene was prepared by successive modifications of GO nanosheets with magnetic nanoparticles, carboxymethyl cellulose and iminodiacetic acid, and then chelated with copper ions. This composite exhibited high adsorption selectivity toward histidine-rich proteins, which was utilized for isolating hemoglobin from human whole blood, and also for isolating a polyhistidine-tagged recombinant protein from *Escherichia coli lysate*, namely *Staphylococcus aureus* enterotoxin B [77]. For macromolecular separation, cellulose acetate nanocomposite ultrafiltration membranes were fabricated using 2D layered nanosheets, e.g., GO and exfoliated molybdenum disulfide (MoS_2), and were successfully tested using macromolecular bovine serum albumin [79].

Nanocellulose/graphene composites are also important components of tissue engineering scaffolds, improving their mechanical properties and their bioactivity. In the studies by Pal et al. (2017), mentioned above, a PLA/CNC/rGO nanocomposite film showed antibacterial activity against Gram-positive *Staphylococcus aureus* and against Gram-negative *Escherichia coli*. At the same time, this film exhibited negligible cytotoxicity against a mouse NIH-3T3 fibroblast cell line, as revealed by an MTT assay of the activity of mitochondrial oxidoreductase enzymes [87]. Nanocomposites of CNCs and rGO, incorporated into PLA matrix through the melt-mixing method, were noncytotoxic and cytocompatible with epithelial human embryonic kidney 293 (HEK293) cells [68]. PLA incorporated with rGO and TEMPO-oxidized CNCs, grafted with poly(ethylene glycol) (PEG), displayed radical scavenging activity and negligible toxicity and cytocompatibility to mouse embryonic C3H10T1/2 cells [69]. A composite film consisting of hydrophilic bacterial cellulose nanofibers and hydrophobic rGO, prepared from GO using a bacterial reduction method, supported the adhesion, viability and proliferation of human bone marrow mesenchymal stem cells in a similar way to standard cell culture polystyrene, and better than pure rGO films [37]. Incorporating GO into electrospun cellulose acetate nanofibrous scaffolds enhanced the adhesion and growth of human umbilical cord mesenchymal stem cells. It also enhanced osteogenic differentiation of these cells, manifested by the activity of alkaline phosphatase, and biomineralization of the scaffolds in a simulated body fluid [59]. Nanofibrous composites of bacterial nanocellulose, a conductive poly(3,4-ethylene dioxythiophene) (PEDOT) polymer and GO, mimicking the native extracellular matrix and allowing electrical stimulation of neural PC12 cells, induced specific orientation and differentiation of these cells [105]. Polyvinyl alcohol/carboxymethyl cellulose (PVA/CMC) scaffolds loaded with rGO nanoparticles, prepared by lyophilization, enhanced the proliferation of EA.hy926 endothelial cells in vitro and angiogenesis in vivo using a chick chorioallantoic membrane model [107]. Polyacrylamide-sodium carboxymethylcellulose hybrid hydrogels reinforced with GO and/or CNCs also have potential for tissue engineering applications due to their tunable mechanical properties [135]. Genetically modified hydrophobin, a fungal cysteine-rich protein, was used to connect nanofibrillated cellulose of wood origin and graphene flakes in order to construct biomimetic mechanically-resistant materials similar to nacre and combining high toughness, strength and stiffness [136].

Nanocellulose/graphene composites also have great potential for the fabrication of antibacterial textiles and for advanced wound dressing. Antibacterial textiles were prepared by electrospinning a mixture containing cellulose acetate, TiO_2 and GO sheets. These textiles showed high antibacterial activity with an inhibition rate higher than 95% against *Bacillus subtilis* and *Bacillus cereus* [137]. Bacterial cellulose is considered as one of the most suitable materials for advanced wound dressing, due to its appropriate mechanical properties, such as strength, Young's modulus, elasticity and conformability, and also due to its great capacity to retain moisture in the wound (for a review, see [2]). These favorable properties can be further enhanced by adding graphene-based materials and by crosslinking with synthetic polymers, such as poly (ethylene glycol), poly (vinyl alcohol), poly (acrylic acid) and poly (acrylamide). In a study by Chen et al. (2019), a bacterial nanocellulose-grafted poly (acrylic acid)/GO composite hydrogel was prepared as a potential wound dressing. The inclusion of GO improved the attachment and proliferation of human dermal fibroblasts in cultures on the composites [108]. Similarly, hydroxypropyl cellulose matrix incorporated with GO and silver-coated ZnO nanoparticles showed improved tensile strength, and also anti-ultraviolet, antibacterial and immunostimulatory effects, which promoted wound healing in an in vivo mouse model [109].

4. Nanocellulose/Carbon Nanotube Composites

4.1. Characterization of Carbon Nanotubes

Carbon nanotubes (CNTs; Figure 3a) are tubular structures formed by a single cylindrical graphene sheet (single-walled carbon nanotubes, referred to as SWCNTs or SWNTs) or several graphene sheets arranged concentrically (multiwalled carbon nanotubes, referred to as MWCNTs or MWNTs, which also include double-walled CNTs (DWCNTs [16]), and few-walled CNTs (FWCNTs [17]). Carbon nanotubes were discovered as a by-product of fullerene synthesis, and were first described by Iijima et al. (1991) [138]. CNTs have a high aspect ratio (i.e., length to diameter ratio) and thus a relatively large surface area. Their diameter is on the nanometer scale (e.g., from 0.4 nm to 2–3 nm in single-walled nanotubes), but their length can reach several micrometers or even centimeters. Due to these properties, CNTs are suitable candidates for hydrogen storage, for the removal of contaminants from water and air, and also for drug delivery. CNTs have excellent mechanical properties, mainly due to the sp^2 bonds. The tensile strength of SWCNTs has been reported to be almost 100 times higher than that of steel, while their specific weight is about six times lower. CNTs can therefore be used for reinforcing various synthetic and natural polymers for industrial and biomedical applications, e.g., for hard tissue engineering. When added to a polymer matrix, CNTs can resemble inorganic mineral nanoparticles in the bone tissue, and they can form nanoscale irregularities on the surface of 2D materials and in the pores of 3D materials, which improve the cell adhesion and growth. CNTs are electrically conductive and enable electrical stimulation of cells, which further improves the adhesion, growth and differentiation of cells (for a review, see [10–14,18,139]). However, free CNTs can be cytotoxic, which is attributed to their ability to cause oxidative damage, and also to their contamination with transition metals (e.g., Fe, Ni, Y), which serve as catalysts during CNT preparation. Methods for producing metal-free CNTs have therefore been developed, e.g., arc-discharge evaporation of graphite rods [139].

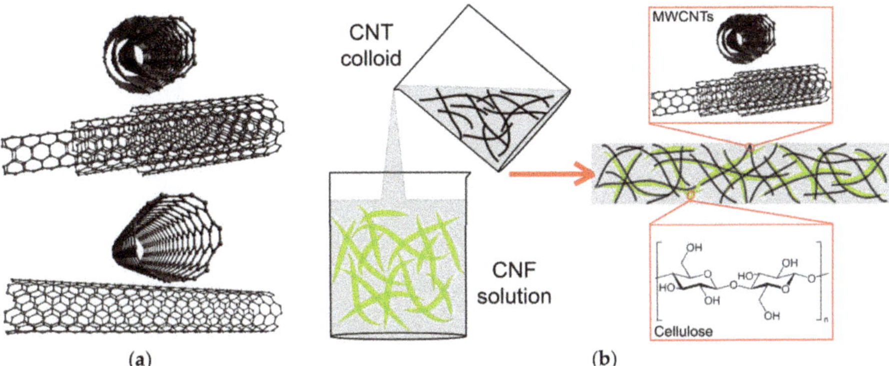

Figure 3. Scheme of multi-walled and single-walled carbon nanotubes (**a**) and of the preparation and structure of nanocellulose/carbon nanotube composites (**b**).

CNTs also resemble CNFs from the point of view of their morphology and their mechanical properties. For example, highly crystalline, thick CNFs derived from tunicates exhibited mean strength of 3–6 GPa, which was comparable with commercially available MWCNTs. However, the mean strength of other types of CNFs is lower; for example, in wood-derived CNFs the mean strength ranged from 1.6 to 3 GPa [140]. CNTs therefore improve the mechanical strength of nanocellulose/CNT composites, and endow them with electrical conductivity, similarly as graphene. As a result, nanocellulose/CNT composites are used in similar industrial and biomedical applications as nanocellulose/graphene composites, e.g., water purification, energy generation, storage and conversion, filling polymeric materials, constructing sensors and biosensors, drug delivery, cancer treatment, electrical stimulation of tissues, and tissue engineering.

4.2. Preparation and Industrial Application of Nanocellulose/CNT Composites

The nanocellulose component in the nanocellulose/CNT composites is used in the form of nanofibrils and nanocrystals, and carbon nanotubes usually in the form of SWCNTs or MWCNTs. Similarly as in composites containing graphene and other carbon allotropes, cellulose nanoparticles facilitate the homogeneous dispersion of CNTs in aqueous environments (Figure 3b), where the two types of nanoparticles are linked by noncovalent interactions, e.g., hydrophobic and electrostatic interactions [42,46]. The dispersion of CNTs can be further facilitated by TEMPO-mediated oxidation of cellulose nanoparticles, which endows them with abundant anionic carboxyl groups [141]. Other ways include the functionalization of CNTs with self-assembling amphiphilic glycosylated proteins [142] or the use of oil-in-water Pickering emulsions of cellulose nanoparticles [143]. From the aqueous dispersions, 2D and 3D nanocellulose/CNT composites can be formed, e.g., by vacuum filtration, centrifugal cast molding, foam forming, casting and printing [39,45,50,141]. Pickering emulsions of cellulose nanocrystals and SWCNTs or MWCNTs were used for fabricating aerogels and foams by freeze-drying [143]. Similarly as in nanocellulose/graphene composites, CNTs can be added to cultures of bacteria producing nanocellulose, and incorporated into the bacterial nanocellulose during its growth [53,57]. Composite nanocellulose/SWCNT films can be transparent or semitransparent [39, 45,141], and can transmit radiant energy [144].

Like nanocellulose/graphene composites, nanocellulose/CNT composites can be combined with various atoms, molecules and nanoparticles in order to enhance their properties for specific applications. For example, Ag nanoparticles attached to the surface of MWCNTs influenced the electrochemical properties of CNT-based films developed on a bacterial nanocellulose membrane [36]. Nanocellulose and CNTs can be also used as additives to various hybrid materials. For example, a hybrid material, created by combination of poly (3,4-ethylenedioxythiophene)-poly (styrenesulfonate) (PEDOT:PSS), silver nanoparticles (AgNPs), CNTs and a nanocellulose layer, was used for constructing a tactile sensor [103]. Incorporation of polypyrrole-coated CNTs into chemically cross-linked CNC aerogels created promising materials for flexible 3D supercapacitors [145]. A combination of cellulose acetate, chitosan and SWCNTs with Fe_3O_4 and TiO_2 in electrospun nanofibers enables combined removal of Cr^{6+}, As^{5+}, methylene blue and Congo red from aqueous solutions via the adsorption and photocatalytic reduction processes [146].

Energy-related applications of nanocellulose/CNT composites include biofuel cells, varactors, supercapacitors, and electrodes for lithium batteries, thermoelectric generators for heat-to-electricity conversion or for constructing heating elements [50]. A biofuel cell comprising electrodes based on supercapacitive materials, i.e., on CNTs and a nanocellulose/polypyrrole composite, was utilized to power an oxygen biosensor. Laccase, immobilized on naphthylated MWCNTs, and fructose dehydrogenase, adsorbed on a porous polypyrrole matrix, were used as cathode and anode bioelectrocatalysts, respectively [84]. Another biofuel cell was based on a conductive MWCNT network, developed on a bacterial nanocellulose film, and functionalized with redox enzymes, including pyroquinoline quinone glucose dehydrogenase (anodic catalyst) and bilirubin oxidase (cathodic catalyst). This system generated electrical power via the oxidation of glucose and the reduction of molecular oxygen [44]. Microelectromechanical system varactors, i.e., voltage-controlled capacitors, consisted of a freestanding SWCNT film, which was employed as a movable component, and a flexible nanocellulose aerogel filling [85]. Supercapacitors with high physical flexibility, desirable electrochemical properties and excellent mechanical integrity were realized by rationally exploiting the unique properties of bacterial nanocellulose, CNTs, and ionic liquid-based polymer gel electrolytes [147]. Other flexible 3D supercapacitor devices were fabricated by incorporating polypyrrole nanofibers, polypyrrole-coated CNTs, and manganese dioxide (MnO_2) nanoparticles in chemically cross-linked cellulose nanocrystal aerogels [145]. Electrospun core-shell nanofibrous membranes, containing CNTs stabilized with cellulose nanocrystals, were developed for use as high-performance flexible supercapacitor electrodes with enhanced water resistance, thermal stability and mechanical toughness [40]. Electrodes for lithium batteries were based on freestanding

LiCoO$_2$/MWCNT/cellulose nanofibril composites, fabricated by a vacuum filtration technique [148], or on freestanding CNT-nanocrystalline cellulose composite films [41]. Thermoelectric generators for heat-to-electricity conversion were based on large-area bacterial nanocellulose films with an embedded/dispersed CNT percolation network, incorporated into the films during nanocellulose production by bacteria in culture [57]. Electrical energy can also be converted into thermal energy, as demonstrated by the composite structure of wood-derived nanocellulose, MWCNTs and pulp, designed for a heating element application [50].

4.3. Biomedical Application of Nanocellulose/CNT Composites

Biomedical applications of composites of cellulose and CNTs are summarized in Table 1. These composites are important systems for drug delivery. The CNTs in the composites can be conjugated with many therapeutics, usually anticancer drugs, but also other types of drugs. For example, an osmotic pump tablet system coated with cellulose acetate membrane containing MWCNTs was developed for delivery of indomethacin (for a review, see [16,65]). Cancer cells can be killed by nanocellulose-CNT dispersions even without the presence of anticancer drugs. For example, nonmercerized type-II cellulose nanocrystals in dispersions with SWCNTs displayed cytotoxicity for human epithelial colorectal adenocarcinoma Caco-2 cells, but they enhanced the mitochondrial metabolism of normal cells [149].

Construction of sensors and biosensors is another important application of nanocellulose/CNT composites useful for (bio) technology and medicine. A tactile piezoresistance and thermoelectric-based sensor, mentioned above, which is capable of simultaneously sensing temperature and pressure, is fabricated from TEMPO-oxidized cellulose, PEDOT:PSS, AgNPs and CNTs [103]. Another pressure sensor was developed using aerogels consisting of plant cellulose nanofibers and functionalized few-walled CNTs [17]. Highly conductive and flexible membranes with a semi-interpenetrating network structure, fabricated from MWCNTs and cellulose nanofibers, showed the electrical features of capacitive pressure sensors and were promising for various electronics applications, e.g., touch screens [150]. Advanced flexible strain sensors for controlling the human body motion were fabricated by pumping hybrid fillers consisting of CNTs/CNCs into porous electrospun thermoplastic polyurethane membranes [91]. Other strain sensors were fabricated by a facile latex assembly approach, in which CNCs played a key role in tailoring the percolating network of conductive natural rubber/CNT composites [114]. A water-responsive shape memory hybrid polymer, based on a thermoplastic polyurethane matrix crosslinked with hydroxyethyl cotton cellulose nanofibers and MWCNTs, was also developed for constructing a strain sensor [90]. A flexible and highly sensitive humidity sensor, capable of monitoring human breath, was based on TEMPO-oxidized nanofibrillated cellulose and CNTs [104]. An electrochemical biosensor for three adenosine triphosphate (ATP) metabolites, namely uric acid, xanthine and hypoxanthine, was based on a composite of NH$_2$-MWCNT/black phosphorene/AgNPs, dispersed in carboxymethyl cellulose [102]. Another electrochemical molecularly-imprinted sensor was based on a nanofibrous membrane prepared by the electrospinning technique from cellulose acetate, MWCNTs and polyvinylpyrrolidone, and was used for determining ascorbic acid [151]. An oxygen biosensor powered by a biofuel cell containing MWNCTs, a nanocellulose/polypyrrole composite, laccase and fructose dehydrogenase, was mentioned above [84]. Versatile wearable textile sensors, e.g., for gas sensing, were produced from cellulose nanofibers extracted from tunicates, homogeneously composited with SWCNTs, by wet spinning in an aligned direction [60].

Other important biomedical applications of nanocellulose/CNT composites are in electrical stimulation of cells and tissues in order to improve their regeneration and function, and in tissue engineering. For example, stretchable, flexible and electrically conductive biopatches for restorating conduction in damaged cardiac regions and for preventing arrhythmias were prepared. These patches were based on nanofibrillated cellulose/SWCNT ink three-dimensionally printed onto bacterial nanocellulose. They restored cardiac conduction after its disruption by a surgical incision made in the ventricular part of the heart in experimental dogs [52]. For neural tissue stimulation, multiblock conductive nerve scaffolds with self-powered electrical stimulation were prepared. These scaffolds were based on polypyrrole/bacterial nanocellulose composites with platinum nanoparticles on the anode side for glucose oxidation, and nitrogen-doped CNTs on the cathode side for oxygen reduction. These scaffolds enhanced the elongation of neurites outgrowing from rat dorsal root ganglions in vitro and stimulated nerve regeneration in a rat sciatic nerve gap model in vivo in comparison with composites containing only polypyrrole and bacterial nanocellulose. These scaffolds could replace the metal needles that are currently used for external electrical stimulation of neural tissue, which may cause pain and a risk of infection [106]. 3D printing was also used for creating scaffolds based on a conductive ink composed of wood-derived CNFs and SWCNTs. These scaffolds were intended for neural tissue engineering for experimental brain studies, and supported the attachment, growth and viability of human neuroblastoma SH-SHY5Y cells [51]. Other scaffolds for tissue engineering consisted of electrospun cellulose acetate nanofibers, assembled with positively-charged chitosan and negatively-charged MWCNTs via an LbL technique. These scaffolds promoted the adsorption of cell adhesion-mediating molecules from the serum supplement of the culture medium and the adhesion and growth of mouse subcutaneous L929 fibroblasts [117]. Our own results related to the potential application of nanocellulose/CNT composites as scaffolds for tissue engineering are reported in Appendix A.

5. Nanocellulose/Nanodiamond Composites

5.1. Characterization of Nanodiamond

Diamond is an allotrope of carbon, consisting of carbon atoms arranged in a cubic crystal structure covalently bonded in sp^3 hybridization (Figure 4a). Like all nanostructured materials, nanodiamonds or diamond nanoparticles are defined as features not exceeding 100 nm in at least one dimension, although some larger diamond particles, i.e., 125–210 nm, are still referred to as nanodiamonds (for a review, see [152]). At the same time, the size of ultrananocrystalline diamond particles is 3–5 nm [153,154]. Diamond nanoparticles can be prepared by various methods. The most widely used techniques are detonation of carbon-containing explosives in an oxygen-deficit environment and microwave-enhanced plasma chemical vapor deposition (MECVD). Other techniques include the radiofrequency plasma-assisted chemical vapor deposition (PACVD) method, milling of diamond microcrystals, hydrothermal synthesis, ion bombardment, laser bombardment, ultrasound synthesis and electrochemical synthesis (for a review, see [10,152–156]). Nanodiamonds are considered to be the most advanced carbon materials in the world. This is due to their excellent mechanical, optical, electrical, thermal and chemical properties. The mechanical properties of nanodiamonds include the highest hardness of all materials on earth, a high Young's modulus, high fracture toughness, high pressure resistance and a low friction coefficient. Their optical properties include transparency, high optical dispersion, and their ability to display various colors and to emit intrinsic luminescence (fluorescence), which is due to defects in the diamond lattice or contamination of the lattice with foreign atoms, such as N, B, H, Ni, Co, Cr or Si. Regarding their electrical properties, nanodiamonds can act as good insulators in their pristine state and as semiconductors after doping, usually with boron. Their thermal properties include superior thermal conductivity and low thermal expansion. The chemical properties of nanodiamonds include low chemical reactivity and resistance to liquid- and gas-phase oxidations. However, nanodiamonds can be doped with various atoms, and their surface can be functionalized by various atoms, chemical groups and (bio)molecules ([152,153,157,158]; for a review, see [10–15,155]).

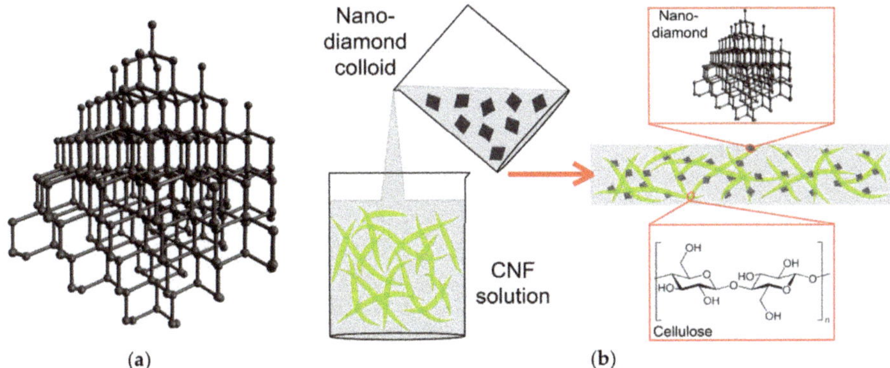

Figure 4. Scheme of a nanodiamond (**a**) and of the preparation and structure of nanocellulose/nanodiamond composites (**b**).

5.2. Preparation and (Bio)Application of Nanocellulose/Nanodiamond Composites

There is greater use of diamond nanoparticles than of fullerenes in nanocellulose/nanocarbon composites, but diamond nanoparticles are used less than graphene and carbon nanotubes. This may be because a nanodiamond is more expensive and is electrically nonconductive in its pristine state. In composites with nanofibrillated cellulose (Figure 4b), a nanodiamond was used for constructing highly thermally conductive, mechanically resistant and optically transparent films with potential application as lateral heat spreaders for portable electronic equipment [70]. A highly mechanically resistant and optically transparent nanopaper was made of cationic CNFs and anionic nanodiamond particles by filtration from a hydrocolloid and subsequent drying [159]. Moreover, a diamond can be rendered electrically semiconductive by doping it with boron, and then can be used for constructing biosensors. For example, a sensor for biotin was developed by the adsorption of captavidin, a nitrated avidin with moderate affinity to biotin, on a carboxymethylcellulose layer stabilized on a boron-doped diamond electrode by a Nafion film. This biosensor was used for analyzing biotin in blood plasma [66] (Table 1).

The reinforcing effect of diamond nanoparticles, coupled with their optical transparency, has also been used advantageously for other biomedical applications, particularly for wound dressing. Nanocellulose/nanodiamond composites are more mechanically resistant than purely nanocellulose-based materials, but they retain their flexibility and stretchability. In addition, their optical transparency enables direct inspection of wounds without the need to remove the dressing. For example, incorporating diamond nanoparticles in a concentration of 2 wt % into chitosan/bacterial nanocellulose composite films resulted in a 3.5-fold increase in the elastic modulus of these films. These composite films were transparent, but their transparency can be modulated by the concentration of diamond nanoparticles, turning them gray and semitransparent at higher nanodiamond concentrations (3 and 4 wt %). The viability of mouse subcutaneous L929 fibroblasts in cultures on these films, evaluated by an MTT test of the activity of cell mitochondrial enzymes, was more than 90% at 24 h after seeding. However, at 48 h, it had dropped to about 75%, which indicated that diamond nanoparticles are slightly cytotoxic [62]. A similar result was obtained on L929 fibroblasts grown on electrospun composite nanofibrous mats containing chitosan, bacterial cellulose and 1–3 wt % medical-grade nanodiamonds [58]. The viability of these cells, estimated by the MTT assay, dropped from approx. 90% on day 1 to approx. 75% on day 3. Nevertheless, the addition of nanodiamonds facilitated the electrospinning process, reduced the diameter of the nanofibers in the mats, regulated the water vapor permeability of the mats, enhanced their hydrophilicity and improved their mechanical properties to a similar level as in native skin [58].

Adding diamond nanoparticles per se did not significantly increase the antibacterial activity of chitosan/bacterial nanocellulose composites [62]. This activity can be further enhanced, e.g., by adding silver nanoparticles [160]. Nanocellulose/nanodiamond composites can also act as a suitable platform for drug delivery. This was demonstrated on transparent doxorubicin-loaded carboxylated nanodiamonds/cellulose nanocomposite membranes, which are promising candidates for wound dressings. These membranes are porous, transparent, with appropriate mechanical properties, and without doxorubicin they are noncytotoxic for HeLa cells [97].

6. Composites of Nanocellulose with Other Carbon Nanoparticles

In addition to fullerenes, graphene, nanotubes and nanodiamonds, other important carbon nanoparticles used in industrial, biotechnological and biomedical applications include carbon nanofibers, carbon quantum dots and nanostructures formed by activated carbon and carbon black. All these nanomaterials can be used in nanocellulose/nanocarbon composites. The biomedical applications of these composites are summarized in Table 1.

6.1. Composites of Nanocellulose and Carbon Nanofibers

Carbon nanofibers can be created by carbonization of cellulose nanofibers originating from bacterial nanocellulose [21,23,161], urea [161], filter paper [162] or plant-derived cellulose [22,24]. Another method of preparing carbon nanofibers is chemical vapor deposition (CVD; [25]). These carbon nanofibers can be further combined with other carbon nanoparticles, mainly graphene. For example, a composite paper consisting of nitrogen-doped carbon nanofibers, reduced graphene oxide (rGO) and bacterial cellulose was designed as a high-performance, mechanically tough, and bendable electrode for a supercapacitor. The bacterial nanocellulose in this paper is exploited both as a biomass precursor for the creation of carbon nanofibers by pyrolysis and as a supporting substrate for the newly-created material [21]. In another study, highly conductive freestanding cross-linked carbon nanofibers, derived from bacterial cellulose in a rapid plasma pyrolysis process, were used as substrates for the growth of vertically-oriented graphene sheets for constructing alternating current filtering supercapacitors [23]. A small amount of rGO can also act as an effective initiator of carbonization of cellulose nanofibers through microwave treatment [24]. Carbonization of aerogels, prepared from a mixture of PVA, cellulose nanofibers and GO by freeze-drying, enhanced the hydrophobic properties, the specific surface area and the adsorption capacity of these aerogels. These materials then became suitable candidates for oil-water separation and environmental protection [22]. In addition to graphene, cellulose-derived carbon nanofibers can be combined with various other nanoparticles and nanostructures, such as Pt nanoparticles for methanol oxidation reaction [161], TiO_2 films and Fe_3O_4 nanoparticles for lithium ion batteries [162], tin oxide (SnO) nanoparticles for lithium-sulfur batteries [163] or $NiCo_2S_4$ nanoparticles for hydrogen evolution reaction [164]. Carbon nanofibers are also promising for biomedical applications, particularly bone tissue engineering. Their nanoscale diameter produced a nanoscale surface roughness of their compacts or of their composite with poly-lactic-*co*-glycolic acid (PLGA). This nanoroughness promoted preferential adhesion of osteoblasts from other cell types, particularly fibroblasts, which could prevent fibrous encapsulation of bone implants [25].

6.2. Composites of Nanocellulose and Carbon Quantum Dots

Carbon quantum dots (CQDs) are quasispherical carbon nanoparticles (less than 10 nm in diameter) with a chemical structure and physical properties similar to those of graphene oxide. These nanoparticles emit a strong wavelength-dependent fluorescence. By changing the CQD size, the color of the emitted light can be tuned from deep ultraviolet to visible and near-infrared light. In addition, the fluorescence of CQDs, and also their water solubility, can be further modulated by functionalizing their surface with various atoms, chemical functional groups and molecules, such as metals, carboxyl groups, organic dyes and polymers. CQDs present good photostability, low photobleaching and relatively low cytotoxicity, and they are therefore considered to be suitable for biomedical applications such as bioimaging, biosensing, photodynamic and photothermal therapy of cancer, and drug delivery [165].

In hybrid materials with nanocellulose, CQDs were applied for constructing biosensors and drug delivery systems, and also for water purification. An optical sensor for visual discrimination of biothiols was based on a bacterial cellulose nanopaper substrate with ratiometric fluorescent sensing elements. These elements included N-acetyl l-cysteine capped green cadmium telluride (CdTe) quantum dots-rhodamine B and red CdTe quantum dots-carbon dots [26]. Hybrid materials containing carbon quantum dots and cellulose are also promising carriers for drug delivery. Composite core/shell chitosan-poly (ethylene oxide)-carbon quantum dots/carboxymethyl cellulose-poly(vinyl alcohol) nanofibers were prepared through coaxial electrospinning as a biodegradable implant for local delivery of temozolomide (TMZ), an anticancer drug. When tested in vitro, the antitumor activity of TMZ conjugated with carbon quantum dots against the tumor U251 cell lines was higher than the activity of the free drug [28]. Last but not least, carbon quantum dots, homogeneously dispersed together with magnetic Fe_3O_4 nanoparticles in electrospun cellulose nanofibers, were promising for the removal of Hg(II) ions from water [27].

6.3. Composites of Nanocellulose and Activated Carbon

Activated carbon is a form of carbon processed to have small, low-volume pores that increase the surface area, which is then available for the adsorption and removal of various toxic contaminants and microorganisms. Composite membranes consisting of a bilayer of porous activated carbon and TEMPO-oxidized plant-derived CNFs showed high capability for removing *Escherichia coli* from water [29]. Activated carbon was also a component of a wound dressing material consisting of a polyvinyl alcohol and cellulose acetate phthalate polymeric composite film, reinforced with Cu/Zn bimetal-dispersed activated carbon micro/nanofibers. This material suppressed the growth of *Pseudomonas aeruginosa*, the most prevalent bacteria in infected wounds caused by burns, surgery and traumatic injuries [30].

6.4. Composites of Nanocellulose and Carbon Black

Carbon black is a form of paracrystalline carbon, produced industrially by partial combustion or thermal decomposition of gaseous or liquid hydrocarbons under controlled conditions. Carbon black has a high surface-area-to-volume ratio, though not so high as that of activated carbon. Although it is considered to have low toxicity, the International Agency for Research on Cancer has classified it as possibly carcinogenic to humans. In addition, as a component of environmental pollution, carbon black can cause oxidative damage and an inflammatory reaction, which further mediate genotoxicity, reproductive toxicity, neurotoxicity and diseases of the respiratory and cardiovascular systems [166,167]. Nevertheless, carbon black is currently used as a filler in tires and in other rubber products, and as a pigment in inks, paints and plastics.

Composites of nanocellulose and carbon black have been used mainly for constructing biosensors, particularly wearable sensors for strain and human body motion, e.g., motion of the fingers, the elbow joint and the throat. A strain-sensing device with excellent waterproof, self-cleaning and anticorrosion properties was based on a superhydrophobic electrically conductive paper. This paper fabricated by dip-coating a printing paper into a carbon black/carbon nanotube/methyl cellulose suspension and into a hydrophobic fumed silica suspension [33]. Another strain-sensing device was fabricated by printing carbon black conductive nanostructures on cellulose acetate paper. At the same time, this material had electrochemical properties promising for the detection of hydrogen peroxide [31]. An electrochemical aptasensor for detecting *Staphylococcus aureus*, e.g., in human blood serum, was designed as a nanocomposite of Au nanoparticles, carbon black nanoparticles and cellulose nanofibers, and was endowed with a thiolated specific *S. aureus* aptamer as a sensing element [32].

7. Potential Cytotoxicity and Immunogenicity of Nanocellulose/Nanocarbon Composites

The vast majority of studies dealing with potential biomedical applications of nanocellulose/nanocarbon composites have reported no cytotoxicity or negligible cytotoxicity of these composites, namely of nanocellulose/fullerene composites [35], nanocellulose/graphene composites [68,69,72,87,107], nanocellulose/CNT composites [51,149] and nanocellulose/nanodiamond composites [58,62,97]. Composites containing other carbon nanoparticles, such as activated carbon nanoparticles [30] or carbon quantum dots [28] have also shown no significant cytotoxic effects. In the mentioned studies, cytotoxicity was mainly tested in vitro on various cell types, such as fibroblasts, epithelial and endothelial cells, and mainly on cell lines, including tumor cell lines. The cell viability and proliferation were usually evaluated by tests of the activity of mitochondrial enzymes, such as MTT assay, resazurin (Alamar Blue) assay, or by a direct microscopic examination of the cells. Some composites have also been tested in vivo, e.g., in a rat model (nanocellulose/CNT composites; [106]), a canine model (nanocellulose/CNT composites; [52]), or using a chick chorioallantoic membrane model (nanocellulose/graphene composites, [107]), without adverse effects.

However, the individual components of nanocellulose/nanocarbon composites, particularly carbon nanoparticles, can act as cytotoxic, if they are not bound to any matrix and are free to move. Graphene and graphene-based carbon nanomaterials, such as fullerenes and nanotubes, are hydrophobic in their pristine state, and can enter into hydrophobic interactions with cholesterol in the cell membrane, which can be extracted from the membrane. In this manner, carbon nanoparticles can damage cells even without penetrating them. Another mechanism of cell membrane damage is the generation of reactive oxygen species (ROS) by carbon nanoparticles. In addition, the nanoparticles can penetrate the cell membrane, and can cause oxidative damage to mitochondria, and can also enter the cell nucleus and act as genotoxic agents (for a review, see [10,12,19,20]). Nanodiamonds have been considered to be relatively nontoxic in comparison with other carbon nanoparticles. However, as shown in our earlier studies, hydrophobic, hydrogen-terminated and positively-charged diamond nanoparticles can enter the cells, impair their growth and cause cell death [152,156]. The mechanism of cell damage by nanodiamonds is by generating ROS, and by excessive delivery of sodium ions adsorbed on the nanodiamond surface [168]. Last but not least, carbon nanoparticles can be immunogenic, i.e., they can activate inflammatory reactions, which can be, as has been demonstrated on carbon black, the main pathogenic mechanism of respiratory, cardiovascular and other serious diseases [166,167].

Cellulose nanoparticles, which are generally considered to be biocompatible [34,114] and of a low ecological toxicity [169], can also act as cytotoxic and immunogenic. It has even been speculated that, due to their high aspect ratio and stiffness, CNCs may cause similar pulmonary toxicity as carbon nanotubes and asbestos [170]. In a mouse model, cellulose nanocrystals induced oxidative stress, caused pulmonary inflammation and damage, increased levels of collagen and transforming growth factor-beta (TGF-β) in lungs, and impaired pulmonary functions [170]. In addition, these effects were markedly more pronounced in female mice than in male mice. The immunogenicity of CNCs was also proven in vitro. CNCs and their cationic derivatives CNC-aminoethylmethacrylate and CNC-aminoethylmethacrylamide evoked an inflammatory response in mouse macrophage J774A.1 cells and in peripheral blood mononuclear cells by increasing the level of ROS in mitochondria, the release of ATP from mitochondria and by stimulating the secretion of interleukin-1beta (IL-1β) [171]. The cytotoxicity and immunogenicity of CNCs depend on the preparation conditions and are increased under harsh and caustic conditions, e.g., the so-called mercerization process, i.e., an alkali treatment [149]. CNFs can also cause cytotoxicity and oxidative damage, which can be even more pronounced than in the case of CNCs, and can evoke an inflammatory response (for a review, see [2,172]). The potential cytotoxicity and immunogenicity of nanocellulose, nanocarbon and their composites should therefore be taken into account when they are for use in biomedical applications.

8. Conclusions

Nanocellulose/nanocarbon composites and other hybrid materials containing cellulose nanoparticles (nanofibrils or nanocrystals) and carbon nanoparticles (fullerenes, graphene, carbon nanotubes, nanodiamonds and other carbon nanoparticles) are novel materials that are promising for a wide range of applications in industry, (bio)technology and medicine. This is due to their unique properties, such as high mechanical strength coupled with flexibility and stretchability (composites with graphene, carbon nanotubes and nanodiamond), shape memory (composites with graphene and carbon nanotubes), photodynamic and photothermal activity (composites with fullerenes and graphene), electrical conductivity (composites with graphene and carbon nanotubes), semiconductivity (composites with boron-doped diamond), thermal conductivity (composites with graphene and nanodiamonds), tunable optical transparency (composites with single-walled carbon nanotubes and nanodiamonds), intrinsic fluorescence and luminescence (composites with graphene quantum dots and carbon quantum dots), and high adsorption and filtration capacity (composites with graphene, carbon nanotubes and carbon quantum dots). These properties arise mainly from the advantageous combination of nanocellulose and nanocarbon, which associates and enhances the desirable effects of each of these components. These materials can be prepared relatively easily from a water-based suspension, which is advantageous particularly for biomedical applications. These applications include drug delivery, biosensorics, isolation of various biomolecules, electrical stimulation of damaged tissues, and particularly tissue engineering (bone, neural and vascular) and wound dressing. Our results have proven supportive effects of nanocellulose/carbon nanotube composites on the adhesion and growth of human and porcine adipose tissue-derived stem cells, particularly under dynamic cultivation in a pressure-generating lab-made bioreactor (see Appendix A). However, it should be pointed out that the biomedical applications of nanocellulose/nanocarbon composites are associated with the risk of their potential cytotoxicity and immunogenicity, although this risk appears to be lower than for the single components of these materials.

Author Contributions: All authors have read and agree to the published version of the manuscript. Conceptualization, L.B., S.S. and A.S.; methodology, J.P., M.T., R.M., J.S., S.P., A.S., S.S.; software, R.M., J.S., A.B.; validation, L.B., R.M.; formal analysis, L.B., A.B., A.S., S.S.; investigation, J.P., M.T., R.M., J.S., S.P., A.S., S.S.; resources, L.B., P.K.; data curation, J.P., R.M., A.B., A.S.; writing—original draft preparation, L.B., J.P., R.M.; writing—review and editing, A.B., A.S., S.S., P.K.; visualization, J.P., M.T., R.M., A.B.; supervision, L.B., P.K.; project administration, L.B.; funding acquisition, L.B.

Funding: This research was funded by the Czech Science Foundation (grant no. 17-00885S). Another support was provided by the BIOCEV – Biotechnology and Biomedicine Centre of the Academy of Sciences and Charles University project (CZ.1.05/1.1.00/02.0109), funded by the European Regional Development Fund.

Acknowledgments: Robin Healey (Czech Technical University, Prague) is gratefully acknowledged for his language revision of the manuscript. Panu Lahtinen from VTT (Technical Research Center of Finland, Espoo, Finland) is acknowledged for providing nanocellulose for collaborative work between Tampere University of Technology and Institute of Physiology of the Czech Academy of Sciences.

Conflicts of Interest: The authors declare no conflict of interest. The funding sponsors had no role in the design of the study; in the collection, analyses, or interpretation of data; in the writing of the manuscript, and in the decision to publish the results.

Appendix A

In our own experiments, we have contributed to the knowledge on potential biomedical applications of nanocellulose/carbon nanotube composites, namely regarding their application in tissue engineering. In these experiments, a PurCotton® cellulose mesh (Winner Industrial Park, Shenzhen, China) was modified with an aqueous dispersion of positively-charged (i.e., cationic) wood-derived CNFs, described in our earlier review article [2], and MWCNTs. Two types of composite samples were prepared, namely the samples with "thick" and "thin" coating of the fibers in the cellulose mesh. For thick coating, square-shaped samples of the cellulose mesh were fully immersed in an aqueous suspension of CNFs+MWCNTs, and during this immersion, the samples were homogeneously impregnated with the nanoparticles. For thin coating, only one corner of the cellulose samples was submerged into the CNF+MWCNT suspension, which resulted in infiltration of the suspension throughout the cellulose mesh by capillary forces. Both types of samples were then dried for 5 h at 60 °C. Both types of samples displayed a grayish color, which was more intense in samples with a thick coating. A pure CNF coating was prepared similarly to the thick coatings, and the meshes without coating were used as a control.

The samples were sterilized by UV light (20 min for each side), fixed into CellCrowns™ inserts (Scaffdex Ltd., Tampere, Finland), placed into 24-well cell culture plates (TPP, Trasadingen, Switzerland) and seeded with human adipose tissue-derived stem cells (hASC) or porcine adipose tissue-derived stem cells (pASC). Human ASC were isolated from subcutaneous fat tissue, obtained by liposuction from the abdominal region of healthy female donors after their informed consent, under ethical approval issued by the Ethics Committee of Hospital "Na Bulovce" in Prague, and in compliance with the tenets of the Declaration of Helsinki on experiments involving human tissues. The isolation was described in more details in our earlier studies [173,174]. Porcine ASC were isolated by enzymatic disintegration of subcutaneous fat tissue samples obtained by excision from laboratory pigs in collaboration with the Institute of Clinical and Experimental Medicine (IKEM) in Prague, Czech Republic. Characterization of cells by flow cytometry revealed the positivity of cells for standard surface markers of ASCs, namely CD105, CD90, CD73 and CD29 in hASC, and CD105, CD90, CD29 and CD44 in pASCs. The ASCs from both species were negative or almost negative for hematopoietic markers, such as CD34 and CD45, and for CD31, an endothelial marker [174,175].

The cells were seeded on the material samples at a density of 50,000 cells per well into 1.5 mL of the culture media. Human ASCs were cultivated in Dulbecco's modified Eagle's Medium (DMEM; Life Technologies, Gibco, Carlsbad, CA, USA) with 10% of fetal bovine serum (FBS; Life Technologies, Gibco), 40 µg/mL of gentamicin (LEK, Ljubljana, Slovenia) and 10 ng/mL of recombinant human fibroblast growth factor basic (FGF2; GenScript, Piscataway, NJ, USA). Porcine ASCs were cultivated in Dulbecco's modified Eagle's Medium (Low glucose, Sigma-Aldrich Co., St. Louis, MO, USA) and Ham's Nutrient Mixture F12 medium (DMEM/F 12, Sigma-Aldrich Co.) in a ratio of 1:1 with 10% of fetal bovine serum (FBS; Life Technologies, Gibco), 1% Antibiotic Antimycotic solution (Sigma-Aldrich Co.) and 10 ng/mL of recombinant human fibroblast growth factor basic (FGF2; GenScript).

After one or seven days of cultivation, the cells were fixed with 4% paraformaldehyde (Sigma-Aldrich Co.) for 20 min, and 0.1% Triton X-100 (Sigma-Aldrich Co.) diluted in phosphate-buffered saline (PBS) was applied for 20 min at room temperature in order to permeabilize the cell membranes. Nonspecific binding sites for antibodies were then blocked by a solution of 1% bovine serum albumin and 0.1% Tween 20 in PBS (all Sigma-Aldrich Co.). Vinculin, a protein of focal adhesion plaques associated with integrin adhesion receptors, was visualized by treating the samples for 1 h at 37 °C with primary antibody against human vinculin (V9131, monoclonal mouse antibody, clone hVIN-1, Sigma-Aldrich Co.), diluted in the blocking solution (1% albumin and 0.1% Tween 20 in PBS) in a ratio of 1:200. After washing with PBS, the samples were incubated for 1 h at room temperature in the dark with a secondary antibody, i.e., goat anti-mouse F(ab')2 fragments of IgG (H + L), conjugated with Alexa Fluor® 488 (A11017; Molecular Probes, Eugene, OR, USA; Thermo Fisher Scientific, Waltham, MA, USA), diluted in PBS to a ratio of 1:400. Finally, cytoskeletal F-actin filaments were stained by phalloidin conjugated with tetramethylrhodamine isothiocynate (TRITC) fluorescent dye (Sigma-Aldrich Co.), diluted in PBS to a final concentration of 5 µg/mL, for 1 h at room temperature in the dark. Microscopy images were acquired using spinning disk confocal system Dragonfly 503 (Andor, Belfast, UK) with Zyla 4.2 PLUS sCMOS camera (Andor, Belfast, UK) mounted on microscope Leica DMi8 (Leica Microsystems, Wetzlar, Germany) with objective HC PL APO 20x/0.75 IMM CORR CS2; Free Working Distance = 0.66 mm or HC PL APO 40x/1.10 W CORR CS2; Free Working Distance = 0.65 mm.

We found that the initial adhesion and subsequent growth of cells, evaluated by the cell number and spreading on days 1 and 7, were similar on all coated samples. There was no apparent difference between samples coated with thick and thin layers of CNFs + MWCNTs and samples coated only with CNFs. However, all types of coatings markedly improved the adhesion and growth of cells in comparison with a pure uncoated cellulose mesh (Figure A1). In general, the cell growth was relatively slow in all tested samples. On day 7, the cell number in all tested samples was only slightly higher than on day 1. In addition, hASCs grew slightly better than pASCs, particularly on samples coated with CNFs+MWCNTs. Therefore, we decided to cultivate pASCs on the tested samples under dynamic conditions, which are known to improve the growth of cells by their mechanical stimulation, by a better supply of nutrients and oxygen and by quicker waste removal.

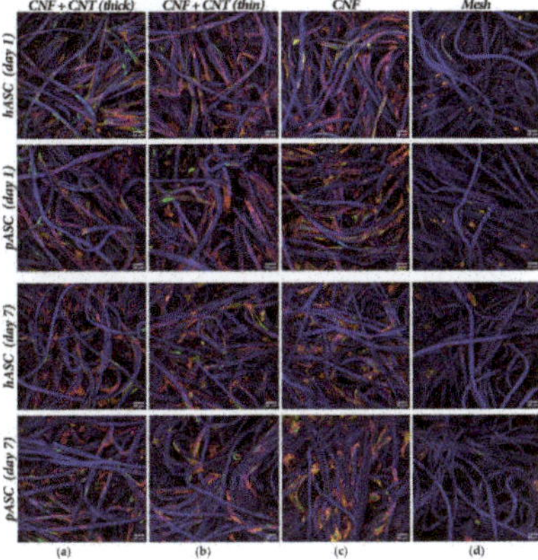

Figure A1. Human adipose tissue-derived stem cells (hASC) and porcine adipose tissue-derived stem cells (pASC) on days 1 and 7 after seeding on a cellulose mesh with thick or thin CNF+MWCNT coating (column (**a**) and (**b**), respectively), with CNF coating (column (**c**)), and without any coating (column (**d**)). Cells were stained by immunofluorescence for vinculin (green), with TRITC-conjugated phalloidin for F-actin (red) and with Hoechst #33258 for the nuclei (blue). Cellulose mesh had autofluorescence in the blue channel. Dragonfly 503 spinning disk confocal microscope with a Zyla 4.2 PLUS sCMOS camera, objective HC PL APO 20x/0.75 IMM CORR CS2. Scale bar: 50 μm.

The dynamic cultivation was held in a unique lab-made cultivation chamber (Figure A2). This chamber allows for fixing a standard well plate with tested substrates and its hermetical sealing to maintain the desired pressure. This chamber was connected to a custom pressure stimulator. This stimulator consists of a servo-controlled linear stage with piston pump and special controlling software. This software allows for setting stimulation parameters that include high/low pressure, motor speed, pulsatile frequency and the shape of the pressure wave.

Porcine ASCs were seeded on CellCrown-fixed substrates in 24-well plates at the same number and in the same cultivation medium as mentioned above. Afterwards, the well plate for dynamic cultivation was fixed into the cultivation chamber, and this chamber was sealed and connected to the pressure stimulator. In the initial phase, the cells were left for 24 h without any pressure stimulation in order to allow their adhesion to the materials. The system was opened through a 220-nm filter to atmosphere forced with slow motion of pump piston to equilibrate CO_2 level and pH of medium. After this resting phase, the pressure stimulation was set to 15.9/10.6 kPa (120/80 mmHg) high/low pressure with frequency of 1 Hz (60 pulses per minute) with triangular pulse shape with 1:1 ratio. This dynamic cultivation lasted for 72 h (96 h of cultivation in total including the 24-h rest phase). Static control samples were cultivated for 96 h in a well plate with standard lid in the same CO_2 incubator as the dynamic samples.

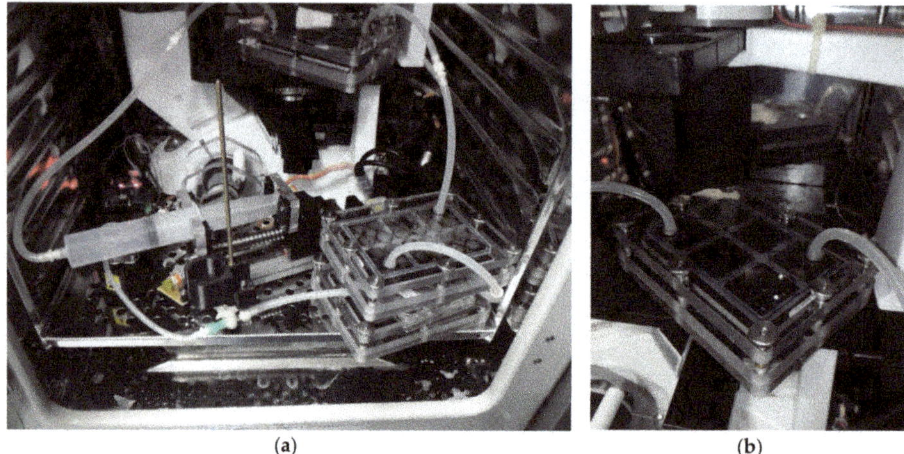

(a) (b)

Figure A2. Lab-made dynamic cultivation system for pressure stimulation of cells on the tested material samples. The whole system in a cell incubator (**a**); detail of a cultivation chamber (**b**).

We found that dynamic cell cultivation markedly improved the adhesion and subsequent growth of pASCs. The cells were better spread and their number after three days of dynamic cultivation (day 4 after seeding) was markedly higher than in the corresponding samples incubated under static conditions for four days (Figure A3), and also for seven days (Figure A1). The improvement in cell colonization by dynamic cultivation was observed particularly on samples with thin CNF + MWCNT coating. On both thick and thin CNF + MWCNT coatings, the cells under dynamic conditions were distributed almost homogeneously, while on the pure CNF coating, the cells tended to form clusters. A similar picture was observed in our earlier study performed on human dermal fibroblasts in four-day-old static cultures on the same cationic CNFs, where the cells were less widespread and distributed less homogeneously than on anionic CNFs [2]. Therefore, it can be concluded that the addition of MWCNTs to cationic CNFs improved the colonization of the material with pASCs under dynamic cell culture conditions.

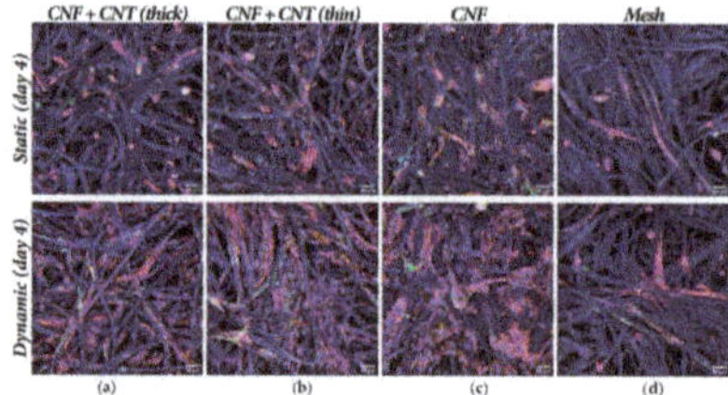

Figure A3. Porcine adipose tissue-derived stem cells (pASC) cultivated in a conventional static cell culture system for four days or in a pressure-generating dynamic cell culture system for three days (after one day of static cultivation). The cells were grown on a cellulose mesh with thick or thin CNF + MWCNT coating (column (**a**) and (**b**), respectively), with CNF coating (column (**c**)), and without any coating (column (**d**)). Cells were stained by immunofluorescence for vinculin (green), with TRITC-conjugated phalloidin for F-actin (red) and with DAPI for the nuclei (blue). Cellulose mesh had autofluorescence in the blue channel. Dragonfly 503 spinning disk confocal microscope with a Zyla 4.2 PLUS sCMOS camera, objective HC PL APO 20x/0.75 IMM CORR CS2. Scale bar: 50 µm.

References

1. Zhang, H.; Dou, C.; Pal, L.; Hubbe, M.A. Review of Electrically Conductive Composites and Films Containing Cellulosic Fibers or Nanocellulose. *Bioresources* **2019**, 14.

2. Bacakova, L.; Pajorova, J.; Bacakova, M.; Skogberg, A.; Kallio, P.; Kolarova, K.; Svorcik, V. Versatile Application of Nanocellulose: From Industry to Skin Tissue Engineering and Wound Healing. *Nanomaterials (Basel)* **2019**, *9*, 164. [CrossRef] [PubMed]

3. Zhang, Y.X.; Nypelo, T.; Salas, C.; Arboleda, J.; Hoeger, I.C.; Rojas, O.J. Cellulose Nanofibrils: From Strong Materials to Bioactive Surfaces. *J Renew Mater* **2013**, *1*, 195–211. [CrossRef]

4. Lin, N.; Dufresne, A. Nanocellulose in biomedicine: Current status and future prospect. *Eur Polym J* **2014**, *59*, 302–325. [CrossRef]

5. Bhattacharya, M.; Malinen, M.M.; Lauren, P.; Lou, Y.R.; Kuisma, S.W.; Kanninen, L.; Lille, M.; Corlu, A.; GuGuen-Guillouzo, C.; Ikkala, O.; et al. Nanofibrillar cellulose hydrogel promotes three-dimensional liver cell culture. *J Control Release* **2012**, *164*, 291–298. [CrossRef]

6. Lou, Y.R.; Kanninen, L.; Kuisma, T.; Niklander, J.; Noon, L.A.; Burks, D.; Urtti, A.; Yliperttula, M. The Use of Nanofibrillar Cellulose Hydrogel As a Flexible Three-Dimensional Model to Culture Human Pluripotent Stem Cells. *Stem Cells Dev* **2014**, *23*, 380–392. [CrossRef]

7. Julkapli, N.M.; Bagheri, S. Nanocellulose as a green and sustainable emerging material in energy applications: a review. *Polym Advan Technol* **2017**, *28*, 1583–1594. [CrossRef]

8. *Nanotechnologies — Standard terms and their definition for cellulose nanomaterial. ISO/TS 20477:2017(E)*, 1st ed.; ISO/TS 20477:2017; ISO: Vernier, Switzerland; Geneva, Switzerland, 2017.

9. Habibi, Y.; Lucia, L.A.; Rojas, O.J. Cellulose Nanocrystals: Chemistry, Self-Assembly, and Applications. *Chem Rev* **2010**, *110*, 3479–3500. [CrossRef] [PubMed]

10. Bacakova, L.; Grausova, L.; Vandrovcova, M.; Vacik, J.; Frazcek, A.; Blazewicz, S.; Kromka, A.; Vanecek, M.; Nesladek, M.; Svorcik, V.; et al. Carbon nanoparticles as substrates for cell adhesion and growth. In *Nanoparticles: New Research*; Lombardi, S.L., Ed.; Nova Science Publishers, Inc.: Hauppauge, NY, USA, 2008; pp. 39–107. ISBN 978-1-60456-704-5.

11. Bacakova, L.; Grausova, L.; Vacik, J.; Kromka, A.; Biederman, H.; Choukourov, A.; Stary, V. Nanocomposite and nanostructured carbon-based films as growth substrates for bone cells. In *Advances in Diverse Industrial Applications of Nanocomposites*; Reddy, B., Ed.; IntechOpen: London, UK, 2011; pp. 399–435. ISBN 978-953-307-202-9.

12. Bacakova, L.; Kopova, I.; Vacik, J.; Lavrentiev, V. Interaction of fullerenes and metal-fullerene composites with cells. In *Fullerenes: Chemistry, Natural Sources and Technological Applications*; Ellis, S.B., Ed.; Nova Science Publishers, Inc.: Hauppauge, NY, USA, 2014; pp. 1–33.

13. Bacakova, L.; Kopova, I.; Stankova, L.; Liskova, J.; Vacik, J.; Lavrentiev, V.; Kromka, A.; Potocky, S.; Stranska, D. Bone cells in cultures on nanocarbon-based materials for potential bone tissue engineering: A review. *Phys Status Solidi A* **2014**, *211*, 2688–2702. [CrossRef]

14. Bacakova, L.; Filova, E.; Liskova, J.; Kopova, I.; Vandrovcova, M.; Havlikova, J. Nanostructured materials as substrates for the adhesion, growth, and osteogenic differentiation of bone cells. *Appl Nanobiomater* **2016**, *4*, 103–153. [CrossRef]

15. Bacakova, L.; Broz, A.; Liskova, J.; Stankova, L.; Potocky, S.; Kromka, A. Application of nanodiamond in biotechnology and tissue engineering. In *Diamond and Carbon Composites and Nanocomposites*; Aliofkhazraei, M., Ed.; IntechOpen: London, UK, 2016; pp. 59–88. ISBN 978-953-51-2453-5. [CrossRef]

16. Wong, B.S.; Yoong, S.L.; Jagusiak, A.; Panczyk, T.; Ho, H.K.; Ang, W.H.; Pastorin, G. Carbon nanotubes for delivery of small molecule drugs. *Adv Drug Deliv Rev* **2013**, *65*, 1964–2015. [CrossRef] [PubMed]

17. Wang, M.; Anoshkin, I.V.; Nasibulin, A.G.; Korhonen, J.T.; Seitsonen, J.; Pere, J.; Kauppinen, E.I.; Ras, R.H.; Ikkala, O. Modifying native nanocellulose aerogels with carbon nanotubes for mechanoresponsive conductivity and pressure sensing. *Adv Mater* **2013**, *25*, 2428–2432. [CrossRef] [PubMed]

18. Yin, R.; Agrawal, T.; Khan, U.; Gupta, G.K.; Rai, V.; Huang, Y.Y.; Hamblin, M.R. Antimicrobial photodynamic inactivation in nanomedicine: small light strides against bad bugs. *Nanomedicine (Lond)* **2015**, *10*, 2379–2404. [CrossRef] [PubMed]

19. Liao, C.; Li, Y.; Tjong, S.C. Graphene Nanomaterials: Synthesis, Biocompatibility, and Cytotoxicity. *Int J Mol Sci* **2018**, *19*, 3564. [CrossRef]

20. Placha, D.; Jampilek, J. Graphenic Materials for Biomedical Applications. *Nanomaterials (Basel)* **2019**, *9*, 1758. [CrossRef]
21. Ma, L.; Liu, R.; Niu, H.; Xing, L.; Liu, L.; Huang, Y. Flexible and Freestanding Supercapacitor Electrodes Based on Nitrogen-Doped Carbon Networks/Graphene/Bacterial Cellulose with Ultrahigh Areal Capacitance. *ACS Appl Mater Interfaces* **2016**, *8*, 33608–33618. [CrossRef]
22. Xu, Z.; Zhou, H.; Tan, S.; Jiang, X.; Wu, W.; Shi, J.; Chen, P. Ultralight super-hydrophobic carbon aerogels based on cellulose nanofibers/poly(vinyl alcohol)/graphene oxide (CNFs/PVA/GO) for highly effective oil-water separation. *Beilstein J Nanotechnol* **2018**, *9*, 508–519. [CrossRef]
23. Li, W.; Islam, N.; Ren, G.; Li, S.; Fan, Z. AC-Filtering Supercapacitors Based on Edge Oriented Vertical Graphene and Cross-Linked Carbon Nanofiber. *Materials (Basel)* **2019**, *12*, 604. [CrossRef]
24. Shi, Q.; Liu, D.; Wang, Y.; Zhao, Y.; Yang, X.; Huang, J. High-Performance Sodium-Ion Battery Anode via Rapid Microwave Carbonization of Natural Cellulose Nanofibers with Graphene Initiator. *Small* **2019**, *15*, e1901724. [CrossRef]
25. Price, R.L.; Ellison, K.; Haberstroh, K.M.; Webster, T.J. Nanometer surface roughness increases select osteoblast adhesion on carbon nanofiber compacts. *J Biomed Mater Res A* **2004**, *70*, 129–138. [CrossRef]
26. Abbasi-Moayed, S.; Golmohammadi, H.; Bigdeli, A.; Hormozi-Nezhad, M.R. A rainbow ratiometric fluorescent sensor array on bacterial nanocellulose for visual discrimination of biothiols. *Analyst* **2018**, *143*, 3415–3424. [CrossRef]
27. Li, L.; Wang, F.; Lv, Y.; Liu, J.; Bian, H.; Wang, W.; Li, Y.; Shao, Z. CQDs-Doped Magnetic Electrospun Nanofibers: Fluorescence Self-Display and Adsorption Removal of Mercury(II). *ACS Omega* **2018**, *3*, 4220–4230. [CrossRef] [PubMed]
28. Shamsipour, M.; Mansouri, A.M.; Moradipour, P. Temozolomide Conjugated Carbon Quantum Dots Embedded in Core/Shell Nanofibers Prepared by Coaxial Electrospinning as an Implantable Delivery System for Cell Imaging and Sustained Drug Release. *AAPS PharmSciTech* **2019**, *20*, 259. [CrossRef] [PubMed]
29. Hassan, M.; Abou-Zeid, R.; Hassan, E.; Berglund, L.; Aitomaki, Y.; Oksman, K. Membranes Based on Cellulose Nanofibers and Activated Carbon for Removal of Escherichia coli Bacteria from Water. *Polymers (Basel)* **2017**, *9*, 335. [CrossRef] [PubMed]
30. Ashfaq, M.; Verma, N.; Khan, S. Highly effective Cu/Zn-carbon micro/nanofiber-polymer nanocomposite-based wound dressing biomaterial against the P. aeruginosa multi- and extensively drug-resistant strains. *Mater Sci Eng C Mater Biol Appl* **2017**, *77*, 630–641. [CrossRef]
31. Santhiago, M.; Correa, C.C.; Bernardes, J.S.; Pereira, M.P.; Oliveira, L.J.M.; Strauss, M.; Bufon, C.C.B. Flexible and Foldable Fully-Printed Carbon Black Conductive Nanostructures on Paper for High-Performance Electronic, Electrochemical, and Wearable Devices. *ACS Appl Mater Interfaces* **2017**, *9*, 24365–24372. [CrossRef]
32. Ranjbar, S.; Shahrokhian, S. Design and fabrication of an electrochemical aptasensor using Au nanoparticles/carbon nanoparticles/cellulose nanofibers nanocomposite for rapid and sensitive detection of Staphylococcus aureus. *Bioelectrochemistry* **2018**, *123*, 70–76. [CrossRef]
33. Li, Q.M.; Liu, H.; Zhang, S.D.; Zhang, D.B.; Liu, X.H.; He, Y.X.; Mi, L.W.; Zhang, J.X.; Liu, C.T.; Shen, C.Y.; et al. Superhydrophobic Electrically Conductive Paper for Ultrasensitive Strain Sensor with Excellent Anticorrosion and Self-Cleaning Property. *Acs Appl Mater Inter* **2019**, *11*, 21904–21914. [CrossRef]
34. Lin, J.; Zhong, Z.; Li, Q.; Tan, Z.; Lin, T.; Quan, Y.; Zhang, D. Facile Low-Temperature Synthesis of Cellulose Nanocrystals Carrying Buckminsterfullerene and Its Radical Scavenging Property in Vitro. *Biomacromolecules* **2017**, *18*, 4034–4040. [CrossRef]
35. Herreros-Lopez, A.; Carini, M.; Da Ros, T.; Carofiglio, T.; Marega, C.; La Parola, V.; Rapozzi, V.; Xodo, L.E.; Alshatwi, A.A.; Hadad, C.; et al. Nanocrystalline cellulose-fullerene: Novel conjugates. *Carbohydr Polym* **2017**, *164*, 92–101. [CrossRef]
36. Kim, Y.; Kim, H.S.; Yun, Y.S.; Bak, H.; Jin, H.J. Ag-doped multiwalled carbon nanotube/polymer composite electrodes. *J Nanosci Nanotechnol* **2010**, *10*, 3571–3575. [CrossRef] [PubMed]
37. Jin, L.; Zeng, Z.; Kuddannaya, S.; Wu, D.; Zhang, Y.; Wang, Z. Biocompatible, Free-Standing Film Composed of Bacterial Cellulose Nanofibers-Graphene Composite. *ACS Appl Mater Interfaces* **2016**, *8*, 1011–1018. [CrossRef] [PubMed]
38. Kim, S.; Xiong, R.; Tsukruk, V.V. Probing Flexural Properties of Cellulose Nanocrystal-Graphene Nanomembranes with Force Spectroscopy and Bulging Test. *Langmuir* **2016**, *32*, 5383–5393. [CrossRef] [PubMed]

39. Siljander, S.; Keinanen, P.; Raty, A.; Ramakrishnan, K.R.; Tuukkanen, S.; Kunnari, V.; Harlin, A.; Vuorinen, J.; Kanerva, M. Effect of Surfactant Type and Sonication Energy on the Electrical Conductivity Properties of Nanocellulose-CNT Nanocomposite Films. *Int J Mol Sci* **2018**, *19*, 1819. [CrossRef] [PubMed]

40. Han, J.; Wang, S.; Zhu, S.; Huang, C.; Yue, Y.; Mei, C.; Xu, X.; Xia, C. Electrospun Core-Shell Nanofibrous Membranes with Nanocellulose-Stabilized Carbon Nanotubes for Use as High-Performance Flexible Supercapacitor Electrodes with Enhanced Water Resistance, Thermal Stability, and Mechanical Toughness. *ACS Appl Mater Interfaces* **2019**, *11*, 44624–44635. [CrossRef]

41. Nguyen, H.K.; Bae, J.; Hur, J.; Park, S.J.; Park, M.S.; Kim, I.T. Tailoring of Aqueous-Based Carbon Nanotube(-)Nanocellulose Films as Self-Standing Flexible Anodes for Lithium-Ion Storage. *Nanomaterials (Basel)* **2019**, *9*, 655. [CrossRef]

42. Jiang, M.; Seney, R.; Bayliss, P.C.; Kitchens, C.L. Carbon Nanotube and Cellulose Nanocrystal Hybrid Films. *Molecules* **2019**, *24*, 2662. [CrossRef]

43. Liu, P.; Zhu, C.; Mathew, A.P. Mechanically robust high flux graphene oxide - nanocellulose membranes for dye removal from water. *J Hazard Mater* **2019**, *371*, 484–493. [CrossRef]

44. Hasan, M.Q.; Yuen, J.; Slaughter, G. Carbon Nanotube-Cellulose Pellicle for Glucose Biofuel Cell. *Conf Proc IEEE Eng Med Biol Soc* **2018**, *2018*, 1–4. [CrossRef]

45. Hamedi, M.M.; Hajian, A.; Fall, A.B.; Hakansson, K.; Salajkova, M.; Lundell, F.; Wagberg, L.; Berglund, L.A. Highly conducting, strong nanocomposites based on nanocellulose-assisted aqueous dispersions of single-wall carbon nanotubes. *ACS Nano* **2014**, *8*, 2467–2476. [CrossRef] [PubMed]

46. Hajian, A.; Lindstrom, S.B.; Pettersson, T.; Hamedi, M.M.; Wagberg, L. Understanding the Dispersive Action of Nanocellulose for Carbon Nanomaterials. *Nano Lett* **2017**, *17*, 1439–1447. [CrossRef] [PubMed]

47. Hamedi, M.; Karabulut, E.; Marais, A.; Herland, A.; Nystrom, G.; Wagberg, L. Nanocellulose aerogels functionalized by rapid layer-by-layer assembly for high charge storage and beyond. *Angew Chem Int Ed Engl* **2013**, *52*, 12038–12042. [CrossRef] [PubMed]

48. Wicklein, B.; Kocjan, A.; Salazar-Alvarez, G.; Carosio, F.; Camino, G.; Antonietti, M.; Bergstrom, L. Thermally insulating and fire-retardant lightweight anisotropic foams based on nanocellulose and graphene oxide. *Nat Nanotechnol* **2015**, *10*, 277–283. [CrossRef] [PubMed]

49. Yousefi, N.; Wong, K.K.W.; Hosseinidoust, Z.; Sorensen, H.O.; Bruns, S.; Zheng, Y.; Tufenkji, N. Hierarchically porous, ultra-strong reduced graphene oxide-cellulose nanocrystal sponges for exceptional adsorption of water contaminants. *Nanoscale* **2018**, *10*, 7171–7184. [CrossRef] [PubMed]

50. Siljander, S.; Keinanen, P.; Ivanova, A.; Lehmonen, J.; Tuukkanen, S.; Kanerva, M.; Bjorkqvist, T. Conductive Cellulose based Foam Formed 3D Shapes-From Innovation to Designed Prototype. *Materials (Basel)* **2019**, *12*, 430. [CrossRef]

51. Kuzmenko, V.; Karabulut, E.; Pernevik, E.; Enoksson, P.; Gatenholm, P. Tailor-made conductive inks from cellulose nanofibrils for 3D printing of neural guidelines. *Carbohydr Polym* **2018**, *189*, 22–30. [CrossRef]

52. Pedrotty, D.M.; Kuzmenko, V.; Karabulut, E.; Sugrue, A.M.; Livia, C.; Vaidya, V.R.; McLeod, C.J.; Asirvatham, S.J.; Gatenholm, P.; Kapa, S. Three-Dimensional Printed Biopatches With Conductive Ink Facilitate Cardiac Conduction When Applied to Disrupted Myocardium. *Circ Arrhythm Electrophysiol* **2019**, *12*, e006920. [CrossRef]

53. Shah, N.; Ul-Islam, M.; Khattak, W.A.; Park, J.K. Overview of bacterial cellulose composites: a multipurpose advanced material. *Carbohydr Polym* **2013**, *98*, 1585–1598. [CrossRef]

54. Xu, T.; Jiang, Q.; Ghim, D.; Liu, K.K.; Sun, H.; Derami, H.G.; Wang, Z.; Tadepalli, S.; Jun, Y.S.; Zhang, Q.; et al. Catalytically Active Bacterial Nanocellulose-Based Ultrafiltration Membrane. *Small* **2018**, *14*, e1704006. [CrossRef]

55. Jun, Y.S.; Wu, X.; Ghim, D.; Jiang, Q.; Cao, S.; Singamaneni, S. Photothermal Membrane Water Treatment for Two Worlds. *Acc Chem Res* **2019**, *52*, 1215–1225. [CrossRef]

56. Jiang, Q.; Ghim, D.; Cao, S.; Tadepalli, S.; Liu, K.K.; Kwon, H.; Luan, J.; Min, Y.; Jun, Y.S.; Singamaneni, S. Photothermally Active Reduced Graphene Oxide/Bacterial Nanocellulose Composites as Biofouling-Resistant Ultrafiltration Membranes. *Environ Sci Technol* **2019**, *53*, 412–421. [CrossRef] [PubMed]

57. Abol-Fotouh, D.; Dorling, B.; Zapata-Arteaga, O.; Rodriguez-Martinez, X.; Gomez, A.; Reparaz, J.S.; Laromaine, A.; Roig, A.; Campoy-Quiles, M. Farming thermoelectric paper. *Energy Environ Sci* **2019**, *12*, 716–726. [CrossRef] [PubMed]

58. Mahdavi, M.; Mahmoudi, N.; Rezaie Anaran, F.; Simchi, A. Electrospinning of Nanodiamond-Modified Polysaccharide Nanofibers with Physico-Mechanical Properties Close to Natural Skins. *Mar Drugs* **2016**, *14*, 128. [CrossRef] [PubMed]

59. Liu, X.; Shen, H.; Song, S.; Chen, W.; Zhang, Z. Accelerated biomineralization of graphene oxide - incorporated cellulose acetate nanofibrous scaffolds for mesenchymal stem cell osteogenesis. *Colloids Surf B Biointerfaces* **2017**, *159*, 251–258. [CrossRef] [PubMed]

60. Cho, S.Y.; Yu, H.; Choi, J.; Kang, H.; Park, S.; Jang, J.S.; Hong, H.J.; Kim, I.D.; Lee, S.K.; Jeong, H.S.; et al. Continuous Meter-Scale Synthesis of Weavable Tunicate Cellulose/Carbon Nanotube Fibers for High-Performance Wearable Sensors. *ACS Nano* **2019**, *13*, 9332–9341. [CrossRef] [PubMed]

61. Zhu, C.; Liu, P.; Mathew, A.P. Self-Assembled TEMPO Cellulose Nanofibers: Graphene Oxide-Based Biohybrids for Water Purification. *ACS Appl Mater Interfaces* **2017**, *9*, 21048–21058. [CrossRef] [PubMed]

62. Ostadhossein, F.; Mahmoudi, N.; Morales-Cid, G.; Tamjid, E.; Navas-Martos, F.J.; Soriano-Cuadrado, B.; Paniza, J.M.L.; Simchi, A. Development of Chitosan/Bacterial Cellulose Composite Films Containing Nanodiamonds as a Potential Flexible Platform for Wound Dressing. *Materials (Basel)* **2015**, *8*, 6401–6418. [CrossRef]

63. Zheng, C.; Yue, Y.; Gan, L.; Xu, X.; Mei, C.; Han, J. Highly Stretchable and Self-Healing Strain Sensors Based on Nanocellulose-Supported Graphene Dispersed in Electro-Conductive Hydrogels. *Nanomaterials (Basel)* **2019**, *9*, 937. [CrossRef]

64. Xing, J.; Tao, P.; Wu, Z.; Xing, C.; Liao, X.; Nie, S. Nanocellulose-graphene composites: A promising nanomaterial for flexible supercapacitors. *Carbohydr Polym* **2019**, *207*, 447–459. [CrossRef]

65. Shi, Z.; Phillips, G.O.; Yang, G. Nanocellulose electroconductive composites. *Nanoscale* **2013**, *5*, 3194–3201. [CrossRef]

66. Buzid, A.; Hayes, P.E.; Glennon, J.D.; Luong, J.H.T. Captavidin as a regenerable biorecognition element on boron-doped diamond for biotin sensing. *Anal Chim Acta* **2019**, *1059*, 42–48. [CrossRef] [PubMed]

67. Li, F.; Yu, H.Y.; Wang, Y.Y.; Zhou, Y.; Zhang, H.; Yao, J.M.; Abdalkarim, S.Y.H.; Tam, K.C. Natural Biodegradable Poly(3-hydroxybutyrate-co-3-hydroxyvalerate) Nanocomposites with Multifunctional Cellulose Nanocrystals/Graphene Oxide Hybrids for High-Performance Food Packaging. *J Agric Food Chem* **2019**, *67*, 10954–10967. [CrossRef] [PubMed]

68. Pal, N.; Banerjee, S.; Roy, P.; Pal, K. Melt-blending of unmodified and modified cellulose nanocrystals with reduced graphene oxide into PLA matrix for biomedical application. *Polym Advan Technol* **2019**, *30*, 3049–3060. [CrossRef]

69. Pal, N.; Banerjee, S.; Roy, P.; Pal, K. Reduced graphene oxide and PEG-grafted TEMPO-oxidized cellulose nanocrystal reinforced poly-lactic acid nanocomposite film for biomedical application. *Mater Sci Eng C Mater Biol Appl* **2019**, *104*, 109956. [CrossRef] [PubMed]

70. Song, N.; Cui, S.; Hou, X.; Ding, P.; Shi, L. Significant Enhancement of Thermal Conductivity in Nanofibrillated Cellulose Films with Low Mass Fraction of Nanodiamond. *ACS Appl Mater Interfaces* **2017**, *9*, 40766–40773. [CrossRef]

71. Ruiz-Palomero, C.; Benitez-Martinez, S.; Soriano, M.L.; Valcarcel, M. Fluorescent nanocellulosic hydrogels based on graphene quantum dots for sensing laccase. *Anal Chim Acta* **2017**, *974*, 93–99. [CrossRef] [PubMed]

72. Javanbakht, S.; Namazi, H. Doxorubicin loaded carboxymethyl cellulose/graphene quantum dot nanocomposite hydrogel films as a potential anticancer drug delivery system. *Mater Sci Eng C Mater Biol Appl* **2018**, *87*, 50–59. [CrossRef]

73. Anirudhan, T.S.; Deepa, J.R. Nano-zinc oxide incorporated graphene oxide/nanocellulose composite for the adsorption and photo catalytic degradation of ciprofloxacin hydrochloride from aqueous solutions. *J Colloid Interface Sci* **2017**, *490*, 343–356. [CrossRef]

74. Xu, Z.; Zhou, H.; Jiang, X.; Li, J.; Huang, F. Facile synthesis of reduced graphene oxide/trimethyl chlorosilane-coated cellulose nanofibres aerogel for oil absorption. *IET Nanobiotechnol* **2017**, *11*, 929–934. [CrossRef]

75. Yao, Q.; Fan, B.; Xiong, Y.; Jin, C.; Sun, Q.; Sheng, C. 3D assembly based on 2D structure of Cellulose Nanofibril/Graphene Oxide Hybrid Aerogel for Adsorptive Removal of Antibiotics in Water. *Sci Rep* **2017**, *7*, 45914. [CrossRef]

76. Alizadehgiashi, M.; Khuu, N.; Khabibullin, A.; Henry, A.; Tebbe, M.; Suzuki, T.; Kumacheva, E. Nanocolloidal Hydrogel for Heavy Metal Scavenging. *ACS Nano* **2018**, *12*, 8160–8168. [CrossRef] [PubMed]

77. Liang, Y.; Liu, J.; Wang, L.; Wan, Y.; Shen, J.; Bai, Q. Metal affinity-carboxymethyl cellulose functionalized magnetic graphene composite for highly selective isolation of histidine-rich proteins. *Talanta* **2019**, *195*, 381–389. [CrossRef] [PubMed]

78. Sun, H.X.; Ma, C.; Wang, T.; Xu, Y.Y.; Yuan, B.B.; Li, P.; Kong, Y. Preparation and Characterization of C60-Filled Ethyl Cellulose Mixed-Matrix Membranes for Gas Separation of Propylene/Propane. *Chem Eng Technol* **2014**, *37*, 611–619. [CrossRef]

79. Vetrivel, S.; Saraswathi, M.S.A.; Rana, D.; Nagendran, A. Fabrication of cellulose acetate nanocomposite membranes using 2D layered nanomaterials for macromolecular separation. *Int J Biol Macromol* **2018**, *107*, 1607–1612. [CrossRef] [PubMed]

80. Blomquist, N.; Wells, T.; Andres, B.; Backstrom, J.; Forsberg, S.; Olin, H. Metal-free supercapacitor with aqueous electrolyte and low-cost carbon materials. *Sci Rep* **2017**, *7*, 39836. [CrossRef]

81. Xu, X.; Hsieh, Y.L. Aqueous exfoliated graphene by amphiphilic nanocellulose and its application in moisture-responsive foldable actuators. *Nanoscale* **2019**, *11*, 11719–11729. [CrossRef]

82. Jiang, Q.; Tian, L.; Liu, K.K.; Tadepalli, S.; Raliya, R.; Biswas, P.; Naik, R.R.; Singamaneni, S. Bilayered Biofoam for Highly Efficient Solar Steam Generation. *Adv Mater* **2016**, *28*, 9400–9407. [CrossRef]

83. Li, X.; Shao, C.; Zhuo, B.; Yang, S.; Zhu, Z.; Su, C.; Yuan, Q. The use of nanofibrillated cellulose to fabricate a homogeneous and flexible graphene-based electric heating membrane. *Int J Biol Macromol* **2019**, *139*, 1103–1116. [CrossRef]

84. Kizling, M.; Draminska, S.; Stolarczyk, K.; Tammela, P.; Wang, Z.; Nyholm, L.; Bilewicz, R. Biosupercapacitors for powering oxygen sensing devices. *Bioelectrochemistry* **2015**, *106*, 34–40. [CrossRef]

85. Generalov, A.A.; Anoshkin, I.V.; Erdmanis, M.; Lioubtchenko, D.V.; Ovchinnikov, V.; Nasibulin, A.G.; Raisanen, A.V. Carbon nanotube network varactor. *Nanotechnology* **2015**, *26*, 045201. [CrossRef]

86. Asmat, S.; Husain, Q. Exquisite stability and catalytic performance of immobilized lipase on novel fabricated nanocellulose fused polypyrrole/graphene oxide nanocomposite: Characterization and application. *Int J Biol Macromol* **2018**, *117*, 331–341. [CrossRef] [PubMed]

87. Pal, N.; Dubey, P.; Gopinath, P.; Pal, K. Combined effect of cellulose nanocrystal and reduced graphene oxide into poly-lactic acid matrix nanocomposite as a scaffold and its anti-bacterial activity. *Int J Biol Macromol* **2017**, *95*, 94–105. [CrossRef] [PubMed]

88. Valentini, L.; Cardinali, M.; Fortunati, E.; Kenny, J.M. Nonvolatile memory behavior of nanocrystalline cellulose/graphene oxide composite films. *Appl Phys Lett* **2014**, *105*. [CrossRef]

89. Song, L.; Li, Y.; Xiong, Z.; Pan, L.; Luo, Q.; Xu, X.; Lu, S. Water-Induced shape memory effect of nanocellulose papers from sisal cellulose nanofibers with graphene oxide. *Carbohydr Polym* **2018**, *179*, 110–117. [CrossRef]

90. Wu, G.; Gu, Y.; Hou, X.; Li, R.; Ke, H.; Xiao, X. Hybrid Nanocomposites of Cellulose/Carbon-Nanotubes/Polyurethane with Rapidly Water Sensitive Shape Memory Effect and Strain Sensing Performance. *Polymers (Basel)* **2019**, *11*, 1586. [CrossRef] [PubMed]

91. Zhu, L.; Zhou, X.; Liu, Y.; Fu, Q. Highly Sensitive, Ultrastretchable Strain Sensors Prepared by Pumping Hybrid Fillers of Carbon Nanotubes/Cellulose Nanocrystal into Electrospun Polyurethane Membranes. *ACS Appl Mater Interfaces* **2019**, *11*, 12968–12977. [CrossRef]

92. Awan, F.; Bulger, E.; Berry, R.M.; Tam, K.C. Enhanced radical scavenging activity of polyhydroxylated C-60 functionalized cellulose nanocrystals. *Cellulose* **2016**, *23*, 3589–3599. [CrossRef]

93. Luo, J.; Deng, W.; Yang, F.; Wu, Z.; Huang, M.; Gu, M. Gold nanoparticles decorated graphene oxide/nanocellulose paper for NIR laser-induced photothermal ablation of pathogenic bacteria. *Carbohydr Polym* **2018**, *198*, 206–214. [CrossRef]

94. Anirudhan, T.S.; Sekhar, V.C.; Shainy, F.; Thomas, J.P. Effect of dual stimuli responsive dextran/nanocellulose polyelectrolyte complexes for chemophotothermal synergistic cancer therapy. *International Journal of Biological Macromolecules* **2019**, *135*, 776–789. [CrossRef]

95. Rasoulzadeh, M.; Namazi, H. Carboxymethyl cellulose/graphene oxide bio-nanocomposite hydrogel beads as anticancer drug carrier agent. *Carbohydr Polym* **2017**, *168*, 320–326. [CrossRef]

96. Wang, X.D.; Yu, K.X.; An, R.; Han, L.L.; Zhang, Y.L.; Shi, L.Y.; Ran, R. Self-assembling GO/modified HEC hybrid stabilized pickering emulsions and template polymerization for biomedical hydrogels. *Carbohyd Polym* **2019**, *207*, 694–703. [CrossRef] [PubMed]

97. Luo, X.; Zhang, H.; Cao, Z.; Cai, N.; Xue, Y.; Yu, F. A simple route to develop transparent doxorubicin-loaded nanodiamonds/cellulose nanocomposite membranes as potential wound dressings. *Carbohydr Polym* **2016**, *143*, 231–238. [CrossRef] [PubMed]

98. Anirudhan, T.S.; Deepa, J.R.; Binussreejayan. Electrochemical sensing of cholesterol by molecularly imprinted polymer of silylated graphene oxide and chemically modified nanocellulose polymer. *Mater Sci Eng C Mater Biol Appl* **2018**, *92*, 942–956. [CrossRef]

99. Fu, W.; Dai, Y.; Meng, X.; Xu, W.; Zhou, J.; Liu, Z.; Lu, W.; Wang, S.; Huang, C.; Sun, Y. Electronic textiles based on aligned electrospun belt-like cellulose acetate nanofibers and graphene sheets: portable, scalable and eco-friendly strain sensor. *Nanotechnology* **2019**, *30*, 045602. [CrossRef] [PubMed]

100. Zou, Y.; Zhang, Y.; Xu, Y.; Chen, Y.; Huang, S.; Lyu, Y.; Duan, H.; Chen, Z.; Tan, W. Portable and Label-Free Detection of Blood Bilirubin with Graphene-Isolated-Au-Nanocrystals Paper Strip. *Anal Chem* **2018**, *90*, 13687–13694. [CrossRef] [PubMed]

101. Jia, Y.; Yu, H.; Zhang, Y.; Dong, F.; Li, Z. Cellulose acetate nanofibers coated layer-by-layer with polyethylenimine and graphene oxide on a quartz crystal microbalance for use as a highly sensitive ammonia sensor. *Colloids Surf B Biointerfaces* **2016**, *148*, 263–269. [CrossRef] [PubMed]

102. Xue, T.; Sheng, Y.Y.; Xu, J.K.; Li, Y.Y.; Lu, X.Y.; Zhu, Y.F.; Duan, X.M.; Wen, Y.P. In-situ reduction of Ag+ on black phosphorene and its NH2-MWCNT nanohybrid with high stability and dispersibility as nanozyme sensor for three ATP metabolites. *Biosens Bioelectron* **2019**, *145*. [CrossRef]

103. Jung, M.; Kim, K.; Kim, B.; Lee, K.J.; Kang, J.W.; Jeon, S. Vertically stacked nanocellulose tactile sensor. *Nanoscale* **2017**, *9*, 17212–17219. [CrossRef]

104. Zhu, P.; Liu, Y.; Fang, Z.; Kuang, Y.; Zhang, Y.; Peng, C.; Chen, G. Flexible and Highly Sensitive Humidity Sensor Based on Cellulose Nanofibers and Carbon Nanotube Composite Film. *Langmuir* **2019**, *35*, 4834–4842. [CrossRef]

105. Chen, C.; Zhang, T.; Zhang, Q.; Chen, X.; Zhu, C.; Xu, Y.; Yang, J.; Liu, J.; Sun, D. Biointerface by Cell Growth on Graphene Oxide Doped Bacterial Cellulose/Poly(3,4-ethylenedioxythiophene) Nanofibers. *ACS Appl Mater Interfaces* **2016**, *8*, 10183–10192. [CrossRef]

106. Sun, Y.; Quan, Q.; Meng, H.; Zheng, Y.; Peng, J.; Hu, Y.; Feng, Z.; Sang, X.; Qiao, K.; He, W.; et al. Enhanced Neurite Outgrowth on a Multiblock Conductive Nerve Scaffold with Self-Powered Electrical Stimulation. *Adv Healthc Mater* **2019**, *8*, e1900127. [CrossRef] [PubMed]

107. Chakraborty, S.; Ponrasu, T.; Chandel, S.; Dixit, M.; Muthuvijayan, V. Reduced graphene oxide-loaded nanocomposite scaffolds for enhancing angiogenesis in tissue engineering applications. *R Soc Open Sci* **2018**, *5*, 172017. [CrossRef] [PubMed]

108. Chen, X.Y.; Low, H.R.; Loi, X.Y.; Merel, L.; Mohd Cairul Iqbal, M.A. Fabrication and evaluation of bacterial nanocellulose/poly(acrylic acid)/graphene oxide composite hydrogel: Characterizations and biocompatibility studies for wound dressing. *J Biomed Mater Res B Appl Biomater* **2019**, *107*, 2140–2151. [CrossRef] [PubMed]

109. Wang, Y.; Shi, L.; Wu, H.; Li, Q.; Hu, W.; Zhang, Z.; Huang, L.; Zhang, J.; Chen, D.; Deng, S.; et al. Graphene Oxide-IPDI-Ag/ZnO@Hydroxypropyl Cellulose Nanocomposite Films for Biological Wound-Dressing Applications. *ACS Omega* **2019**, *4*, 15373–15381. [CrossRef] [PubMed]

110. Burrs, S.L.; Bhargava, M.; Sidhu, R.; Kiernan-Lewis, J.; Gomes, C.; Claussen, J.C.; McLamore, E.S. A paper based graphene-nanocauliflower hybrid composite for point of care biosensing. *Biosens Bioelectron* **2016**, *85*, 479–487. [CrossRef] [PubMed]

111. Liu, C.; Dong, J.; Waterhouse, G.I.N.; Cheng, Z.Q.; Ai, S.Y. Electrochemical immunosensor with nanocellulose-Au composite assisted multiple signal amplification for detection of avian leukosis virus subgroup J. *Biosens Bioelectron* **2018**, *101*, 110–115. [CrossRef] [PubMed]

112. Cao, J.; Zhang, X.X.; Wu, X.D.; Wang, S.M.; Lu, C.H. Cellulose nanocrystals mediated assembly of graphene in rubber composites for chemical sensing applications. *Carbohyd Polym* **2016**, *140*, 88–95. [CrossRef]

113. Yan, C.Y.; Wang, J.X.; Kang, W.B.; Cui, M.Q.; Wang, X.; Foo, C.Y.; Chee, K.J.; Lee, P.S. Highly Stretchable Piezoresistive Graphene-Nanocellulose Nanopaper for Strain Sensors. *Advanced Materials* **2014**, *26*, 2022–2027. [CrossRef]

114. Wang, S.; Zhang, X.; Wu, X.; Lu, C. Tailoring percolating conductive networks of natural rubber composites for flexible strain sensors via a cellulose nanocrystal templated assembly. *Soft Matter* **2016**, *12*, 845–852. [CrossRef]

115. Baleizao, C.; Nagl, S.; Schaferling, M.; Berberan-Santos, M.N.; Wolfbeis, O.S. Dual fluorescence sensor for trace oxygen and temperature with unmatched range and sensitivity. *Anal Chem* **2008**, *80*, 6449–6457. [CrossRef]

116. Kochmann, S.; Baleizao, C.; Berberan-Santos, M.N.; Wolfbeis, O.S. Sensing and imaging of oxygen with parts per billion limits of detection and based on the quenching of the delayed fluorescence of (13)C70 fullerene in polymer hosts. *Anal Chem* **2013**, *85*, 1300–1304. [CrossRef] [PubMed]

117. Luo, Y.; Wang, S.; Shen, M.; Qi, R.; Fang, Y.; Guo, R.; Cai, H.; Cao, X.; Tomas, H.; Zhu, M.; et al. Carbon nanotube-incorporated multilayered cellulose acetate nanofibers for tissue engineering applications. *Carbohydr Polym* **2013**, *91*, 419–427. [CrossRef] [PubMed]

118. Duri, S.; Harkins, A.L.; Frazier, A.J.; Tran, C.D. Composites Containing Fullerenes and Polysaccharides: Green and Facile Synthesis, Biocompatibility, and Antimicrobial Activity. *Acs Sustain Chem Eng* **2017**, *5*, 5408–5417. [CrossRef]

119. Kopova, I.; Bacakova, L.; Lavrentiev, V.; Vacik, J. Growth and potential damage of human bone-derived cells on fresh and aged fullerene c60 films. *Int J Mol Sci* **2013**, *14*, 9182–9204. [CrossRef]

120. Kopova, I.; Lavrentiev, V.; Vacik, J.; Bacakova, L. Growth and potential damage of human bone-derived cells cultured on fresh and aged C60/Ti films. *PLoS One* **2015**, *10*, e0123680. [CrossRef]

121. Sheka, E.F. *Chapter 1: Concepts and grounds. In Fullerenes: Nanochemistry, Nanomagnetism, Nanomedicine, Nanophotonics*; Sheka, E.F., Ed.; CRC Press Taylor and Francis Group: Boca Raton, FL, USA, 2011; pp. 1–14. ISBN 9781439806425.

122. Sheka, E.F. Chapter 9: Nanomedicine of fullerene C60. In *Fullerenes: Nanochemistry, Nanomagnetism, Nanomedicine, Nanophotonics*, 1st ed.; Sheka, E.F., Ed.; CRC Press Taylor and Francis Group: Boca Raton, FL, USA, 2011; pp. 175–191. ISBN 9781439806425.

123. Li, J.; Kee, C.D.; Vadahanambi, S.; Oh, I.K. A Novel Biocompatible Actuator based on Electrospun Cellulose Acetate. *Adv Mater Res-Switz* **2011**, *214*, 359. [CrossRef]

124. Alekseeva, O.V.; Bagrovskaya, N.A.; Noskov, A.V. The Sorption Activity of a Cellulose-Fullerene Composite Relative to Heavy Metal Ions. *Prot Met Phys Chem+* **2019**, *55*, 15–20. [CrossRef]

125. Shams, S.S.; Zhang, R.Y.; Zhu, J. Graphene synthesis: a Review. *Mater Sci-Poland* **2015**, *33*, 566–578. [CrossRef]

126. Coros, M.; Pogacean, F.; Magerusan, L.; Socaci, C.; Pruneanu, S. A brief overview on synthesis and applications of graphene and graphene-based nanomaterials. *Front Mater Sci* **2019**, *13*, 23–32. [CrossRef]

127. Malho, J.M.; Laaksonen, P.; Walther, A.; Ikkala, O.; Linder, M.B. Facile Method for Stiff, Tough, and Strong Nanocomposites by Direct Exfoliation of Multilayered Graphene into Native Nanocellulose Matrix. *Biomacromolecules* **2012**, *13*, 1093–1099. [CrossRef]

128. Zhang, X.F.; Lu, Z.X.; Zhao, J.Q.; Li, Q.Y.; Zhang, W.; Lu, C.H. Exfoliation/dispersion of low-temperature expandable graphite in nanocellulose matrix by wet co-milling. *Carbohyd Polym* **2017**, *157*, 1434–1441. [CrossRef] [PubMed]

129. Zhang, G.Q.; Lv, J.L.; Yang, F.L. Optimized anti-biofouling performance of bactericides/cellulose nanocrystals composites modified PVDF ultrafiltration membrane for micro-polluted source water purification. *Water Sci Technol* **2019**, *79*, 1437–1446. [CrossRef]

130. Yuan, H.; Pan, H.; Meng, X.; Zhu, C.L.; Liu, S.Y.; Chen, Z.X.; Ma, J.; Zhu, S.M. Assembly of MnO/CNC/rGO fibers from colloidal liquid crystal for flexible supercapacitors via a continuous one-process method. *Nanotechnology* **2019**, *30*. [CrossRef]

131. Dhar, P.; Gaur, S.S.; Kumar, A.; Katiyar, V. Cellulose Nanocrystal Templated Graphene Nanoscrolls for High Performance Supercapacitors and Hydrogen Storage: An Experimental and Molecular Simulation Study. *Sci Rep-Uk* **2018**, *8*. [CrossRef]

132. Li, G.X.; Yu, J.Y.; Zhou, Z.Q.; Li, R.K.; Xiang, Z.H.; Cao, Q.; Zhao, L.L.; Peng, X.W.; Liu, H.; Zhou, W.J. N-Doped Mo2C Nanobelts/Graphene Nanosheets Bonded with Hydroxy Nanocellulose as Flexible and Editable Electrode for Hydrogen Evolution Reaction. *Iscience* **2019**, *19*, 1090. [CrossRef]

133. Zhou, X.M.; Liu, Y.; Du, C.Y.; Ren, Y.; Li, X.L.; Zuo, P.J.; Yin, G.P.; Ma, Y.L.; Cheng, X.Q.; Gao, Y.Z. Free-Standing Sandwich-Type Graphene/Nanocellulose/Silicon Laminar Anode for Flexible Rechargeable Lithium Ion Batteries. *Acs Appl Mater Inter* **2018**, *10*, 29638–29646. [CrossRef]

134. Wang, Z.H.; Tammela, P.; Stromme, M.; Nyholm, L. Nanocellulose coupled flexible polypyrrole@graphene oxide composite paper electrodes with high volumetric capacitance. *Nanoscale* **2015**, *7*, 3418–3423. [CrossRef]

135. Kumar, A.; Rao, K.M.; Han, S.S. Mechanically viscoelastic nanoreinforced hybrid hydrogels composed of polyacrylamide, sodium carboxymethylcellulose, graphene oxide, and cellulose nanocrystals. *Carbohyd Polym* **2018**, *193*, 228–238. [CrossRef]

136. Laaksonen, P.; Walther, A.; Malho, J.M.; Kainlauri, M.; Ikkala, O.; Linder, M.B. Genetic Engineering of Biomimetic Nanocomposites: Diblock Proteins, Graphene, and Nanofibrillated Cellulose. *Angew Chem Int Edit* **2011**, *50*, 8688–8691. [CrossRef]

137. Jia, L.L.; Huang, X.Y.; Liang, H.E.; Tao, Q. Enhanced hydrophilic and antibacterial efficiencies by the synergetic effect TiO2 nanofiber and graphene oxide in cellulose acetate nanofibers. *International Journal of Biological Macromolecules* **2019**, *132*, 1039–1043. [CrossRef]

138. Iijima, S. Helical Microtubules of Graphitic Carbon. *Nature* **1991**, *354*, 56–58. [CrossRef]

139. Stankova, L.; Fraczek-Szczypta, A.; Blazewicz, M.; Filova, E.; Blazewicz, S.; Lisa, V.; Bacakova, L. Human osteoblast-like MG 63 cells on polysulfone modified with carbon nanotubes or carbon nanohorns. *Carbon* **2014**, *67*, 578–591. [CrossRef]

140. Saito, T.; Kuramae, R.; Wohlert, J.; Berglund, L.A.; Isogai, A. An Ultrastrong Nanofibrillar Biomaterial: The Strength of Single Cellulose Nanofibrils Revealed via Sonication-Induced Fragmentation. *Biomacromolecules* **2013**, *14*, 248–253. [CrossRef]

141. Koga, H.; Saito, T.; Kitaoka, T.; Nogi, M.; Suganuma, K.; Isogai, A. Transparent, Conductive, and Printable Composites Consisting of TEMPO-Oxidized Nanocellulose and Carbon Nanotube. *Biomacromolecules* **2013**, *14*, 1160–1165. [CrossRef]

142. Fang, W.; Linder, M.B.; Laaksonen, P. Modification of carbon nanotubes by amphiphilic glycosylated proteins. *J Colloid Interf Sci* **2018**, *512*, 318–324. [CrossRef]

143. Mougel, J.B.; Bertoncini, P.; Cathala, B.; Chauvet, O.; Capron, I. Macroporous hybrid Pickering foams based on carbon nanotubes and cellulose nanocrystals. *J Colloid Interf Sci* **2019**, *544*, 78–87. [CrossRef]

144. Trigueiro, J.P.C.; Silva, G.G.; Pereira, F.V.; Lavall, R.L. Layer-by-layer assembled films of multi-walled carbon nanotubes with chitosan and cellulose nanocrystals. *J Colloid Interf Sci* **2014**, *432*, 214–220. [CrossRef]

145. Yang, X.; Shi, K.Y.; Zhitomirsky, I.; Cranston, E.D. Cellulose Nanocrystal Aerogels as Universal 3D Lightweight Substrates for Supercapacitor Materials. *Advanced Materials* **2015**, *27*, 6104–6109. [CrossRef]

146. ZabihiSahebi, A.; Koushkbaghi, S.; Pishnamazi, M.; Askari, A.; Khosravi, R.; Irani, M. Synthesis of cellulose acetate/chitosan/SWCNT/Fe3O4/TiO2 composite nanofibers for the removal of Cr(VI), As(V), Methylene blue and Congo red from aqueous solutions. *Int J Biol Macromol* **2019**, *140*, 1296–1304. [CrossRef]

147. Kang, Y.J.; Chun, S.J.; Lee, S.S.; Kim, B.Y.; Kim, J.H.; Chung, H.; Lee, S.Y.; Kim, W. All-solid-state flexible supercapacitors fabricated with bacterial nanocellulose papers, carbon nanotubes, and triblock-copolymer ion gels. *ACS Nano* **2012**, *6*, 6400–6406. [CrossRef]

148. Zhou, Y.; Lee, Y.; Sun, H.; Wallas, J.M.; George, S.M.; Xie, M. Coating Solution for High-Voltage Cathode: AlF3 Atomic Layer Deposition for Freestanding LiCoO2 Electrodes with High Energy Density and Excellent Flexibility. *ACS Appl Mater Interfaces* **2017**, *9*, 9614–9619. [CrossRef]

149. Gonzalez-Dominguez, J.M.; Anson-Casaos, A.; Grasa, L.; Abenia, L.; Salvador, A.; Colom, E.; Mesonero, J.E.; Garcia-Bordeje, J.E.; Benito, A.M.; Maser, W.K. Unique Properties and Behavior of Nonmercerized Type-II Cellulose Nanocrystals as Carbon Nanotube Biocompatible Dispersants. *Biomacromolecules* **2019**, *20*, 3147–3160. [CrossRef]

150. Zhang, H.; Sun, X.; Hubbe, M.A.; Pal, L. Highly conductive carbon nanotubes and flexible cellulose nanofibers composite membranes with semi-interpenetrating networks structure. *Carbohydr Polym* **2019**, *222*, 115013. [CrossRef]

151. Zhai, Y.; Wang, D.; Liu, H.; Zeng, Y.; Yin, Z.; Li, L. Electrochemical Molecular Imprinted Sensors Based on Electrospun Nanofiber and Determination of Ascorbic Acid. *Anal Sci* **2015**, *31*, 793–798. [CrossRef]

152. Broz, A.; Bacakova, L.; Stenclova, P.; Kromka, A.; Potocky, S. Uptake and intracellular accumulation of diamond nanoparticles - a metabolic and cytotoxic study. *Beilstein J Nanotechnol* **2017**, *8*, 1649–1657. [CrossRef]

153. Williams, O.A. Ultrananocrystalline diamond for electronic applications. *Semicond Sci Tech* **2006**, *21*, R49–R56. [CrossRef]

154. Shenderova, O.A.; Gruen, D.M. *Ultrananocrystalline Diamond: Synthesis, Properties and Applications of UNCD*, 2nd ed.; Shenderova, O.A., Gruen, D.M., Eds.; William Andrew Publishing: Oxford, UK, 2012; p. 584. ISBN 9781437734652.

155. Mochalin, V.N.; Shenderova, O.; Ho, D.; Gogotsi, Y. The properties and applications of nanodiamonds. *Nat Nanotechnol* **2011**, *7*, 11–23. [CrossRef]

156. Stankova, L.; Musilkova, J.; Broz, A.; Potocky, S.; Kromka, A.; Kozak, H.; Izak, T.; Artemenko, A.; Stranska, D.; Bacakova, L. Alterations to the adhesion, growth and osteogenic differentiation of human osteoblast-like cells on nanofibrous polylactide scaffolds with diamond nanoparticles. *Diam Relat Mater* **2019**, *97*. [CrossRef]

157. Grausova, L.; Kromka, A.; Burdikova, Z.; Eckhardt, A.; Rezek, B.; Vacik, J.; Haenen, K.; Lisa, V.; Bacakova, L. Enhanced Growth and Osteogenic Differentiation of Human Osteoblast-Like Cells on Boron-Doped Nanocrystalline Diamond Thin Films. *PLoS ONE* **2011**, *6*. [CrossRef]

158. Liskova, J.; Babchenko, O.; Varga, M.; Kromka, A.; Hadraba, D.; Svindrych, Z.; Burdikova, Z.; Bacakova, L. Osteogenic cell differentiation on H-terminated and O-terminated nanocrystalline diamond films. *Int J Nanomed* **2015**, *10*, 869–884. [CrossRef]

159. Morimune-Moriya, S.; Salajkova, M.; Zhou, Q.; Nishino, T.; Berglund, L.A. Reinforcement Effects from Nanodiamond in Cellulose Nanofibril Films. *Biomacromolecules* **2018**, *19*, 2423–2431. [CrossRef]

160. Juknius, T.; Ruzauskas, M.; Tamulevicius, T.; Siugzdiniene, R.; Jukniene, I.; Vasiliauskas, A.; Jurkeviciute, A.; Tamulevicius, S. Antimicrobial Properties of Diamond-Like Carbon/Silver Nanocomposite Thin Films Deposited on Textiles: Towards Smart Bandages. *Materials* **2016**, *9*, 371. [CrossRef]

161. Yuan, F.S.; Huang, Y.; Fan, M.M.; Chen, C.T.; Qian, J.S.; Hao, Q.L.; Yang, J.Z.; Sun, D.P. N-Doped Carbon Nanofibrous Network Derived from Bacterial Cellulose for the Loading of Pt Nanoparticles for Methanol Oxidation Reaction. *Chem-Eur J* **2018**, *24*, 1844–1852. [CrossRef]

162. Li, S.; Wang, M.Y.; Luo, Y.; Huang, J.G. Bio-Inspired Hierarchical Nanofibrous Fe3O4-TiO2-Carbon Composite as a High-Performance Anode Material for Lithium-Ion Batteries. *Acs Appl Mater Inter* **2016**, *8*, 17343–17351. [CrossRef]

163. Celik, K.B.; Cengiz, E.C.; Sar, T.; Dursun, B.; Ozturk, O.; Akbas, M.Y.; Demir-Cakan, R. In-situ wrapping of tin oxide nanoparticles by bacterial cellulose derived carbon nanofibers and its application as freestanding interlayer in lithium sulfide based lithium-sulfur batteries. *J Colloid Interf Sci* **2018**, *530*, 137–145. [CrossRef]

164. Xu, J.C.; Rong, J.; Qiu, F.X.; Zhu, Y.; Mao, K.L.; Fang, Y.Y.; Yang, D.Y.; Zhang, T. Highly dispersive NiCo2S4 nanoparticles anchored on nitrogen-doped carbon nanofibers for efficient hydrogen evolution reaction. *J Colloid Interf Sci* **2019**, *555*, 294–303. [CrossRef]

165. d'Amora, M.; Giordani, S. Carbon Nanomaterials for Nanomedicine. In *Micro and Nano Technologies*; Ciofani, G., Ed.; Elsevier: Amsterdam, The Netherlands, 2018; pp. 103–113. ISBN 978-0-12-814156-4. [CrossRef]

166. Chaudhuri, I.; Fruijtier-Polloth, C.; Ngiewih, Y.; Levy, L. Evaluating the evidence on genotoxicity and reproductive toxicity of carbon black: a critical review. *Crit Rev Toxicol* **2018**, *48*, 143–169. [CrossRef]

167. Niranjan, R.; Thakur, A.K. The Toxicological Mechanisms of Environmental Soot (Black Carbon) and Carbon Black: Focus on Oxidative Stress and Inflammatory Pathways. *Front Immunol* **2017**, *8*. [CrossRef]

168. Zhu, Y.; Li, W.X.; Zhang, Y.; Li, J.; Liang, L.; Zhang, X.Z.; Chen, N.; Sun, Y.H.; Chen, W.; Tai, R.Z.; et al. Excessive Sodium Ions Delivered into Cells by Nanodiamonds: Implications for Tumor Therapy. *Small* **2012**, *8*, 1771–1779. [CrossRef]

169. Kovacs, T.; Naish, V.; O'Connor, B.; Blaise, C.; Gagne, F.; Hall, L.; Trudeau, V.; Martel, P. An ecotoxicological characterization of nanocrystalline cellulose (NCC). *Nanotoxicology* **2010**, *4*, 255–270. [CrossRef]

170. Shvedova, A.A.; Kisin, E.R.; Yanamala, N.; Farcas, M.T.; Menas, A.L.; Williams, A.; Fournier, P.M.; Reynolds, J.S.; Gutkin, D.W.; Star, A.; et al. Gender differences in murine pulmonary responses elicited by cellulose nanocrystals. *Part Fibre Toxicol* **2016**, *13*. [CrossRef] [PubMed]

171. Sunasee, R.; Araoye, E.; Pyram, D.; Hemraz, U.D.; Boluk, Y.; Ckless, K. Cellulose nanocrystal cationic derivative induces NLRP3 inflammasome-dependent IL-1beta secretion associated with mitochondrial ROS production. *Biochem Biophys Rep* **2015**, *4*, 1–9. [CrossRef] [PubMed]

172. Ede, J.D.; Ong, K.J.; Goergen, M.; Rudie, A.; Pomeroy-Carter, C.A.; Shatkin, J.A. Risk Analysis of Cellulose Nanomaterials by Inhalation: Current State of Science. *Nanomaterials (Basel)* **2019**, *9*, 337. [CrossRef]

173. Przekora, A.; Vandrovcova, M.; Travnickova, M.; Pajorova, J.; Molitor, M.; Ginalska, G.; Bacakova, L. Evaluation of the potential of chitosan/beta-1,3-glucan/hydroxyapatite material as a scaffold for living bone graft production in vitro by comparison of ADSC and BMDSC behaviour on its surface. *Biomed Mater* **2017**, *12*, 015030. [CrossRef]

174. Bacakova, L.; Zarubova, J.; Travnickova, M.; Musilkova, J.; Pajorova, J.; Slepicka, P.; Kasalkova, N.S.; Svorcik, V.; Kolska, Z.; Motarjemi, H.; et al. Stem cells: their source, potency and use in regenerative therapies with focus on adipose-derived stem cells - a review. *Biotechnol Adv* **2018**, *36*, 1111–1126. [CrossRef] [PubMed]
175. Filova, E.; Bullett, N.A.; Bacakova, L.; Grausova, L.; Haycock, J.W.; Hlucilova, J.; Klima, J.; Shard, A. Regionally-selective cell colonization of micropatterned surfaces prepared by plasma polymerization of acrylic acid and 1,7-octadiene. *Physiol Res* **2009**, *58*, 669–684.

 *nanomaterials*

Article

Cellulose Nanofibril/Carbon Nanomaterial Hybrid Aerogels for Adsorption Removal of Cationic and Anionic Organic Dyes

Zhencheng Yu [1], Chuanshuang Hu [1,*], Anthony B. Dichiara [2], Weihui Jiang [1] and Jin Gu [1,*]

[1] College of Materials and Energy, South China Agricultural University, Guangzhou 510642, China; yuzansing2017@stu.scau.edu.cn (Z.Y.); 201820117385@mail.scut.edu.cn (W.J.)
[2] School of Environmental and Forest Sciences, University of Washington, Seattle, WA 98195, USA; abdichia@uw.edu
* Correspondence: cshu@scau.edu.cn (C.H.); gujin@scau.edu.cn (J.G.); Tel.: +86-20-85282568 (C.H.); +86-20-85280319 (J.G.)

Received: 30 November 2019; Accepted: 15 January 2020; Published: 19 January 2020

Abstract: Advances in nanoscale science and engineering are providing new opportunities to develop promising adsorbents for environmental remediation. Here, hybrid aerogels are assembled from cellulose nanofibrils (CNFs) and carbon nanomaterials to remove cationic dye methylene blue (MB) and anionic dye Congo red (CR) in single and binary systems. Two classes of carbon nanomaterials, carbon nanotubes (CNTs) and graphene nanoplates (GnPs), are incorporated into CNFs with various amounts, respectively. The adsorption, mechanics and structure properties of the hybrid aerogels are investigated and compared among different combinations. The results demonstrate CNF–GnP 3:1 hybrid exhibits the best performance among all composites. Regarding a single dye system, both dye adsorptions follow a pseudo-second-order adsorption kinetic and monolayer Langmuir adsorption isotherm. The maximal adsorption capacities of CNF–GnP aerogels for MB and CR are 1178.5 mg g^{-1} and 585.3 mg g^{-1}, respectively. CNF–GnP hybrid show a superior binary dye adsorption capacity than pristine CNF or GnP. Furthermore, nearly 80% of MB or CR can be desorbed from CNF–GNP using ethanol as the desorption agent, indicating the reusability of this hybrid material. Hence, the CNF–GnP aerogels show great promise as adsorption materials for wastewater treatment.

Keywords: cellulose nanofibrils; graphene nanoplates; carbon nanotubes; aerogel; organic dyes; adsorption

1. Introduction

Dyes are colored organic chemicals typically classified as anionic (acid, reactive, and direct dyes), cationic (all basic dyes), and non-ionic (dispersed dyes) based on their charge upon dissolution in aqueous solutions [1]. These complex molecules are widely used in many industrial fields, such as textile, paper, leather tanning, food processing, plastics, cosmetics, rubber, and printing. The contamination of the hydrosphere with dyes raises serious environmental and sanitary concerns due to their ubiquity, toxicity and deleterious effects on photosynthetic activity in aquatic life due to decreased sunlight penetration [2]. Particularly, methylene blue (MB), one of the most widely used basic dyes in the printing and textile industries, can cause a variety of harmful effects, such as eye burns, gastrointestinal tract and skin irritation [3]. As a typical direct azo dye, Congo red (CR) is mainly applied in a relatively large dosage for dyeing biological samples, and can increase the risk of cancer if absorbed into the human body [4]. Currently, there are various industrial methods for treating waste dye solutions, including adsorption, microbial treatment, chemical oxidation or reduction, flocculation precipitation, ozone oxidization, chemical precipitation, nanofiltration, catalytic degradation etcetera [5–7]. Due to its simple operation and low cost, adsorption is widely used as a commercial way to remove organic dyes from aqueous solutions [8].

Recently, carbon nanomaterials have attracted a broad interest in pollutant adsorption. Among them, carbon nanotubes (CNTs) are comprised of one or several sheets of hexagonally packed carbon atoms rolled into concentric seamless cylinders; these exhibit strong mechanical properties, a high aspect ratio, excellent chemical stability and a large specific surface area, which make them desirable for dye adsorption [9–11]. The adsorption mechanism of organic compounds on carbon nanomaterials can be described as an interplay between different intermolecular forces (i.e., hydrophobic, van der Waals forces, π–π bonding, hydrogen bonding, and electrostatic interactions), whose contributions depend on the adsorbate nature and the surface chemistry of the nano-adsorbent [12]. Another type of carbon nanomaterial, graphene, is a planar honeycomb-shaped nanomaterial consisting of six-member rings with a large specific surface area (theoretical value of 2630 m^2 g^{-1}) [13]. It is more easily prepared than CNT and shows an extensive application prospect in the removal of organic and inorganic pollutants [14] and dyes [15,16]. Graphene could be modified to adsorb pollutants via π–π, electrostatic interaction and hydrogen bonding. However, the cost of a carbon-based nano-adsorbent is high and the regeneration treatment of spent materials is often challenging [17]. Due to strong π–π bonds and van der Waals forces, graphene sheets tend to condense and restack. CNTs often group into bundles and tangle due to the same reason. Therefore, the available specific surface area of these carbon nanomaterial is much lower than the theoretical value, greatly restricting their practical applications [18]. Thus, providing a recycling function and reducing the aggregation potential of the carbon nanomaterials are vital ways to design high-efficiency adsorbents.

Nanocelluloses are a class of renewable nanomaterials derived from the most abundant natural polymer on earth. Their high specific surface area, bioavailability and surface reactivity make nanocelluloses interesting materials for pollutant adsorbents [19]. Since unmodified nanocelluloses expose abundant surface hydroxyl groups and perhaps other negatively charged groups such as sulfate or carboxylate groups depending on the preparation procedure [20,21], they are well-suited for the adsorption of cationic molecules. Recently, high surface-area cellulose nanofibril (CNF) aerogels prepared via 2,2,6,6-tetramethylpyperidine-1-oxyl (TEMPO) oxidation were shown to absorb cationic malachite green (MG) dye (212.7 mg g^{-1}) due to electrostatic interactions between MG and negatively charged oxygen moieties on the CNF surface [22]. To make nanocellulose adsorbents for anionic dye, chemical modifications, such as introduction of positively charged amino groups, are usually required [23]. However, these positively charged nanocelluloses are no longer suitable for cationic dye adsorption.

While the above studies demonstrate the individual efficacy of carbon nanomaterials and nanocellulose adsorbents, improved adsorption performance and cost optimization may be realized by combining these nanomaterials together. Since carbon nanomaterials and nanocellulose exhibit different affinities for given molecules; hence, their combination can increase the variety of pollutants that may be adsorbed. Recently, CNFs were shown to improve the dispersion of carbon nanotubes [24,25] and graphene [26] in aqueous solutions. Hajian et al. [26] suggested the charges on the TEMPO oxidized CNFs induced an electrostatic stabilization of the CNF–carbon nanomaterial complexes to prevent aggregation. When oxygen-containing carbon nanomaterials, such as graphene oxide (GO) or oxidized CNTs, were used, there was a strong interaction between the oxygen-containing groups of carbon nanomaterials and hydroxyl groups of nanocellulose [27]. Recently, Wei et al. [27] synthesized GO/microcrystalline cellulose (MCC) aerogels in an LiBr aqueous solution. Hybrid GO/MCC aerogel had higher adsorption capacity of MB per unit mass of GO (2630 mg g^{-1}) than pure GO when the content of GO was low (0.3 wt%). Hussain et al. [28] fabricated GO/CNF monoliths based on a urea-assisted self-assembly method. The maximum adsorption capacity of these hybrid monoliths to MB achieved 227.27 mg g^{-1}. Wu et al. [29] exploited cellulose nanofiber as a cross-linker to interweave between reduced graphene oxide (rGO) layers and obtained hybridized monolith aerogels by a hydrothermal method. The hybridized monolith was able to adsorb not only hydrophilic dyes, but also hydrophobic organic oil. To our knowledge, the influence of carbon nanomaterials features, such as their morphology (nanotubes or nanoplates), on the adsorption behaviors of the CNF/carbon nanomaterial hybrids have not been explored.

Concerning industrial wastewater, different types of dyes could be found. Most of the previous works focus on single solute adsorption in pure water and are not representative of real-world wastewater effluents. During this study, two types of carbon nanomaterials, CNTs and graphene nanoplates (GnPs), are dispersed in water with CNFs using different mass ratios, respectively, to prepare hybrid aerogels through a simple freeze-drying procedure without the assistance of other agents. The aim is to understand the interactions of CNT and GnP with CNF and to explore the adsorption capacity of these hybrid materials to both anionic and cationic dyes in single and binary systems. The preparation, characterization, and adsorption assessments of CNF–CNT and CNF–GnP aerogels for MB and CR dyes are reported. Adsorption behaviors of dyes are inspected by kinetic models and adsorption isothermal models. The adsorption mechanism, contact time, and concentration of MB and CR dye to hybrid aerogels are investigated. Furthermore, desorption of both dyes from the adsorbents are studied.

2. Materials and Methods

2.1. Materials

Rice straw was harvested at the South China Agricultural University in 2016. Benzene (99.5%, AR, Damao Chemical Reagent Factory, Tianjin, China), ethanol (99.7%, AR, Guangzhou Chemical Reagent Factory, Guangzhou, China), sodium chlorite ($NaClO_2$, 80%, Aladdin), acetic acid glacial (CH_3COOH, 99.5%, AR, Guangzhou Chemical Reagent Factory), potassium hydroxide (KOH, 85%, AR, Shanghai RichJoint Chemical Reagents, Shanghai, China), hydrochloric acid (HCl, 1 N, certified, Guangzhou Chemical Reagent Factory), sodium hydroxide (NaOH, 96%, Guangzhou Chemical Reagent Factory), sodium hypochlorite (NaClO, 8.0%, AR, Damao Chemical Reagent Factory), 2,2,6,6-tetramethylpyperdine-1-oxyl (TEMPO, 98%, Aladdin), sodium bromide (NaBr, 99.9%, AR, Tianjin Fuchen Chemical Reagent Factory, Tianjin, China), methylene blue (MB, AR, Tianjin Fuchen Chemical Reagent Factory), Congo red (CR, Tianjin Fuchen Chemical Reagent Factory), graphene nanoplates (GnPs, Aldrich, grade C-750, thickness: a few nm, particle size: <2 μm, specific surface area: 750 $m^2\ g^{-1}$), and OH-functionalized multi-walled carbon nanotubes (CNTs, Cheap Tubes, external diameter: 50–80 nm, internal diameter: 5–10 nm, length: 10–20 μm, OH content: 5.5%, specific surface area: 60 $m^2\ g^{-1}$ [30]) were used as received. The chemical formula, maximum absorption wavelength and electrical property under neutral pH of MB and CR are shown in Table S1. The deionized (DI) water used was purified by an AIKE Advanced-II-OS water purification system (Cheng Du Kangning Science and Technology Development Company, Chengdu, China).

2.2. Preparation of Cellulose Nanofibrils

Pure cellulose was prepared from rice straw by extracting wax and dissolving lignin, hemicellulose and silica [31]. Cellulose nanofibrils (CNFs) were prepared by TEMPO mediated oxidation employing 5 mmol NaClO per gram of cellulose followed by mechanical blending at 37,000 rpm for 30 min [32]. TEMPO-mediated oxidation converted the C6 primary hydroxyls on the surface of the cellulose into carboxyls [33]. The successive mechanical treatment disintegrated oxidized cellulose fibers into individual nanofibrils that were 1–5 nm in width and hundreds of nanometers to 2 μm in length [32]. The carboxylate content of the CNF was measured using a conductometric titration (Oakton CON 6+, Cole-Parmer Instrument Company, Vernon Hills, IL, USA) following a previous method (Gu and Hsieh, 2015) (Figure S1). The carboxylate ($COOH + COO^-$) content was determined to be 1.36 mmol/g at neutral pH (COOH content: 0.137 mmol/g).

2.3. Preparation of CNF, CNF–CNT and CNF–GnP Aerogels

CNTs or GnPs were added to CNF dispersions with various mass ratios of CNF to carbon nanomaterial (1:0, 3:1, 1:1, 1:3 and 0:1). The final concentration of all the mixtures was controlled to be 6 mg mL^{-1}. The mixtures were sonicated by ultrasonic cell disruptor (SCIENTZ, IID, Ningbo, China) at 600 W for 30 min in an ice bath. Each sample was frozen at −21 °C for 4 h in a 15 mL polypropylene centrifuge tube followed by freeze-drying at −50 °C. The CNF–CNT and CNF–GnP hybrid aerogels were obtained. Pure CNF aerogel also was obtained via the same method from a CNF suspension of 6 mg mL^{-1}.

2.4. Characterization

The dispersion quality of aqueous CNT/CNF and GnP/CNF mixtures after sonication was examined by depositing a drop of each suspension on a glass slide for optical microscope observations. The size of the aggregates was measured and averaged over 150–200 particles. Regarding the compression test, the aerogel samples were transferred into a chamber with a temperature of 25 ± 2 °C and a relative humidity of 65 ± 2%. The compressive strength of the samples was measured 24 h later with a compression ratio of 1 mm min^{-1} on a universal mechanical testing machine (CMT1000, SUST, Zhuhai, China). The morphology of CNF, CNF–CNT, CNF–GnP aerogels and CNT, GnP powder was obtained by a field emission scanning electron microscope (FE-SEM, SU-70, Hitachi, Chiyoda, Japan) after sputtering coating the samples with gold at 15 mA for 2 min under vacuum conditions (Hitachi, E-1010 Ion Sputtering System, Japan). Imaging of the samples was performed at 0.5–30 kV acceleration voltage and 1–2 nA current intensity at magnifications of 20–800,000 times. The chemical structures of CNF, CNF–CNT, CNF–GnP aerogels and CNT, GnP powder were characterized by Fourier transform infrared spectroscopy (FTIR, PerkinElmer, Spectrum 100, Waltham, MA, USA) scanning from 4000 to 400 cm^{-1} with a resolution of 4 cm^{-1}. Before FTIR measurements, 2 mg dry sample was ground into powder with 200 mg KBr and pressed into pellets. The specific surface area (SSA) of the samples was calculated using Equation (1) [34].

$$\text{SSA} = \frac{N_A A_{\text{MB}}(C_o - C_e)V}{M_{\text{MB}} M_S} \tag{1}$$

where N_A is Avogadro's number (6.023 × 10^{23} mol^{-1}), A_{MB} is the covered area per MB molecules (typically assumed to be 1.35 nm^2), C_o and C_e are the initial and equilibrium concentration of MB, respectively, V is the volume of the MB solution, M_{MB} is the relative molecular mass of MB, and M_S is the mass of the sample.

2.5. Adsorption of Dyes in Single Systems

Anionic dye CR and cationic dye MB were used in the adsorption experiment. CNF, CNF–CNT, CNF–GnP were aerogels, while pristine CNT and GnP were in the form of powder. Aqueous phase adsorption studies were conducted at 25 °C by submerging 5 mg of a specific nano adsorbent into 20 mL of each dye solution at neutral pH. The mixture was continuously agitated on an orbital shaker at 120 rpm and the amount of residual dye in the solution was determined by ultraviolet-visible spectroscopy (UV-Vis, Thermo Scientific, Evolution 201, Waltham, MA, USA) at the maximum absorption wavelength using a measured extinction coefficient from a Beer's law analysis for each solution. The adsorption capacity of MB, CR on each adsorbent was calculated using Equation (2).

$$q_t = \frac{(C_0 - C_t)V}{m} \tag{2}$$

where q_t is the amount of dye adsorbed at a given time (mg g^{-1}), C_0 is the initial dye concentration (mg L^{-1}), C_t is the residual dye concentration at a given time (mg L^{-1}), V is the solution volume (L), and m is the mass of the adsorbent (mg).

The effect of contact time (0–240 min) was examined using 5 mg of adsorbent and an initial dye concentration of 10, 250 and 500 mg L^{-1} for MB and 100, 600 and 2000 mg L^{-1} for CR. The effect of initial dye concentration on the final adsorption capacity was investigated in a range of dye concentrations (MB: 10, 50, 100, 150, 200, 250, 300, 400, 500, 600, 800 and 1000 mg L^{-1}; CR: 10, 50, 100, 150, 200, 250, 300, 400, 500, 600, 800, 1000, 1500 and 2000 mg L^{-1}) using an adsorbent of 5 mg at 25 °C and 120 rpm for 16 h.

2.6. Adsorption of Dyes in Binary Systems

MB and CR were added into 20 mL DI water with various mass ratios at a total concentration of 200 mg L^{-1}. The mass ratio of MB to CR was 3:1, 1:1 or 1:3. An adsorbent of 5 mg was added, and the mixture was agitated at 25 °C and 120 rpm for 16 h. Regarding a binary system, dye A (MB) and dye B (CR) concentrations were calculated as follows:

$$C_A = \frac{k_{B2}d_1 - k_{B1}d_2}{k_{A1}k_{B2} - k_{A2}k_{B1}} \tag{3}$$

$$C_B = \frac{k_{A1}d_2 - k_{A2}d_1}{k_{A1}k_{B2} - k_{A2}k_{B1}} \tag{4}$$

where d_1 and d_2 are the mixture optical densities measured at λ_1 (664 nm) and λ_2 (498 nm), respectively. k_{A1}, k_{B1}, k_{A2}, and k_{B2} are the calibration constants for components A and B at wavelengths of λ_1 and λ_2, respectively [35,36].

2.7. Desorption of Dyes

Desorption of MB and CR from CNF, GnP and CNF–GnP 3:1 was performed in ethanol, acetonitrile, acetone or 400 mM NaCl at 25 °C. First, the adsorbents were added into 20 mL of 100 mg L^{-1} dye solutions for adsorption at 120 rpm and then taken out from the dye solutions after 16 h. These adsorbents were transferred to 20 mL of desorption agents. After 1h, the adsorbents were removed and the dye concentration in the solution was measured. The adsorbents might again have been transferred to a fresh desorption agent and the desorption step was repeated for several cycles. Dye percentage removal (%) was calculated by Equation (5):

$$\text{Removal} \ (\%) \ = \ \frac{D_t}{C_0 - C_e} \times 100 \tag{5}$$

where D_t is the concentration of dye in the desorption (mg L^{-1}), C_0 and C_e are the initial and final concentrations of dye in the adsorption (mg L^{-1}), respectively.

3. Results and Discussion

3.1. Effect of Adsorbent Structure

Figure 1 shows the uptake of MB or CR after 16 h with constant agitation on different nanosorbents. The initial MB and CR concentrations were 500 mg L^{-1} and 2000 mg L^{-1}, respectively. Pure CNF adsorbed the highest amount of cationic MB per unit mass (1207.2 mg g^{-1}) due to the large quantity of negatively charged carboxyl groups present on its surface. However, the amount of CR adsorbed on pure CNF was low (175.5 mg g^{-1}) possibly due to Coulombic repulsions between negatively charged CNF and anionic CR. Pure GnP powder (1491.7 mg g^{-1}) was able to adsorb the highest amount of CR followed by pure CNT powder (777.9 mg g^{-1}). The ability of GnP to remove MB was also superior to CNT, which can be attributed to the larger surface area of GnP with more sites available for adsorption. Compared to nanocellulose, carbon nanomaterials could adsorb both cationic and anionic dyes mainly though π–π interaction [37]. Among the hybrid aerogels, CNF–GnP aerogels were generally better than CNF–CNT aerogels in the ability to remove both types of dyes from water at the same CNF to carbon nanomaterial ratio. Moreover, the adsorption capacity of MB onto CNF–GnP 3:1 (1166.1 mg g^{-1}) was very close to that of pure CNF. As the quantity of carbon nanomaterials in the hybrid increased, the MB uptake gradually decreased because CNF was displaced by nano adsorbents with lower affinity for MB. Interestingly, among all hybrid aerogels, CNF–GnP 3:1 also exhibited the highest CR uptake (507.1 mg g^{-1}), almost two times greater than pure CNF. Similar results were obtained for CNF–CNT hybrids, with CNF–CNT 3:1 having superior MB and CR uptakes than other CNF–CNT combinations. This may suggest that high CNF concentrations can reduce carbon nanomaterial aggregation and improve dispersion quality to promote more effective contact between the solute and the sorbent. *t*-test results indicated the adsorption of MB onto CNF–GnP 3:1 and CNF–CNT 3:1 had no significant difference ($p > 0.01$), but CNF–GnP 3:1 was superior to CNF–CNT 3:1 in the adsorption of CR ($p < 0.01$) (Tables S2–S5). Based on these observations, hybrid aerogels comprising a CNF to carbon nanomaterial ratio of 3:1 were selected for further dispersion, mechanics, morphology, and chemical structure analysis. The CNF–CNT 3:1 and CNF–GnP 3:1 sorbents are reported henceforth as CNF–CNT and CNF–GnP, respectively.

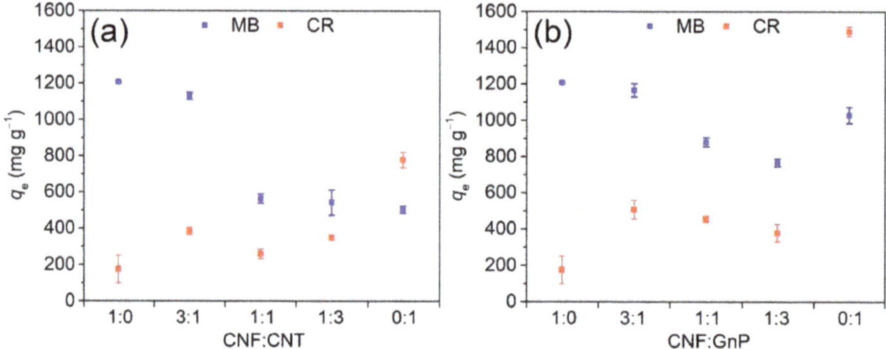

Figure 1. The final adsorption of MB and CR onto CNF–CNT aerogels with the mass ratio of 1:0, 3:1, 1:1, 1:3 and 0:1 (**a**), CNF–GnP aerogels with the mass ratio of 1:0, 3:1, 1:1, 1:3 and 0:1 (**b**). Initial MB concentration was 500 mg L^{-1}. Initial CR concentration was 2000 mg L^{-1}.

The dispersion states of CNT and GnP aqueous suspensions in the presence of CNFs were investigated by optical microscopy after probe sonication for 5 min and 30 min (Figure S2). After 5 min of ultrasonic treatment, the average particle sizes of CNTs and GnPs were 8.8 ± 13.1 µm (Figure S2a) and 6.6 ± 11.0 µm (Figure S2c), respectively. After 30 min of ultrasonic treatment, large particle aggregations disappeared in each case (Figure S2b,d). The average particle sizes of CNTs and GnPs were 2.1 ± 0.5 µm and 3.1 ± 0.9 µm, respectively. The hydrophobicity of carbon nanomaterials and their tendency to readily form aggregates by hexagonal packing of individual particles with high van der Waals binding energy can greatly reduce the surface area available for adsorption. Ultrasonication provided sufficient energy to break the aggregates of carbon nanomaterials. While TEMPO-oxidized CNFs had negatively charged surface carboxyls, the counterions on the surface of the CNFs induced dipoles in the sp^2 carbon lattice of the carbon nanomaterials. Then, the charges on the CNFs induced electrostatic stabilization between CNF and CNT/GnP that prevented the carbon nanomaterials from reaggregation [26]. Van der Waals interactions also may occur between CNF and CNT/GnP. CNTs contained few hydroxyl groups and were able to form hydrogen bonds with CNFs. The homogeneous CNF/CNT and CNF/GnP suspensions remained stable for at least 12 h, which was long enough for the preparation of the hybrid aerogels.

Pure CNF aerogel was white with a porous external structure and turned black with the incorporation of carbon nanomaterials (Figure 2, inset). The aerogels formed inside centrifuge tubes during freeze-drying remained intact as cylindrical blocks and could be easily cut into slices using a sharp blade with no apparent deformation. The compression stress-strain curves of the three aerogels are shown in Figure 2. All curves exhibited "slow slope type" before the strain reached 80%. The initial strength of CNF and CNF–GnP aerogels was similar and higher than that of CNF–CNT. After the strains reached 80%, the curves were "steep type". Inflection points were observed when the strains were approximately 80%. Under the strains of 80%, the compression strengths of CNF, CNF–CNT and CNF–GnP were 0.064, 0.014 and 0.036 MPa, respectively. When immersed in water after compression testing, CNF, CNF–CNT and CNF–GnP aerogels exhibited a water activated shape recovery property (Supplementary Movies S1–S3). The aerogels absorbed water and restored the deformation. Most of the water could be easily squeezed out with tweezers and the compact aerogels could reabsorb water and return to their original size and shape. The fact that the aerogel cylinders were easily squeezed to ~10% of their length and quickly regained the same dimensions with a complete recovery, indicate that the aerogels were mechanically strong, and their open structure allowed the liquid solution to rapidly and freely flow in and out.

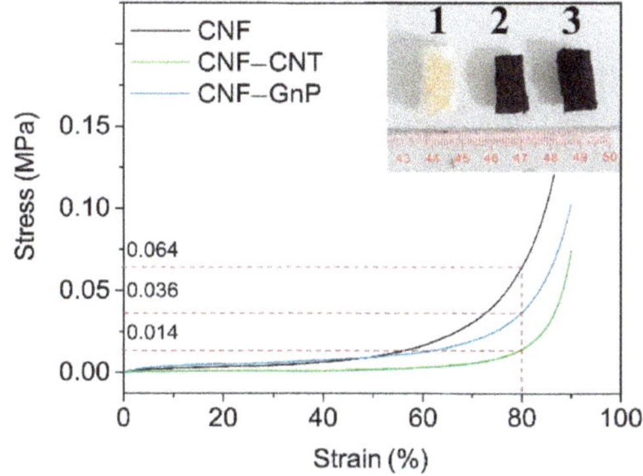

Figure 2. Stress–strain curves of CNF, CNF–CNT and CNF–GnP. The inset is a photograph of the freeze-dried aerogels. 1: CNF, 2: CNF–GnP and 3: CNF–CNT.

The morphologies of the CNF, CNT, GnP, CNF–CNT and CNF–GnP adsorbents have been characterized by SEM (Figure 3a–h). Pure CNF aerogel exhibited a three-dimensional structure with an intercalation of flat and folded sheets and contained pores of various shapes, which may be ascribed to the high suspension concentration (i.e., 6 mg mL^{-1}) used before freeze-drying. Pristine CNT and GnP powders revealed the presence of bundles and stacked aggregates due to strong attractive forces between individual particles (Figure 3c,d). The combination between CNT and CNF affected the formation of the CNF sheet structure (Figure 3e). This is the reason that CNF–CNT aerogel produces debris after compression performance testing (Supplementary Movie 2). CNT dispersed better in the CNF matrix (Figure 3f). However, at higher magnification (Figure 3f, inset), some CNT aggregates could be observed still in the matrix. Since both CNFs and CNTs were anisotropic rods, and CNTs were much longer than CNFs, individual CNT partly uncovered by CNFs may re-associate with each other during the freeze-drying process. Thus, the CNT surface active sites available for dye adsorption were reduced. CNF–GnP formed porous aerogels with the main framework still being composed of CNF, while granular GnPs were evenly distributed in the CNFs matrix after ultrasonication and freeze-drying (Figure 3g,h). This can be attributed to the strong interactions between CNF and GnP, preventing GnP stacking and improving the hydrophilicity of GnPs [38]. The graphene platelets were well separated by rod-like CNFs. To contrast with the irregular and aggregated structure of CNT and GnP powders, the hybrid aerogels exhibited an open pore network that can facilitate fast molecular diffusion, hence promoting the accessibility of adsorption sites to relatively large dye molecules. Noteworthy, the morphology of CNF–GnP aerogels was quite different from the curly morphology of pure graphene aerogel in a previous study [39].

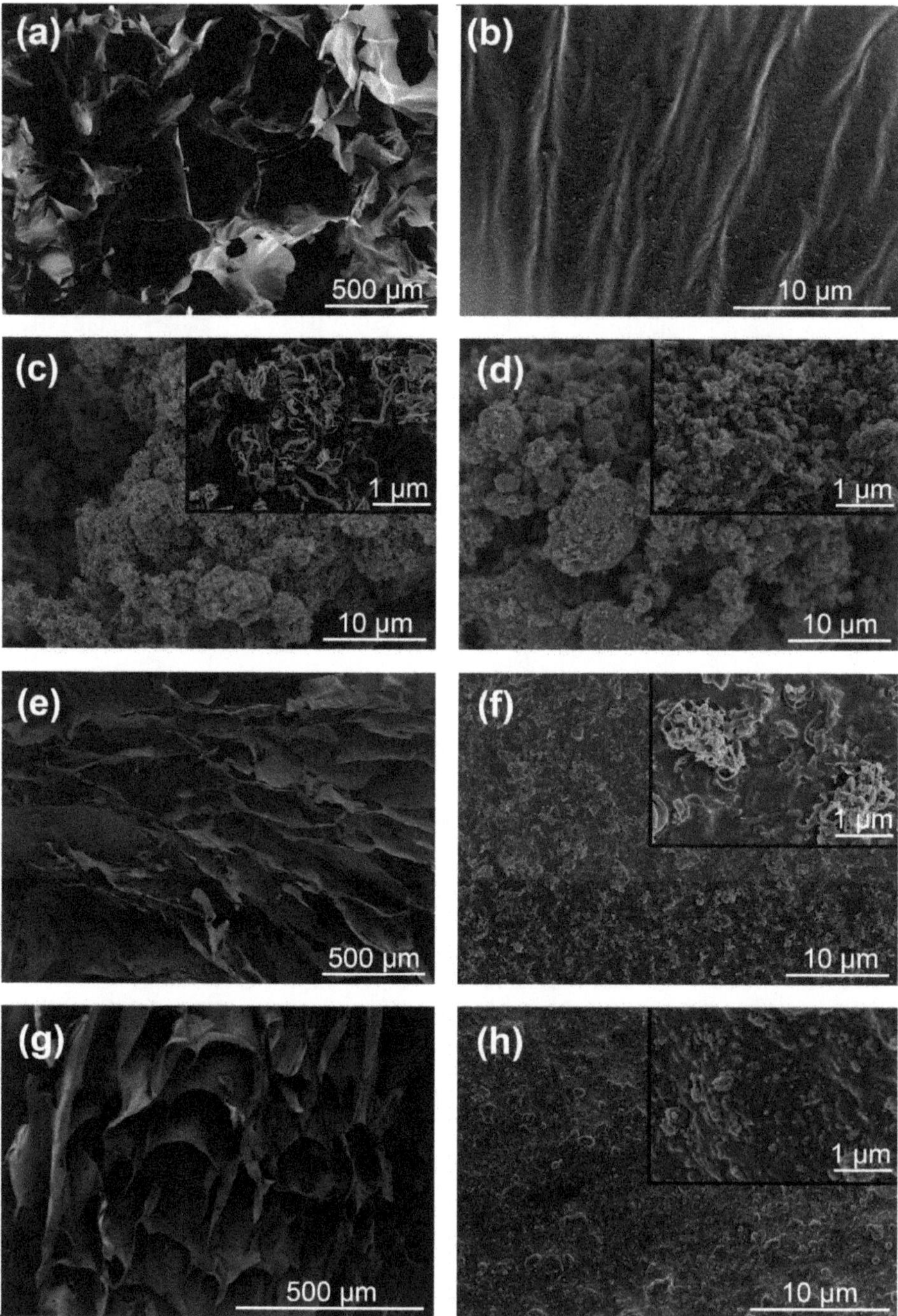

Figure 3. FE–SEM images of CNF aerogel (**a**,**b**), CNT (**c**), GnP (**d**), CNF–CNT aerogel (**e**,**f**) and CNF–GnP aerogel (**g**,**h**). The insets in (**c**,**d**,**f**,**h**) are FE–SEM images of CNT, GnP, CNF–CNT and CNF–GnP at a high magnification, respectively.

The FTIR spectra of CNF, CNF–CNT, CNF–GnP, CNT and GnP are shown in Figure 4. The CNF spectrum showed common cellulose peaks: broad hydroxyl stretching at 3360 cm^{-1} and bending at 1610 cm^{-1}, predominant C–O peaks at 1168, 1112, and 1062 cm^{-1}, and a C–H stretching peak at 2900 cm^{-1}, respectively. The small shoulder at 1712 cm^{-1} was associated with the carbonyl stretching of the carboxylic acid, confirming C6 primary hydroxyl conversion to carboxyls from TEMPO oxidation [21]. The CNT and GnP spectra were nearly featureless. A small bump at 1570 cm^{-1} was assigned to C=C groups in graphene. Bending vibrations of C–O–C at 1210 cm^{-1} and C–O at 1038 cm^{-1}, respectively, indicated epoxide or C–OH structure existing in CNT and graphene. These weak vibration peaks confirmed that the degree of oxidation in CNT and GnP were low. The cellulose characteristic peaks also were observed in the hybrid aerogels containing 25% CNT/GnP. The change of wavenumber for O–H in the hybrids indicated the existence of hydrogen bonding between CNT/GnP and CNF [40]. Since CNT and GnP only had few oxygen containing groups, van der Waals forces and, perhaps, hydrophobic interactions also contributed to the combination of the carbon nanomaterials and CNFs.

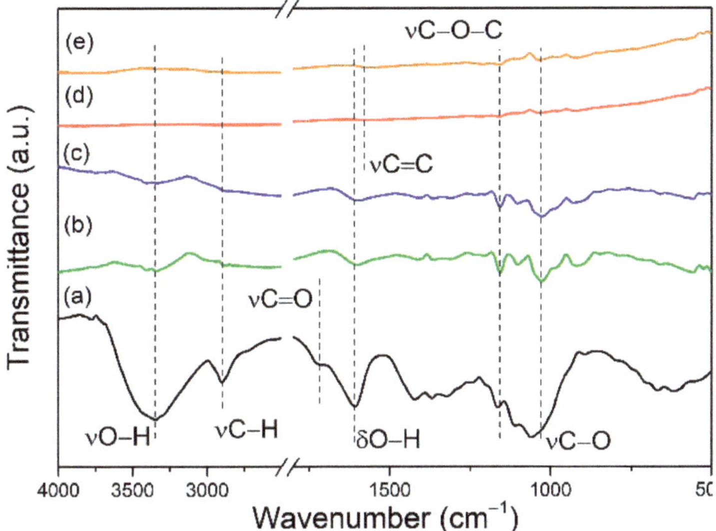

Figure 4. FTIR spectra of CNF (**a**), CNF–CNT (**b**), CNF–GnP (**c**), CNT (**d**) and GnP (**e**). To optimize the representation, the region of 2500–1800 cm^{-1} is omitted.

Based on adsorption, dispersion, mechanics, morphology and IR results mentioned above, the best performance of the CNF–GnP 3:1 hybrid aerogel perhaps resulted from the plate structure and large surface area of GnPs (i.e., The specific surface areas of GnP and CNT were 750 and 60 m^2, respectively). A sufficient amount of CNFs prevented GnPs from stacking and improved the hydrophilicity of GnPs. However, CNTs still tangled with each other in the CNF matrix. Dye molecules adsorbed to the GnP portion mainly through π–π and hydrophobic interactions, while cationic MB was able to adsorb to the negatively charged CNF portion by electrostatic interactions. Thus, CNF–GnP 3:1 aerogel was selected for future adsorption kinetics and isothermal modeling.

3.2. Effect of Contact Time and Adsorption Kinetics

The effect of CNF, CNF–GnP and GnP contact time (25 °C, 120 rpm) on dye removal was studied at low, medium and high initial dye concentrations (Figure 5). The initial MB concentrations were 10, 250 and 500 mg L^{-1} and the initial CR concentrations were 10, 600 and 2000 mg L^{-1}. The adsorption of MB dye onto CNF, CNF–GnP and GnP occurred rapidly during the first 30 min, then leveled beyond

60 min at all initial MB dye concentrations. The adsorption of CR dye onto CNF, CNF–GnP and GnP occurred at a lower speed compared to that of MB. The adsorption onto CNF and CNF–GnP slowed with adsorption time and reached a plateau beyond 90 min at all initial CR concentrations. However, the adsorption onto GnP still increased very slowly even after 120 min at all initial CR concentrations.

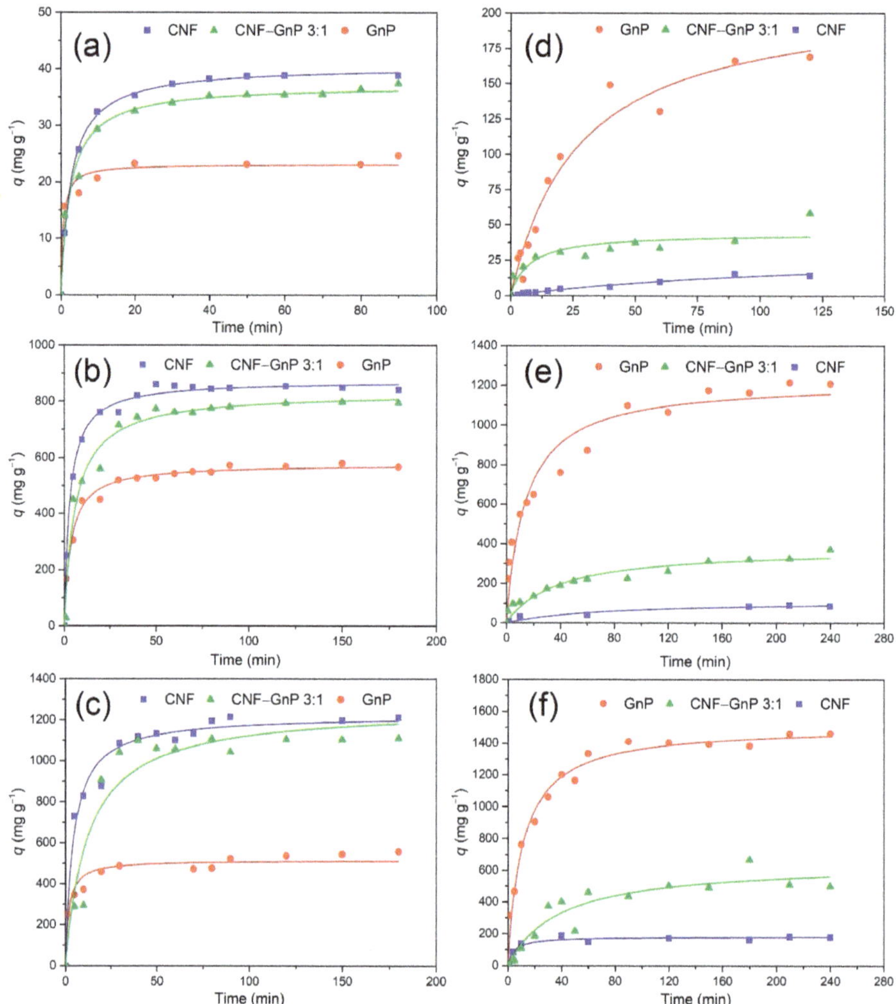

Figure 5. Effects of contact time on the dye removal efficiencies of MB and CR using CNF, CNF–GnP and GnP. Initial MB dye concentration: 10 mg L^{-1} (**a**), 250 mg L^{-1} (**b**), 500 mg L^{-1} (**c**). Initial CR concentration: 100 mg L^{-1} (**d**), 600 mg L^{-1} (**e**), 2000 mg L^{-1} (**f**). Pseudo-second-order adsorption kinetics was applied for all conditions in (**a–f**).

Adsorption kinetics models can be employed to predict the equilibrium adsorption capacity and elucidate the adsorption mechanism. During the adsorption process, the dye molecules migrated from the aqueous solution onto the surface of the adsorbent. MB molecules were adsorbed through electrostatic interactions. The electrostatic interactions occurred when the cationic dye MB was close enough to the adsorption sites ($-COO^-$, $-OH$) on the adsorbent surface. CR molecules were adsorbed mainly through π–π bonding and hydrophobic interactions with carbon nanomaterials.

Accompanying the increase in contact time, the accumulation of dye molecules on the adsorbent surface gradually increased and eventually reached equilibrium. The adsorption kinetics of MB and CR on different absorbents was evaluated using both the Lagergren's pseudo-first-order and the Ho's pseudo-second-order models. The Lagergren's pseudo-first-order kinetics is expressed as Equation (6) [41]:

$$q_t = q_e(1 - e^{-k_1 t}) \tag{6}$$

where k_1 is the rate constant (min^{-1}), q_t is the amounts of dye absorbed at a given time (mg g^{-1}), and q_e is the amount of dye adsorbed at equilibrium (mg g^{-1}). Nonlinear regression analysis was used to assess the values of q_e, k_1. The Ho's pseudo-second-order kinetics was expressed as Equation (7) [42]:

$$q_t = \frac{q_e^2 k_2 t}{1 + q_e k_2 t} \tag{7}$$

where k_2 is the pseudo-second-order rate constant (g mg^{-1} min^{-1}). Nonlinear regression analysis was used to assess the values of q_e, k_2. The initial adsorption rate v_0 at $t = 0$ could be calculated using Equation (8):

$$v_0 = k_2 \times q_e^2 \tag{8}$$

Regarding all adsorbents, the pseudo-second-order model was generally more applicable for describing the adsorption of MB, as demonstrated by the higher correlation coefficients (R^2), compared to the first-order kinetics (Table 1). This result was consistent with previous reports of cationic dyes adsorbed onto pure CNF [22] and pure graphene [43]. The initial MB adsorption rate (v_0) and MB dye adsorption capacity (q_e) increased rapidly with increasing original dye concentrations from 10 to 250 mg L^{-1} for all adsorbents. Further increasing of the original MB concentration from 250 mg L^{-1} to 500 mg L^{-1} resulted in an increase of v_0 for pure GnP, while v_0 did not change significantly in the cases of pure CNF and CNF–GnP hybrid. This phenomenon may be attributed to the limited amount of negatively charged adsorption sites on the CNF surface at longer contact times and higher MB concentrations. Although the adsorption of MB onto CNF–GnP sorbent was relatively slower than that on each component alone, the hybrid aerogel exhibited the highest theoretical MB adsorption capacity ($q_e = 1264.5$ mg g^{-1}) at a high initial MB concentration (i.e., 500 mg L^{-1}).

The pseudo-second order kinetic model was also clearly a better fit for the adsorption of CR onto pure GnP, which is consistent with a previous study [44]. However, the sorption of CR onto pure CNF could be described as either pseudo first-order kinetics or pseudo second-order kinetics, as indicated by the similar R^2 (0.94–0.98) for both models. Occurring at neutral pH, both CR and CNF were negatively charged. Adsorption of CR onto CNF was relatively low and possibly resulted from hydrophobic interaction. The adsorption of CR onto the CNF–GnP hybrid also could be represented by either the first-order or second-order model and exhibited much higher uptake values than pure CNF. This result possibly indicates that in the hybrid material, CNF did not significantly affect the adsorption kinetics of GnP, even though the GnP content was relatively small. Both the CR adsorption rate (v_0) and CR adsorption capacity (q_e) increased with increasing the initial dye concentrations from 100 to 2000 mg L^{-1} for all adsorbents. The augmentation of the initial concentration provided a greater driving force for the mass transfer and subsequent adsorption on the nanomaterials [45]. The theoretical q_e value for CNF–GnP reached 648.5 mg g^{-1} at a high initial CR concentration, 2.5 times higher than pristine CNF (182.4 mg g^{-1}).

Table 1. Estimated kinetic parameters of the two adsorption models for methylene blue (MB) and Congo red (CR).

Sample		MB Concentration (mg L^{-1})			CR Concentration (mg L^{-1})		
GnP	Param.	10	250	500	100	600	2000
pseudo-first order	q_e (mg g^{-1})	22.2	544.2	496.7	167.0	1113.7	1383.5
	k_1 (min^{-1})	1.192	0.170	0.282	0.040	0.049	0.059
	R^2	0.927	0.955	0.856	0.971	0.912	0.960
pseudo-second order	q_e (mg g^{-1})	23.2	578.3	515.9	213.2	1223.5	1510.4
	k_2 (g mg^{-1} min^{-1})	7.1×10^{-2}	4.7×10^{-4}	1.1×10^{-3}	1.7×10^{-4}	5.9×10^{-5}	6.1×10^{-5}
	v_0 (mg g^{-1} min^{-1})	38.0	157.1	282.2	7.76	87.9	138.4
	R^2	0.966	0.985	0.937	0.968	0.954	0.981
CNF							
pseudo-first order	q_e (mg g^{-1})	37.8	831.0	1140.0	20.7	92.5	173.4
	k_1 (min^{-1})	0.232	0.198	0.145	0.011	0.012	0.153
	R^2	0.986	0.972	0.933	0.982	0.942	0.960
pseudo-second order	q_e (mg g^{-1})	40.6	874.8	1225.0	33.0	114.5	182.4
	k_2 (g mg^{-1} min^{-1})	8.9×10^{-3}	3.8×10^{-4}	1.9×10^{-4}	2.2×10^{-4}	1.20×10^{-4}	1.2×10^{-3}
	v_0 (mg g^{-1} min^{-1})	14.7	294.0	278.1	0.241	1.57	38.6
	R^2	0.999	0.994	0.976	0.978	0.948	0.952
CNF–GnP 3:1							
pseudo-first order	q_e (mg g^{-1})	35.2	767.2	1114.0	37.9	321.4	538.8
	k_1 (min^{-1})	0.210	0.112	0.063	0.128	0.022	0.024
	R^2	0.944	0.961	0.951	0.915	0.931	0.992
pseudo-second order	q_e (mg g^{-1})	37.2	834.3	1264.5	44.1	379.4	648.5
	k_2 (g mg^{-1} min^{-1})	1.0×10^{-2}	2.0×10^{-4}	6.3×10^{-5}	3.0×10^{-3}	7.07×10^{-5}	4.0×10^{-5}
	v_0 (mg g^{-1} min^{-1})	14.1	137.2	100.2	5.92	10.2	16.9
	R^2	0.978	0.977	0.910	0.940	0.955	0.972

3.3. Effect of Initial Dye Concentration and Adsorption Isotherm

The effect of initial dye concentration on the adsorption capacity was investigated in the 10–1000 mg L^{-1} and 10–2000 mg L^{-1} ranges for MB and CR, respectively. The systems were mixed at 25 °C and 120 rpm for 16 h and equilibrium was declared when there was no appreciable change in solution concentration with additional contact time. Concerning all adsorbents, at low dye concentration, adsorption increased dramatically with increasing concentration (Figure 6). The adsorption of MB reached a plateau when residual MB concentration was above 200 mg L^{-1} in all cases. The final MB adsorption at equilibrium for CNF–GnP was 1207.5 mg g^{-1} at the initial MB concentration of 1000 mg L^{-1}. The adsorption of CR increased relatively slowly when the residual CR concentration was above 400 mg L^{-1} in all cases. The final CR adsorption at equilibrium of CNF–GnP was 507.1 mg g^{-1} at the initial CR concentration of 2000 mg L^{-1}.

To further understand the mechanism of adsorption, the adsorbed quantities and residual dyes in the solution at equilibrium were fitted with isothermal models. The adsorption isotherm models describe the interaction between the adsorbate and adsorbent. Three models, Langmuir, Freundlich and Sips, were used to obtain the isotherm parameters for adsorption of dyes onto CNF, GnP, CNF–GnP.

The Langmuir isotherm equation is expressed as follows [46]:

$$q_e = \frac{q_{max} K_L C_e}{1 + K_L C_e} \tag{9}$$

where q_e is the equilibrium adsorption amount per unit weight of the adsorbent (mg g^{-1}), C_e is the equilibrium concentration of adsorbate in the solution (mg L^{-1}), q_{max} is the maximum amount of the dyes adsorbed per unit weight of the adsorbent (mg g^{-1}), which describes the complete single-layer coverage on the surface of the dye at a high equilibrium concentration of dyes, and K_L is the Langmuir adsorption equilibrium constant related to binding site affinity (L g^{-1}), representing the bonding energy of adsorbent and adsorption product. The Langmuir isotherm model is based on the monolayer sorption on a surface with a finite number of identical sites and uniform adsorption energies.

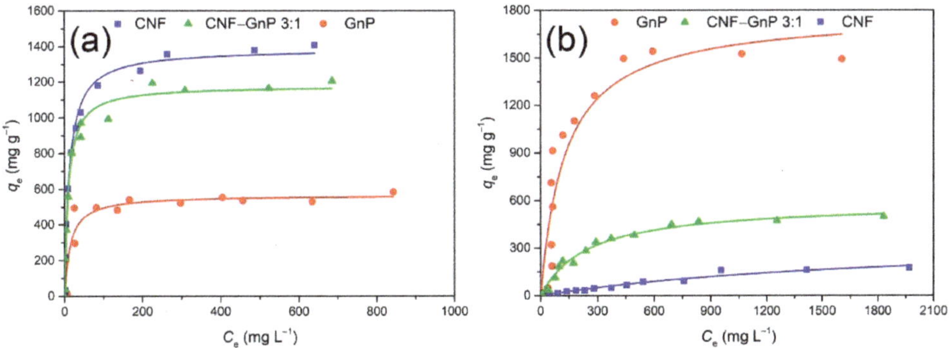

Figure 6. Adsorption of MB (**a**) and CR (**b**) to CNF, CNF–GnP and GnP. The Langmuir adsorption model was applied for data fitting.

The Freundlich isotherm equation is expressed as below [47]:

$$q_e = K_F \times C_e^{\frac{1}{n}} \tag{10}$$

where K_F (mg g^{-1}) and n are the Freundlich constants. The Freundlich isotherm model is an empirical equation for understanding the adsorption of heterogeneous surfaces with multiple adsorption layers. K_F and n are related to adsorption capacity, adsorption strength and spontaneity, respectively. When the value of n is within the range of $1 < n < 10$, it indicates a good adsorption process. The larger n value, the better the adsorption effect.

The Sips isotherm equation is given as follows [48]:

$$\frac{1}{q_e} = \frac{1}{q_{max}K_S}\left(\frac{1}{C_e}\right)^{\frac{1}{n}} + \frac{1}{q_{max}} \tag{11}$$

where q_{max} is the Sips constant related to maximum adsorption capacity (mg g^{-1}), K_S is the isotherm constant of Sips related to adsorption energy (L g^{-1}), and n is the heterogeneity factor. The Sips model is a combination of the Langmuir and Freundlich isotherms. As K_S approaches 0, the Sips isotherm equation follows the Freundlich model. When n approaches or equals 1, the Sips isotherm equation is reduced to the Langmuir isotherm.

The fitting parameters of each model are listed in Table 2. According to the correlation coefficients (R^2), both the Langmuir and Sips adsorption models could adequately describe the dye adsorption on each adsorbent, while the Freundlich isotherm model was the least suitable. Since the Sips model is derived from the Langmuir equation, employs one more fitting parameter, and yields similar correlation coefficients, it could be concluded that the Langmuir model is more appropriate to describe the adsorption behavior. The Langmuir fitting curves for the adsorption of MB and CR on the different adsorbents are shown in Figure 6. The binding constant K_L is related to the adsorption energy between the adsorbent and the dye. MB displayed higher binding constants (GnP: 7.0×10^{-2} L g^{-1}, CNF: 8.3×10^{-2} L g^{-1}, CNF–GnP: 1.1×10^{-1} L g^{-1}) than CR (GnP: 7.1×10^{-3} L g^{-1}, CNF: 8.6×10^{-4} L g^{-1},

CNF–GnP: 3.8×10^{-3} L g^{-1}) regardless of the adsorbent nature, indicating a higher binding affinity to MB. Compared to the other dye-adsorbent complexes, the uptake of MB on CNF–GnP was the most favorable. The Langmuir model revealed that both MB and CR adsorbed as a monolayer on the CNF–GnP surfaces, with maximum adsorption capacities of 1178.5 mg g^{-1} and 585.3 mg g^{-1}, respectively. According to Equation (1) and the monolayer adsorption, the specific surface areas (SSA) of CNF and CNF–GnP were determined to be 3220.4 and 3036.1 m^2 g^{-1}, respectively. Theoretically, if CNF is assumed to be a perfect cylinder with a 1 nm diameter (cellulose density: 1.5 g/cm^3), the surface area of CNF is 2667 m^2/g. The high SSA determined by MB adsorption indicates CNFs contained a large number of micropores.

Table 2. Estimated adsorption parameters of Langmuir, Freundlich and Sips isotherms at room temperature.

Langmuir Adsorption Model		**GnP**		**CNF**		**CNF–GnP 3:1**	
		MB	**CR**	**MB**	**CR**	**MB**	**CR**
	q_{max} (mg g^{-1})	567.1	1787.3	1387.2	351.7	1178.5	585.3
$q_e = (q_{max}K_L C_e)/(1 + K_L C_e)$	K_L (L g^{-1})	7.0×10^{-2}	7.1×10^{-3}	8.3×10^{-2}	8.6×10^{-4}	1.1×10^{-1}	3.8×10^{-3}
	R^2	0.879	0.858	0.984	0.960	0.985	0.980
Freundlich adsorption model		**GnP**		**CNF**		**CNF–GnP 3:1**	
		MB	**CR**	**MB**	**CR**	**MB**	**CR**
	n	5.431	2.995	4.512	1.428	4.984	1.944
$q_e = K_F \times C_e^{1/n}$	K_F (mg g^{-1})	181.3	156.1	376.5	0.966	364.2	29.9
	R^2	0.773	0.719	0.902	0.935	0.892	0.864
Sips adsorption model		**GnP**		**CNF**		**CNF–GnP 3:1**	
		MB	**CR**	**MB**	**CR**	**MB**	**CR**
	q_{max} (mg g^{-1})	552.4	1515.0	1453.0	235.2	1231.1	517.9
$1/q_e = (1/q_{max}K_s) \times (1/C_e)^{1/n} + (1/q_{max})$	K_s (L g^{-1})	4.1×10^{-2}	2.3×10^{-4}	1.2×10^{-1}	1.0×10^{-4}	1.5×10^{-1}	6.8×10^{-4}
	n	0.819	0.534	1.222	0.727	1.243	0.723
	R^2	0.869	0.885	0.991	0.960	0.988	0.969

Extensive research about the uptake of various dyes on carbon-based and cellulose/polysaccharide-based composites has been reported in the literature. Table 3 presents a comparison of the maximum dye adsorption capacity of different adsorbents. CNF–GnP was able to adsorb both cationic MB and anionic CR and yielded higher uptake values compared to recent studies using cellulose, activated carbon, graphene, and CNT-based composites.

Table 3. Comparison of adsorption capacity of different adsorbents.

Adsorbent	q_{max} (mg $^{-1}$) MB	Ref.	Adsorbent	q_{max} (mg g^{-1}) CR	Ref.
CNF–GnP aerogel	1178.5	this study	CNF–GnP aerogel	585.3	this study
Zeolite—activated carbon composite from oil palm ash	285.71	[49]	Polyaniline@GO-multiwalled carbon nanotube nanocomposite	66.67	[44]
Nickel nanoparticles/porous carbon—carbon nanotube hybrids	312	[50]	Chitosan hydrogel beads impregnated with CNT	450.4	[51]
3D rGO/L–Cys hydrogel	660	[52]	Functionalized multiwalled carbon nanotubes	148	[53]
Graphene/cellulose nanofibers	227.27	[28]	Polyacrylamide grafted quaternized cellulose	349.28	[54]
GO/calcium alginate composites	181.81	[55]	CaCO$_3$–cellulose aerogel	75.81	[56]

3.4. Adsorption of Dyes in Binary Systems

Regarding industrial wastewater, different types of dyes could be found and they compete for the adsorption sites on the surface of the adsorbent. To investigate the adsorption capacity of CNF–GnP in a more practical setting, an MB and CR binary system was prepared and investigated. The mass ratio of MB to CR in the solution was designed to be 3:1, 1:1 and 1:3. The total initial dye concentration was set to be 200 mg L^{-1}. The adsorption of dye in single and binary systems was compared (Figure 7). Shown in Figure 7a, when the mass ratio of MB to CR was 3:1 (i.e., MB = 150 mg L^{-1}, CR = 50 mg L^{-1}), the adsorption of MB onto pure CNF in the binary system was lower than that in the single system, indicating MB and CR competed for adsorption sites on the CNF surface. However, the adsorption of MB onto pure GnP and CNF–GnP was similar in both the single and binary systems. The adsorption of CR onto all adsorbents increased in the binary system compared to that in the single system. This may be attributed to the Coulombic attraction between cationic MB adsorbed on the material surface and anionic CR in solution. Similarly, when the mass ratio of MB to CR was 1:1 (Figure 7b), the adsorption of MB was lower and the CR uptake for pure CNF and CNF–GnP was higher in the binary system than in the single system. However, the adsorption of CR onto pure GnP was strongly affected by the presence of MB when the CR concentration increased. Moreover, when CR became the dominant molecule in the binary system (i.e., MB to CR ratio was 1:3, Figure 7c), the CNF–GnP hybrid was able to adsorb even more CR than pure GnP. Regarding all binary systems, the adsorption capacity of CNF–GnP was the best among the different nano adsorbents, and the presence of one dye had no negative impact on the adsorption of the other dye. The hybrid CNF–GnP adsorbed the highest amount of MB and CR combined per unit mass of adsorbent. Perhaps the cationic MB and anionic CR adsorbed onto different types of adsorption sites on the CNF–GnP surface. The cationic MB mainly adsorbed onto the negatively charged CNF portion, while most of the CR adsorbed onto the GnP portion. MB adsorbed on the material surface also may attract anionic CR in solution via Coulombic attraction.

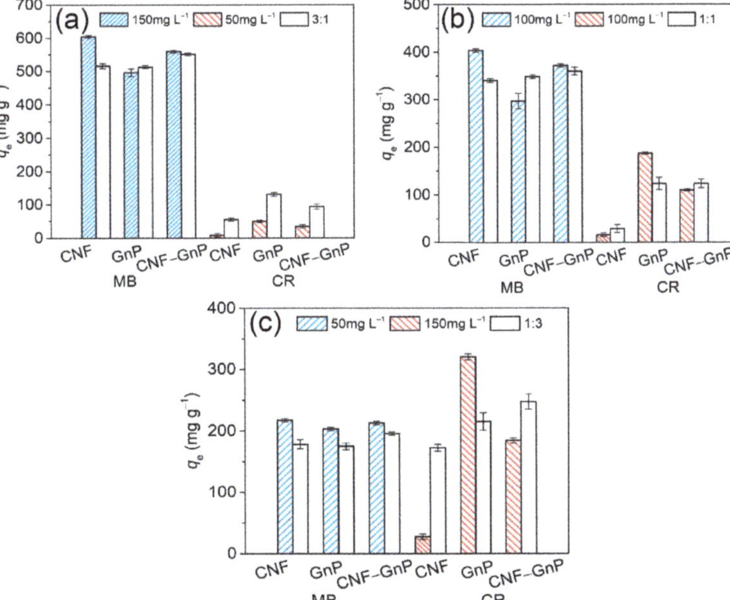

Figure 7. Adsorption of MB and CR in single and binary dye solution. (**a**) mass ratios of MB to CR are 3:1; (**b**) mass ratio of MB to CR are 1:1; (**c**) mass ratio of MB to CR are 1:3. Colored bars represent adsorption in a single dye system. Colorless bars represent adsorption of one dye in a binary dye system.

3.5. Desorption

The desorption of MB and CR from CNF, GnP and CNF–GnP by ethanol, acetonitrile, acetone and 400 mM NaCl was investigated. The desorption of dye adhered onto CNF, GnP and CNF–GnP was not effective using acetonitrile, acetone and NaCl. Upon immersion in acetonitrile, acetone or 400 mM NaCl for 1 h, only 23.6%, 17.2% or 28.3% of pre-adsorbed MB in CNF–GnP hybrid (100 mg L^{-1} MB, 20 mL MB solution, 16 h) was desorbed, respectively. Upon immersion in acetone or 400 mM NaCl, only 6.0% or 20.1% of pre-adsorbed MB in GnP was desorbed. Concerning CNF, only 36.6% MB was desorbed by acetone. Although 400 mM NaCl could desorb 88.6% MB under the same conditions, the pure CNF aerogel could not remain intact after immersion for 1 h. Strong ionic conditions destroyed the hydrogen bonding network in pure CNF. Dyes adhered to CNF, GnP and CNF–GnP were desorbed rapidly by ethanol (Figure 8). Desorption of MB and CR adhered to CNF–GnP by ethanol was relatively more effective. After 1h of immersion, 42.4% of the MB and 51.0% of the CR were desorbed from CNF–GnP by ethanol. Finally, after four rounds of desorption, 79.2% of the MB and 78.3% of the CR were desorbed. Anhydrous ethanol is a protic solvent that contains polarized oxyhydrogen bonds which ionize to form alkoxyl negative ions and protons (hydrogen ions). It can provide lone pair electron interaction with MB (cationic dye) molecules. Concurrently, anhydrous ethanol also can provide protons with CR (anionic dye) molecules to form hydrogen bonds. The rapid desorption of both MB and CR demonstrate that the CNF–GnP hybrid aerogel could be easily regenerated for repeated dye removal applications.

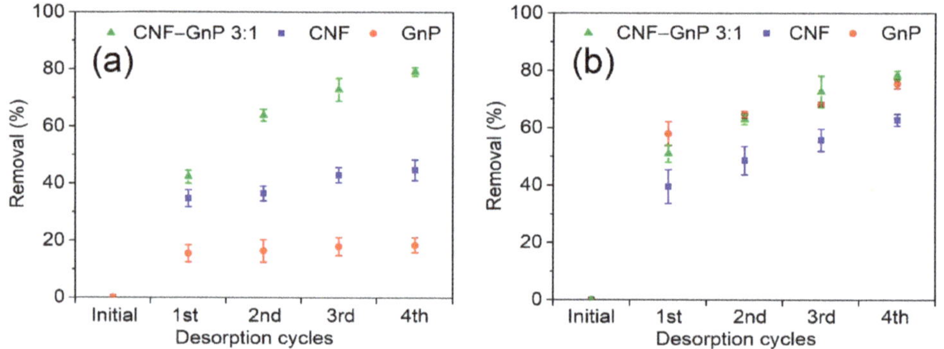

Figure 8. Cyclic desorption of MB (**a**) and CR (**b**) with CNF–GnP, CNF, GnP by ethanol.

4. Conclusions

Hybrid aerogels containing TEMPO oxidized cellulose nanofibrils (CNFs) and carbon nanomaterials (carbon nanotubes (CNTs) or graphene nanoplates (GnPs)), were designed and synthesized by freeze-drying to remove organic dyes from single and binary systems. When the CNF to GnP mass ratio was 3:1, the hybrid aerogel exhibited the most effective adsorption of both methylene blue (MB) and Congo red (CR) among all the hybrid systems tested. The final adsorption capacities of CNF–GnP 3:1 aerogels for MB and CR reached 1166.1 mg g^{-1} and 507.1 mg g^{-1} at initial dye concentrations of 500 mg L^{-1} and 2000 mg L^{-1}, respectively. The CNFs enhanced the dispersion of carbon nanomaterials in an aqueous environment. The hybrid aerogels were mechanically strong and exhibited water-activated shape recovery. Seen in a single dye adsorption system, the adsorption ability measurements demonstrate that the CNF–GnP 3:1 aerogel possessed the adsorption capacity and adsorption rate close to CNF aerogels in cationic MB solutions. The adsorption capacity and adsorption rate of CNF–GnP aerogels was more efficient than the CNF aerogel in an anionic CR solution. Dye adsorptions to CNF–GnP followed a pseudo-second-order adsorption kinetic and the existence of CNF did not affect the adsorption kinetics of GnP. The adsorption followed a monolayer

Langmuir isotherm. Concerning a binary system, the CNF–GnP aerogel removed cationic MB as well as anionic CR at a higher total dye adsorption capacity than pristine CNF or GnP. Moreover, 79.2% and 78.3% of the MB and CR were desorbed from CNF–GnP by using ethanol as the desorption agent, suggesting the reusability of this hybrid material. Results of this study indicate that CNF–GnP show promise as high-potential adsorbents for organic dye removal.

Supplementary Materials: The following are available online at http://www.mdpi.com/2079-4991/10/1/169/s1, Figure S1: Conductometric titration curves of CNF (a) with HCl added and (b) without HCl added., Figure S2: Optical micrographs of (a) CNF/CNT 3:1 after 5 min ultrasonic treatment; (b) CNF/CNT 3:1 after 30 min ultrasonic treatment; and (c) CNF/GnP 3:1 after 5 min of ultrasonic treatment; (d) CNF/GnP 3:1 after 30 min ultrasonic treatment, Table S1: Chemical structure, molecular weight, maximum absorption wavelength and electrical property of methylene blue and Congo red. Table S2: Group statistics of the final adsorption of MB. Table S3: Independent sample tests of the final adsorption of MB onto CNF–GnP 3:1 and CNF–CNT 3:1. Table S4: Group statistics of the final adsorption of CR. Table S5: Independent sample tests of the final adsorption of CR onto CNF–GnP 3:1 and CNF–CNT 3:1. Movie S1: Shape recovery of CNF aerogel in water after compression. Movie S2: Shape recovery of CNF–CNT aerogel in water after compression. Movie S3: Shape recovery of CNF–GnP aerogel in water after compression in water.

Author Contributions: Conceptualization, J.G.; methodology, A.B.D.; investigation, Z.Y. and W.J.; data curation, Z.Y.; writing—original draft preparation, Z.Y.; writing—review and editing, J.G. and A.B.D.; supervision, C.H.; funding acquisition, C.H. and J.G. All authors have read and agreed to the published version of the manuscript.

Funding: This research was funded by Guangzhou Science and Technology Bureau, grant number 201904010308, the Guangdong Government Science and Technology, grant number 2017B020238003 and Bureau of Guangdong Forestry, grant number 2017-LYBZ-012.

Conflicts of Interest: The authors declare no conflict of interest.

References

1. Ahmed, M.J. Application of agricultural based activated carbons by microwave and conventional activations for basic dye adsorption: Review. *J. Environ. Chem. Eng.* **2016**, *4*, 89–99. [CrossRef]

2. Mohammed, N.; Grishkewich, N.; Berry, R.M.; Tam, K.C. Cellulose nanocrystal–alginate hydrogel beads as novel adsorbents for organic dyes in aqueous solutions. *Cellulose* **2015**, *22*, 3725–3738. [CrossRef]

3. Tan, I.A.W.; Ahmad, A.L.; Hameed, B.H. Adsorption of basic dye using activated carbon prepared from oil palm shell: Batch and fixed bed studies. *Desalination* **2008**, *225*, 13–28. [CrossRef]

4. Purkait, M.K.; Maiti, A.; DasGupta, S.; De, S. Removal of congo red using activated carbon and its regeneration. *J. Hazard. Mater.* **2007**, *145*, 287–295. [CrossRef] [PubMed]

5. Singh, N.B.; Nagpal, G.; Agrawal, S. Water purification by using Adsorbents: A Review. *Environ. Technol. Innov.* **2018**, *11*, 187–240. [CrossRef]

6. Gu, J.; Hu, C.; Zhang, W.; Dichiara, A. Reagentless preparation of shape memory cellulose nanofibril aerogels decorated with Pd nanoparticles and their application in dye discoloration. *Appl. Catal. B Environ.* **2018**, *237*, 482–490. [CrossRef]

7. Dante, R.C.; Martín-Ramos, P.; Chamorro-Posada, P.; Meejoo-Smith, S.; Vázquez-Cabo, J.; Rubiños-López, Ó.; Lartundo-Rojas, L.; Sánchez-Árevalo, F.M.; Trakulmututa, J.; Rutto, D.; et al. Comparison of the activities of C2N and BCNO towards Congo red degradation. *Mater. Chem. Phys.* **2019**, *221*, 397–408. [CrossRef]

8. Zhu, X.; Liu, Y.; Zhou, C.; Zhang, S.; Chen, J. Novel and High-Performance Magnetic Carbon Composite Prepared from Waste Hydrochar for Dye Removal. *ACS Sustain. Chem. Eng.* **2014**, *2*, 969–977. [CrossRef]

9. Ma, J.; Yu, F.; Zhou, L.; Jin, L.; Yang, M.; Luan, J.; Tang, Y.; Fan, H.; Yuan, Z.; Chen, J. Enhanced Adsorptive Removal of Methyl Orange and Methylene Blue from Aqueous Solution by Alkali-Activated Multiwalled Carbon Nanotubes. *ACS Appl. Mat. Interfaces* **2012**, *4*, 5749–5760. [CrossRef]

10. Moradi, O. Adsorption Behavior of Basic Red 46 by Single-Walled Carbon Nanotubes Surfaces. *Fuller. Nanotub. Carbon Nanostruct.* **2013**, *21*, 286–301. [CrossRef]

11. Zare, K.; Sadegh, H.; Shahryari-Ghoshekandi, R.; Maazinejad, B.; Ali, V.; Tyagi, I.; Agarwal, S.; Gupta, V.K. Enhanced removal of toxic Congo red dye using multi walled carbon nanotubes: Kinetic, equilibrium studies and its comparison with other adsorbents. *J. Mol. Liq.* **2015**, *212*, 266–271. [CrossRef]

12. Yu, J.; Yu, L.; Yang, H.; Liu, Q.; Chen, X.; Jiang, X.; Chen, X.; Jiao, F. Graphene nanosheets as novel adsorbents in adsorption, preconcentration and removal of gases, organic compounds and metal ions. *Sci. Total Environ.* **2015**, *502*, 70–79. [CrossRef] [PubMed]

13. Bolotin, K.I.; Sikes, K.J.; Jiang, Z.; Klima, M.; Fudenberg, G.; Hone, J.; Kim, P.; Stormer, H.L. Ultrahigh electron mobility in suspended graphene. *Solid State Commun.* **2008**, *146*, 351–355. [CrossRef]

14. Yang, K.; Wang, J.; Chen, X.; Zhao, Q.; Ghaffar, A.; Chen, B. Application of graphene-based materials in water purification: From the nanoscale to specific devices. *Environ. Sci. Nano* **2018**, *5*, 1264–1297. [CrossRef]

15. Pan, X.; Du, Q.; Zhou, Y.; Liu, L.; Xu, G.; Yan, C. Ammonia Borane Promoted Synthesis of Graphene Aerogels as High Efficient Dye Adsorbent. *J. Nanosci. Nanotechnol.* **2018**, *18*, 7231–7240. [CrossRef]

16. Kim, H.; Kang, S.-O.; Park, S.; Park, H.S. Adsorption isotherms and kinetics of cationic and anionic dyes on three-dimensional reduced graphene oxide macrostructure. *J. Ind. Eng. Chem.* **2015**, *21*, 1191–1196. [CrossRef]

17. Dichiara, A.B.; Benton-Smith, J.; Rogers, R.E. Enhanced adsorption of carbon nanocomposites exhausted with 2, 4-dichlorophenoxyacetic acid after regeneration by thermal oxidation and microwave irradiation. *Environ. Sci. Nano* **2014**, *1*, 113–116. [CrossRef]

18. Li, D.; Mueller, M.B.; Gilje, S.; Kaner, R.B.; Wallace, G.G. Processable aqueous dispersions of graphene nanosheets. *Nat. Nanotechnol.* **2008**, *3*, 101–105. [CrossRef]

19. Mahfoudhi, N.; Boufi, S. Nanocellulose as a novel nanostructured adsorbent for environmental remediation: A review. *Cellulose* **2017**, *24*, 1171–1197. [CrossRef]

20. Gu, J.; Catchmark, J.M.; Kaiser, E.Q.; Archibald, D.D. Quantification of cellulose nanowhiskers sulfate esterification levels. *Carbohydr. Polym.* **2013**, *92*, 1809–1816. [CrossRef]

21. Gu, J.; Hsieh, Y.-L. Surface and Structure Characteristics, Self-Assembling, and Solvent Compatibility of Holocellulose Nanofibrils. *ACS Appl. Mater. Interfaces* **2015**, *7*, 4192–4201. [CrossRef] [PubMed]

22. Jiang, F.; Dinh, D.M.; Hsieh, Y.-L. Adsorption and desorption of cationic malachite green dye on cellulose nanofibril aerogels. *Carbohydr. Polym.* **2017**, *173*, 286–294. [CrossRef] [PubMed]

23. Jin, L.; Sun, Q.; Xu, Q.; Xu, Y. Adsorptive removal of anionic dyes from aqueous solutions using microgel based on nanocellulose and polyvinylamine. *Bioresour. Technol.* **2015**, *197*, 348–355. [CrossRef] [PubMed]

24. Hamedi, M.M.; Hajian, A.; Fall, A.B.; Håkansson, K.; Salajkova, M.; Lundell, F.; Wågberg, L.; Berglund, L.A. Highly Conducting, Strong Nanocomposites Based on Nanocellulose-Assisted Aqueous Dispersions of Single-Wall Carbon Nanotubes. *ACS Nano* **2014**, *8*, 2467–2476. [CrossRef] [PubMed]

25. Jing-Quan, H.; Kai-Yue, L.; Yi-Ying, Y.; Chang-Tong, M.; Hui-Xiang, W.; Peng-Bin, Y.; Xin-Wu, X. Synthesis and electrochemical performance of flexible cellulose nanofiber-carbon nanotube/natural rubber composite elastomers as supercapacitor electrodes. *New Carbon Mater.* **2018**, *33*, 341–350.

26. Hajian, A.; Lindström, S.B.; Pettersson, T.; Hamedi, M.M.; Wågberg, L. Understanding the Dispersive Action of Nanocellulose for Carbon Nanomaterials. *Nano Lett.* **2017**, *17*, 1439–1447. [CrossRef]

27. Wei, X.; Huang, T.; Yang, J.-H.; Zhang, N.; Wang, Y.; Zhou, Z.-W. Green synthesis of hybrid graphene oxide/microcrystalline cellulose aerogels and their use as superabsorbents. *J. Hazard. Mater.* **2017**, *335*, 28–38. [CrossRef]

28. Hussain, A.; Li, J.; Wang, J.; Xue, F.; Chen, Y.; Bin Aftab, T.; Li, D. Hybrid Monolith of Graphene/TEMPO-Oxidized Cellulose Nanofiber as Mechanically Robust, Highly Functional, and Recyclable Adsorbent of Methylene Blue Dye. *J. Nanomater.* **2018**, *2018*, 1–12. [CrossRef]

29. Wu, H.; Wang, Z.-M.; Kumagai, A.; Endo, T. Amphiphilic cellulose nanofiber-interwoven graphene aerogel monolith for dyes and silicon oil removal. *Compos. Sci. Technol.* **2019**, *171*, 190–198. [CrossRef]

30. Goodman, S.M.; Ferguson, N.; Dichiara, A.B. Lignin-assisted double acoustic irradiation for concentrated aqueous dispersions of carbon nanotubes. *RSC Adv.* **2017**, *7*, 5488–5496. [CrossRef]

31. Lu, P.; Hsieh, Y.L. Preparation and characterization of cellulose nanocrystals from rice straw. *Carbohydr. Polym.* **2012**, *87*, 564–573. [CrossRef]

32. Gu, J.; Hsieh, Y.-L. Alkaline cellulose nanofibrils from streamlined alkali treated rice straw. *ACS Sustain. Chem. Eng.* **2017**, *5*, 1730–1737. [CrossRef]

33. Isogai, A.; Saito, T.; Fukuzumi, H. TEMPO-oxidized cellulose nanofibers. *Nanoscale* **2011**, *3*, 71–85. [CrossRef] [PubMed]

34. Wan, W.; Zhang, R.; Li, W.; Liu, H.; Lin, Y.; Li, L.; Zhou, Y. Graphene—Carbon nanotube aerogel as an ultra-light, compressible and recyclable highly efficient absorbent for oil and dyes. *Environ. Sci. Nano* **2016**, *3*, 107–113. [CrossRef]

35. Mahmoodi, N.M.; Hayati, B.; Arami, M.; Mazaheri, F. Single and Binary System Dye Removal from Colored Textile Wastewater by a Dendrimer as a Polymeric Nanoarchitecture: Equilibrium and Kinetics. *J. Chem. Eng. Data* **2010**, *55*, 4660–4668. [CrossRef]

36. Mahmoodi, N.M.; Hayati, B.; Arami, M. Textile Dye Removal from Single and Ternary Systems Using Date Stones: Kinetic, Isotherm, and Thermodynamic Studies. *J. Chem. Eng. Data* **2010**, *55*, 4638–4649. [CrossRef]

37. Chen, L.; Han, Q.; Li, W.; Zhou, Z.; Fang, Z.; Xu, Z.; Wang, Z.; Qian, X. Three-Dimensional graphene-Based adsorbents in sewage disposal: A review. *Environ. Sci. Pollut. Res.* **2018**, *25*, 25840–25861. [CrossRef]

38. Zhang, X.; Liu, D.; Sui, G. Superamphiphilic Polyurethane Foams Synergized from Cellulose Nanowhiskers and Graphene Nanoplatelets. *Adv. Mater. Interfaces* **2018**, *5*, 1701094. [CrossRef]

39. Dai, J.; Huang, T.; Tian, S.-Q.; Xiao, Y.-J.; Yang, J.-H.; Zhang, N.; Wang, Y.; Zhou, Z.-W. High structure stability and outstanding adsorption performance of graphene oxide aerogel supported by polyvinyl alcohol for waste water treatment. *Mater. Des.* **2016**, *107*, 187–197. [CrossRef]

40. Dichiara, A.B.; Song, A.; Goodman, S.M.; He, D.; Bai, J. Smart papers comprising carbon nanotubes and cellulose microfibers for multifunctional sensing applications. *J. Mater. Chem. A* **2017**, *5*, 20161–20169. [CrossRef]

41. Lagergren, S. Zur theorie der sogenannten adsorption gelosterstoffeKungliga Svenska Vetenskapsakademiens. *Handlingar* **1898**, *24*, 1–39.

42. Ho, Y.S.; McKay, G. Pseudo-second order model for sorption processes. *Process. Biochem.* **1999**, *5*, 451–465. [CrossRef]

43. Liu, T.; Li, Y.; Du, Q.; Sun, J.; Jiao, Y.; Yang, G.; Wang, Z.; Xia, Y.; Zhang, W.; Wang, K.; et al. Adsorption of methylene blue from aqueous solution by graphene. *Colloids Surf. B* **2012**, *90*, 197–203. [CrossRef] [PubMed]

44. Ansari, M.O.; Kumar, R.; Ansari, S.A.; Ansari, S.P.; Barakat, M.A.; Alshahrie, A.; Cho, M.H. Anion selective pTSA doped polyaniline@graphene oxide-multiwalled carbon nanotube composite for Cr(VI) and Congo red adsorption. *J. Colloid Interface Sci.* **2017**, *496*, 407–415. [CrossRef] [PubMed]

45. Dichiara, A.B.; Sherwood, T.J.; Rogers, R.E. Binder free graphene-single-wall carbon nanotube hybrid papers for the removal of polyaromatic compounds from aqueous systems. *J. Mater. Chem. A* **2013**, *1*, 14480–14483. [CrossRef]

46. Langmuir, I. The adsorption of gases on plane surfaces of glass, mica andplatinum. *J. Am. Chem. Soc.* **1918**, *40*, 42. [CrossRef]

47. Freundlich, H. Veber die adsorption in loesungen (Adsorption in solution). *Z. Phys. Chem.* **1907**, *57*, 385–470.

48. Foo, K.Y.; Hameed, B.H. Insights into the modeling of adsorption isotherm systems. *Chem. Eng. J.* **2010**, *156*, 2–10. [CrossRef]

49. Khanday, W.A.; Marrakchi, F.; Asif, M.; Hameed, B.H. Mesoporous zeolite—Activated carbon composite from oil palm ash as an effective adsorbent for methylene blue. *J. Taiwan Inst. Chem. Eng.* **2017**, *70*, 32–41. [CrossRef]

50. Jin, L.; Zhao, X.; Qian, X.; Dong, M. Nickel nanoparticles encapsulated in porous carbon and carbon nanotube hybrids from bimetallic metal-organic-frameworks for highly efficient adsorption of dyes. *J. Colloid Interface Sci.* **2017**, *509*, S0021979717310226. [CrossRef] [PubMed]

51. Chatterjee, S.; Lee, M.W.; Woo, S.H. Adsorption of congo red by chitosan hydrogel beads impregnated with carbon nanotubes. *Bioresour. Technol.* **2010**, *101*, 1800–1806. [CrossRef] [PubMed]

52. Zhang, X.; Liu, D.; Yang, L.; Zhou, L.; You, T. Self-assembled three-dimensional graphene-based materials for dye adsorption and catalysis. *J. Mater. Chem. A* **2015**, *3*, 10031–10037. [CrossRef]

53. Gupta, V.K.; Kumar, R.; Nayak, A.; Saleh, T.A.; Barakat, M.A. Adsorptive removal of dyes from aqueous solution onto carbon nanotubes: A review. *Adv. Colloid Interface Sci.* **2013**, *193–194*, 24–34. [CrossRef] [PubMed]

54. Wang, Y.; Zhao, L.; Peng, H.; Wu, J.; Liu, Z.; Guo, X. Removal of Anionic Dyes from Aqueous Solutions by Cellulose-Based Adsorbents: Equilibrium, Kinetics, and Thermodynamics. *J. Chem. Eng. Data* **2016**, *61*, 3266–3276. [CrossRef]

55. Li, Y.; Du, Q.; Liu, T.; Sun, J.; Wang, Y.; Wu, S.; Wang, Z.; Xia, Y.; Xia, L. Methylene blue adsorption on graphene oxide/calcium alginate composites. *Carbohydr. Polym.* **2013**, *95*, 501–507. [CrossRef] [PubMed]
56. Chong, K.Y.; Chia, C.H.; Zakaria, S.; Sajab, M.S.; Chook, S.W.; Khiew, P.S. CaCO3-decorated cellulose aerogel for removal of Congo Red from aqueous solution. *Cellulose* **2015**, *22*, 2683–2691. [CrossRef]

 *nanomaterials*

Article

Self-Healable Electro-Conductive Hydrogels Based on Core-Shell Structured Nanocellulose/Carbon Nanotubes Hybrids for Use as Flexible Supercapacitors

Huixiang Wang [1], Subir Kumar Biswas [2], Sailing Zhu [1], Ya Lu [1], Yiying Yue [3], Jingquan Han [1,*], Xinwu Xu [1,*], Qinglin Wu [4] and Huining Xiao [5]

[1] College of Materials Science and Engineering, Joint International Research Lab of Lignocellulosic Functional Materials, Nanjing Forestry University, Nanjing 210037, China; whx9111@163.com (H.W.); zhusailing@njfu.edu.cn (S.Z.); luyajiangsu@163.com (Y.L.)

[2] Laboratory of Active Bio-based Materials, Research Institute for Sustainable Humanosphere, Kyoto University, Uji, Kyoto 611-0011, Japan; subir.biswas.88a@st.kyoto-u.ac.jp

[3] College of Biology and Environment, Nanjing Forestry University, Nanjing 210037, China; yue@njfu.edu.cn

[4] School of Renewable Natural Resources, Louisiana State University, Baton Rouge, LA 70803, USA; wuqing@lsu.edu

[5] Department of Chemical Engineering, University of New Brunswick, Fredericton, NB E3B 5A3, Canada; hxiao@unb.ca

* Correspondence: hjq@njfu.edu.cn (J.H.); xucarpenter@aliyun.com (X.X.)

Received: 8 December 2019; Accepted: 2 January 2020; Published: 6 January 2020

Abstract: Recently, with the development of personal wearable electronic devices, the demand for portable power is miniaturization and flexibility. Electro-conductive hydrogels (ECHs) are considered to have great application prospects in portable energy-storage devices. However, the synergistic properties of self-healability, viscoelasticity, and ideal electrochemistry are key problems. Herein, a novel ECH was synthesized by combining polyvinyl alcohol-borax (PVA) hydrogel matrix and 2,2,6,6-tetramethylpiperidine-1-oxyl (TEMPO)-cellulose nanofibers (TOCNFs), carbon nanotubes (CNTs), and polyaniline (PANI). Among them, CNTs provided excellent electrical conductivity; TOCNFs acted as a dispersant to help CNTs form a stable suspension; PANI enhanced electrochemical performance by forming a "core-shell" structural composite. The freeze-standing composite hydrogel with a hierarchical 3D-network structure possessed the compression stress (~152 kPa) and storage modulus (~18.2 kPa). The composite hydrogel also possessed low density (~1.2 g cm^{-3}), high water-content (~95%), excellent flexibility, self-healing capability, electrical conductivity (15.3 S m^{-1}), and specific capacitance of 226.8 F g^{-1} at 0.4 A g^{-1}. The fabricated solid-state all-in-one supercapacitor device remained capacitance retention (~90%) after 10 cutting/healing cycles and capacitance retention (~85%) after 1000 bending cycles. The novel ECH had potential applications in advanced personalized wearable electronic devices.

Keywords: cellulose nanofibers; carbon nanotube; polyaniline; hydrogels; supercapacitor

---

1. Introduction

Nowadays, with the rapid development of personal wearable electronic devices, miniaturized energy storage devices with mechanical flexibility and even self-healing functions have attracted more and more attention; among them, the supercapacitor is a research hotspot [1,2]. The supercapacitor is a promising new energy storage device, which combines the advantages of high power from double-electric layer capacitors and high energy from batteries [3]. However, traditional supercapacitors are susceptible to various unavoidable mechanical deformations and cannot meet the needs of flexible wearable electronic devices [4]. Therefore, flexibility and self-healing characteristics are requirements for wearable energy storage devices. In order to integrate flexibility, self-healability, and electrochemical performance into a supercapacitor device, searching suitable electrode and electrolyte material has become a key to the manufacture of intelligent devices [5].

Hydrogels are considered as ideal candidates because of their flexible three-dimensional (3D) networks, high deformability, and hydrophilic property [6]. Firstly, hydrogel polymer network with high-water content dissolves ions and provides high ion conductivity, which can be compared with the ionic conductivity of a liquid while maintaining the solid shape and size to avoid liquid leakage during various mechanical deformation. It is an ideal choice for flexible supercapacitor electrolyte and separator materials [7]. Secondly, electrically conductive hydrogel (ECH) is a new functional material that combines flexible 3D hydrogel network with conductive nano-fillers. Due to the inherent porous structure, excellent conductivity, and flexibility of the ECH, it is the perfect choice for flexible electrode materials. The continuous conductive network inside the ECH provides electron transfer pathways. The hierarchical pore structure can ensure sufficient contact between the active material and the electrolyte ions to improve the electrochemical performance [8]. Finally, a supercapacitor assembled from the hydrogel-based electrode and electrolyte, because of the inherent viscoelasticity and flexibility of the hydrogel, will possess an improved interface between the electrode and the electrolyte [9]. However, the related researches on self-healable and flexible hydrogel-based supercapacitors are limited so far [2].

A novel freestanding and moldable hydrogel with excellent self-healing performance was synthesized by incorporating cellulose nanofibers (CNFs) into polyvinyl alcohol-borax (PVA) hydrogel matrix. Conductive materials were incorporated into self-healable PVA hydrogels to form multifunctional ECHs, which had potential applications in flexible and self-healable supercapacitors. In our present study, carbon nanotubes (CNTs) and polyaniline (PANI) were the electrochemically active materials in electrodes. 2,2,6,6-tetramethylpiperidine-1-oxyl (TEMPO)-cellulose nanofibers (TOCNFs) were the bio-dispersant that helped CNTs disperse in water. PANI was coated on the surface of TOCNF-CNT nano-hybrids through in-situ oxidative polymerization to form "core-shell" TOCNF-CNT@PANI composites. These composites built a hierarchically reinforced and conductive 3D network in the self-healable PVA hydrogel matrix. The hydrogel-based electrodes and electrolytes were used to assemble symmetrical solid-state all-in-one supercapacitors with flexible and fast self-healable performances.

2. Materials and Methods

2.1. Materials

Bleached wood pulp was provided by Nippon Paper Chemicals Co., Ltd. (Tokyo, Japan). 2,2,6,6-tetramethylpiperidine-1-oxyl (TEMPO, $C_9H_{18}NO$), sodium bromide (NaBr), sodium hypochlorite solution (NaClO, 6–14% active chlorine), ammonium persulfate (APS, $H_8N_2O_8S_2$), aniline monomers (ANI, C_6H_7N), Sodium hydroxide (NaOH), Potassium hydroxide (KOH), polyvinyl alcohol (PVA, M_w = 124,000–186,000 g mol^{-1}, 99% hydrolyzed), sodium tetraborate decahydrate (Borax, $Na_2B_4O_7 \cdot 10H_2O$, 99.5% purity) were purchased from Aladdin Chemical Reagent Co., Ltd., Shanghai, China. The carbon nanotubes (CNTs) were obtained from Nanotech Port Co., Ltd., Shenzhen, China (the purity >97%; the diameter in the range 10–20 nm; the length in the range 30–100 μm). They were all the analytical grade. Deionized (DI) water was used in all the preparations.

2.2. Preparation of TOCNFs, TOCNF-CNT Nanohybrids, and TOCNF-CNT@PANI Nanohybrids

The bleached wood pulp solution was mixed with TEMPO and NaBr by means of mechanical stirring. Then, NaClO (15 mmol g^{-1} cellulose) solution was added to initiate an oxidation reaction. The value of pH should be maintained at 10.5 by adding 1 M NaOH solution in the reaction process.

After the oxidation reaction, the achieved fibers were washed by DI water and ultrasonicated at 800 W power for 3 min to obtain TOCNFs [10]. The concentration of final homogenized suspension was adjusted to 0.6 wt%.

A total of 0.5 g CNT powers was added to the 62.5 g 0.6 wt% TOCNFs suspensions at the mass ratios m(CNTs):m(TOCNFs) = 1:0.75 while keeping mechanical agitation for 30 min. Then, they were diluted to 500 mL water and ultrasonicated for 10 min at 400 W power in an ice bath to achieve TOCNF-CNT nanohybrids homogenized aqueous dispersion. The sonication was carried out by a high-energy ultrasonic cell disruptor.

TOCNF-CNT@PANI nanohybrids were synthesized based on TOCNF-CNT nanohybrids as a template and APS as oxidant through in-situ polymerization in an acidic water medium, as follows. First, 50 mL 1 M HCl solution containing 0.875 g ANI (0.0094 mol) was added into the 500 mL TOCNF-CNT nanohybrids dispersions and stirred for 10 min in an ice bath. Next, 50 mL 1 M HCl solution containing 2.68 g APS (0.01175 mol, mole ratio of ANI:APS = 1:1.25) was added slowly to the mixture in 30 min. During the process, the uniform mixture was slowly converted to blue tone, indicating polymerization; then, the solution was kept stirring for 6 h and stand for 12 h in an ice bath to complete polymerization. Finally, the obtained TOCNF-CNT@PANI nanohybrids were washed by vacuum filtration, and the excess unreacted ions were removed. The washed product was transferred to dialysis tubes and dialyzed to pH = 7.

A similar method was used to synthesize TOCNF-CNT@PANI nanohybrids containing different PANI content. The mass ratio of ANI monomers to TOCNF-CNT nanohybrids were 0:1, 1:1, 2:1, and 3:1, and the corresponding TOCNF-CNT@PANI nanohybrids samples were denoted TOCNF-CNT, TOCNF-CNT@PANI-1, TOCNF-CNT@PANI-2, and TOCNF-CNT@PANI-3, respectively.

2.3. Preparation of TOCNF-CNT@PANI/PVA Composite Hydrogel

Two grams PVA and 100 mL DI water were stirred for 1 h at 90 °C to achieve a transparent PVA aqueous solution. Then, the solution was added into TOCNF-CNT, TOCNF-CNT@PANI-1, TOCNF-CNT@PANI-2, and TOCNF-CNT@PANI-3 nanohybrids aqueous dispersion and stirred to form homogeneous dispersion, respectively. According to m (PVA):m (borax) = 4:1, borax powders were added slowly into the mixture with stirring at 90 °C until the composite became sticky. The composite showed viscoelastic performance slowly with the decrease of temperature, indicating the formation of the final composite hydrogels. The corresponding hydrogels were named TOCNF-CNT/PVA, TOCNF-CNT@PANI/PVA-1, TOCNF-CNT@PANI/PVA-2, and TOCNF-CNT@PANI/PVA-3, respectively.

2.4. Fabrication of the Self-Healing Solid-State Supercapacitor

The supercapacitor device was a symmetric configuration, which was assembled by two pieces of TOCNF-CNT@PANI/PVA-2 ($5.0 \times 1.5 \times 0.1$ cm^3) and one piece of TOCNF/PVA ($5.0 \times 1.5 \times 0.1$ cm^3). They were immersed into 6 M KOH solution for 12 h before assembling. In symmetrical supercapacitors, the electrolyte and separator were TOCNF/PVA hydrogels, which were sandwiched in TOCNF-CNT@PANI/PVA-2 hydrogels using as electrodes and current collectors. The interfaces between electrodes and separators could be completely glued into a single thin separating layer as excellent self-healing character, thereby manufacturing an integrated solid-state supercapacitor. The electrochemical measurements of supercapacitors were performed on an electrochemical workstation (CHI 760E, Shanghai, China).

2.5. Characterization

The transmission electron microscope (TEM) was carried out by a transmission electron microscope (JEM-1400, Tokyo, Japan) operating at 120 kV. The concentrations of CNTs, TOCNF-CNT, and TOCNF-CNT@PANI water suspensions were 0.05–0.1 wt%. The absorption spectra of CNTs, TOCNF-CNT, and TOCNF-CNT@PANI suspensions were measured at room temperature using an ultraviolet-visible (UV-vis) spectrophotometer (Tu-1810, Purkinje Co., Beijing, China) at a wavelength of 200–1000 nm. Spectra were collected using water as a reference at a scan speed of 0.5 nm s^{-1}. Infrared spectra were performed by Fourier transform infrared (FTIR) spectroscopy (Nicolet iS50, Thermo Fisher Scientific Inc., Madison, WI, USA) with attenuated total reflection (ATR) mode. The wavelength range was from 4000 to 500 cm^{-1} at a resolution of 4 cm^{-1}. X-ray diffraction (XRD) spectra were obtained by an X-ray diffractometer (Ultima IV, Rigaku, Japan) at 40 kV and 30 mA, and the angle range was $2\theta = 5{\sim}40°$ at a scanning rate of 5° min^{-1}. The microstructure of composite hydrogel was observed by a JSM-7600F scanning electron microscope (Nippon electronics Co., Ltd., Tokyo, Japan) at a voltage of 15 kV.

Uniaxial compression measurements were carried out through a universal mechanical testing machine (CMT4304, Shenzhen, China) on a cylindrical sample (diameter of 20 mm; the height of 5 mm) at a compression speed of 8 mm min^{-1} at room temperature in air. The values of stress (σ) and strain (ε) were calculated from the force and deformation of the original size of samples. The compressive elastic modulus (E_e) was calculated from the rake angle ratio of the linear part ($\varepsilon < 20\%$) of the σ-ε curve. The specific stress (σ_s) was obtained by dividing the density (ρ). The energy absorption (E_a) was the integral area of the part under the σ-ε curve.

Tensile stress-strain measurements were performed using a universal mechanical testing machine (CMT4304, Shenzhen, China) at a pulling rate of 20 mm min^{-1}. The hydrogel samples of initial and healed (20 s) in a size of $30 \times 15 \times 5$ mm^3 were nipped to the tensile machine during the testing process. All the tensile measurements were repeated for three times. The healing efficiencies (η_F) were calculated from the break stress (F_{healed}) of the healed hydrogel divided that of the initial one ($F_{original}$). The healing efficiencies (η_K) were defined from that the break strain (K_{healed}) of the healed hydrogel sample divided that of the initial one ($K_{original}$). ($\eta_F = F_{healed}/F_{original} \times 100\%$, $\eta_K = K_{healed}/K_{original} \times 100\%$).

The dynamic rheological properties, including dynamic frequency scanning, dynamic strain scanning, and continuous step strain, were tested by a Rheometer (HAAKE600, Waltham, MA, USA) with a plate diameter of 40 mm and a gap of 500 μm. The dynamic strain range was 0.01 to 100% with angular frequency (ω) of 1 Hz. The linear viscoelastic region (LVR) was decided by storage modulus (G'), and the G' was independent of the strain in the LVR. In the following measurement of each sample, 1% strain (γ) was selected to maintain the dynamic oscillatory deformation within the LVR. In dynamic frequency scanning measurement, the relationship between the shear storage modulus (G'), loss modulus (G''), and angular frequency (ω) were recorded at $\omega = 0.1$–100 rad s^{-1}, $\gamma = 1\%$, and 25 °C. The complex modulus (G^*) was calculated by Equation (1).

$$G = \sqrt{G'^2 + G''^2} \tag{1}$$

Continuous step strain tests were conducted to study the recovery property of the hydrogels under the applied shear stress. A procedure to the program was as follows: 1% (800 s) →80% (800 s) → 1% (800 s) → 80% (800 s) → 1% (800 s), the G' and G'' versus time were measured at $\omega = 1$ Hz and 25 °C.

Conductivity tests of hydrogel electrodes. A square-shaped hydrogel with a size of $1 \times 1 \times 10$ cm^3 was sandwiched by two pieces of platinum electrodes. The resistance (R) values of the hydrogel were decided by current-voltage (I–V) measurement using an electrochemical workstation (CHI 760E, Shanghai, China). The conductivity (σ) was achieved from Equation (2):

$$\sigma = \frac{1 \times d}{R \times S} \tag{2}$$

where σ was the conductivity (S m^{-1}), R was the resistance (Ω), d was the length (m), and S was the cross-sectional area (m^2) of the sample, respectively.

Electrochemical measurements of the hydrogel-based electrodes were performed on a three-electrode system using an electrochemical workstation (CHI 760E, Shanghai, China). The working electrode, the counter electrode, the reference electrode, and electrolyte were the

TOCNF-CNT@PANI/PVA-2 hydrogel (1.5 g), platinum plate electrode, mercury/mercury oxide (Hg/HgO) electrode, and 6 M KOH aqueous solution, respectively. The cyclic voltammetry (CV) test was carried out at scan rates of 40 mV s^{-1} from −0.2–0.8 V, the galvanostatic charge-discharge (G-CD) test was performed over the voltage range of −0.2–0.8 V at a current density of 0.4 A g^{-1}, and electrochemical impedance spectra (EIS) test was measured over the frequency range from 0.01 Hz to 100 kHz at open circuit potential (alternating current perturbation voltage was 5 mV). The specific capacitance (C_s) values were calculated from the G-CD curves using Equation (3):

$$C_s = \frac{I\Delta t}{m\Delta V} \tag{3}$$

where C_s represented the specific capacitance (F g^{-1}), I represented the discharge current (A), Δt represented the time of discharge (s), ΔV represented the voltage of discharge (V), and m represented the mass of active materials (g).

3. Results and Discussion

3.1. Synthesis Process and Mechanism of TOCNF-CNT@PANI/PVA Composite Hydrogels

The preparation process of hierarchical 3D network TOCNF-CNT@PANI/PVA composite hydrogel is illustrated in Figure 1. Firstly, TOCNF suspensions were prepared through a TEMPO oxidation treatment in an aqueous system following an ultrasonication treatment. TEMPO oxidizes the hydroxymethyl group at the glycose C_6 position in the cellulose chain to a more active carboxyl group [11]. Secondly, the homogeneous TOCNF-CNT nanohybrid dispersions were obtained by mixing CNT powders with TOCNF dispersions through ultrasonication. Not only the fluctuation of counter ions on the surface of TOCNF fibers induced the dipoles of the carbon lattice in CNTs, but also the carboxyl groups of TOCNFs produced electrostatic repulsion, which ensured the stabilization of CNTs in water [10]. Thirdly, TOCNF-CNT with excellent dispersibility and high specific surface area were used as nanocarrier of PANI. TOCNF-CNT@PANI nanohybrids were synthesized by in-situ chemical polymerization with APS as oxidant and TOCNF-CNT as a biological template in acid medium. PANI formed a wholly uniform coating layer around TOCNF-CNT nanohybrid bundles (Figure 1d), which would bring enough pseudo-capacitance [12]. Finally, PVA and borax were introduced into the TOCNF-CNT@PANI nanohybrid dispersions to achieve TOCNF-CNT@PANI/PVA composite hydrogel through cross-linking reaction. Borax would decompose into B(OH)$_4^-$ ions in water, and B(OH)$_4^-$ ions would create a reversible connection between the CNF-CNT@PANI composite fibers and the PVA molecular chains, forming a dynamic 3D network structure in the hydrogel [6]. The hierarchical 3D network inside hydrogel is illustrated in Figure 1e; the CNTs provided the fast electron transport path, and the nano-coating layer of PANI-ensured electrons could only pass through a very short distance to the CNT networks with high conductivity, which improved the electrochemistry of TOCNF-CNT@PANI nanohybrid. In addition, the borate ions could combine with the adjacent hydroxyl groups to form dynamic cross-linking between PVA chains and TOCNF-CNT@PANI nanohybrids. The dynamic PVA-borate cross-linking network provided hydrogel with the moldable and self-healing performance. The TOCNF-CNT networks provided an additional platform to improve strength, toughness, and conductivity. The chain entanglement and hydrogen bonding between TOCNF-CNT@PANI nanohybrids and PVA formed a hierarchical 3D network in TOCNF-CNT@PANI/PVA hydrogels.

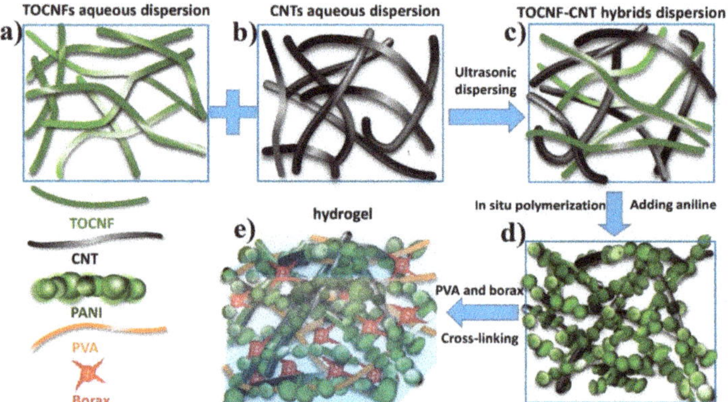

Figure 1. Schematic of the fabrication process of TOCNF-CNT@PANI/PVA composite hydrogel. (**a**) TOCNFs aqueous dispersion, (**b**) CNTs aqueous dispersion, (**c**) TOCNF-CNT nanohybrids aqueous dispersion, (**d**) TOCNF-CNT@PANI nanohybrids aqueous dispersion, (**e**) TOCNF-CNT@PANI/PVA composite hydrogel.

3.2. Dispersion State and Chemical Analysis of TOCNF-CNT and TOCNF-CNT@PANI Nanohybrids

The microstructure morphologies of CNTs, TOCNF-CNT, and TOCNF-CNT@PANI-2 nanohybrids are shown in Figure 2a–c. Due to the hydrophobic surface and strong van der Waals forces between CNT fibers, it was difficult for the pristine CNTs to disperse well in water [13]. It was clear to see that pristine CNTs formed densely agglomeration and precipitation in water (Figure 2a and insert). After introducing TOCNFs into CNT suspension, as shown in Figure 2b, the CNTs (length: 500–700 nm, diameter: 20–30 nm) dispersed in an individual form without aggregation in water with the help of TOCNFs (length: 1 μm, diameter: 20–30 nm).

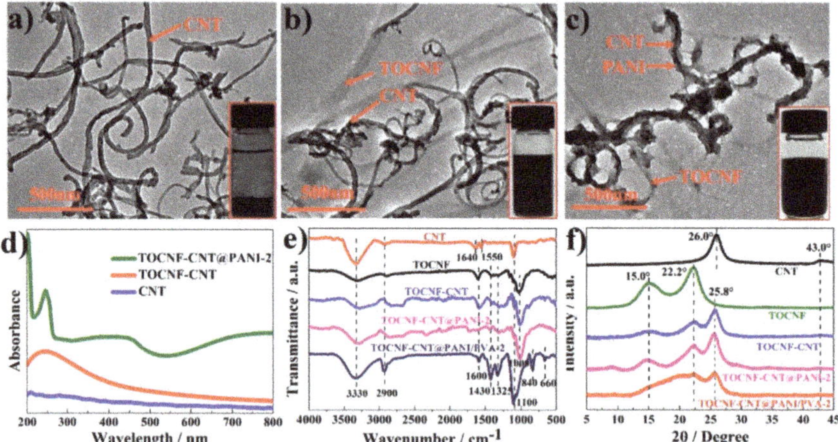

Figure 2. Transmission electron microscope (TEM) images of (**a**) CNTs, (**b**) TOCNF-CNT nanohybrids, (**c**) TOCNF-CNT@PANI nanohybrids, and the inset in each image represents the corresponding dispersion state of each suspension at the concentration of 0.5 wt%; (**d**) Ultraviolet-visible (UV-vis) spectra of neat CNTs, TOCNF-CNT, and TOCNF-CNT@PANI-2 nanohybrids in water. (**e**) Fourier transform infrared (FTIR) spectra and (**f**) X-ray diffraction (XRD) patterns of TOCNFs, CNTs, TOCNF-CNT, TOCNF-CNT@PANI-2 nanohybrids, and TOCNF-CNT@PANI/PVA-2 composite hydrogel.

Figure 2c shows nodular-structure. PANI particles were deposited on the surface of TOCNF-CNT nanohybrids, forming a continuous shell structure with a high aspect ratio. The in-situ deposited aniline fiber could grow a fresh polymer, then initiate the continuous growing process and form a block precipitate, and the initial seed morphology would be transcribed on a long scale [14]. TOCNF-CNT@PANI composite fiber with a "core-shell" structure was the prototype of the 3D hierarchical conductive network. The PANI grew along the TOCNF-CNT templates to form the "shell", which was carried by TOCNF-CNT fibers, and dispersed in water to construct a hierarchical 3D conductive network. It was noted that the continuous conductive networks fabricated by TOCNF-CNT@PANI composite fibers were obligatory to synthesize the ECHs with high conductivity and enhanced mechanical performances [15].

The UV-vis spectroscopy of CNT, TOCNF-CNT, and TOCNF-CNT@PANI-2 complexes are shown in Figure 2d. Due to the poor dispersibility and high aggregation, the absorption peak of pure CNT suspension was much weaker than that of the TOCNF-CNT nanohybrids at the same CNT concentrations (0.015 wt%), suggesting that TOCNFs improved significantly the dispersibility of CNTs in an aqueous medium. The absorption peak of TOCNF-CNT nanohybrids was around 262 nm in the UV-vis spectrum, which could be assigned to the $\pi-\pi$ transitions of CNTs and showed a more uniform dispersion state of CNTs [16]. Two peaks of TOCNF-CNT@PANI-2 at 285 and 475 nm were from the transition of $\pi-\pi$ inter-band and the polaron band of PANI, respectively. The end band (~800 nm) of the TOCNF-CNT@PANI-2 complexes was smooth because of the cross-linking of polymer chains, indicating that it was successful in doping PANI into the conductive state [17].

The FTIR spectra of TOCNFs, CNTs, TOCNF-CNT nanohybrids, TOCNF-CNT@PANI-2 composite, and TOCNF-CNT@PANI/PVA-2 hydrogel is shown in Figure 2e. For all the samples, the broad bands at around 3330 cm^{-1} and sharp bands at around 2900 cm^{-1} were generally due to O-H stretching of the hydrogen bonds and asymmetrically stretching vibration of C–H in the CH_2 group, respectively [18–21]. The peaks around 1000 cm^{-1} and 1600 cm^{-1} were ascribed to the C–O–C vibration and the carbonyl functional groups of cellulose [22–24]. For the spectra of neat CNTs, the characteristic absorption around 1640 cm^{-1} was assigned to the quaking of the carbon skeleton [25,26]. For the spectra of TOCNFs, the peaks at 1430 cm^{-1} and 1325 cm^{-1} were assigned to the hydrogen bonding and CH_2 wagging [27]. The typical absorption peak of TOCNF-CNT-2 nanohybrids was analogous to that of TOCNFs, which was due to the absorption peak of TOCNFs covering the absorption peak of the CNTs in the FTIR spectra, revealing the CNTs were successfully combined with TOCNF bio-templates [28]. Comparing with the spectra of pure CNTs, the C–O–C bending band shifted from 1100 cm^{-1} to 1000 cm^{-1} (TOCNF-CNT-2 nanohybrids), confirming the existence of hydrogen bonding between TOCNFs and CNTs [29].

After in situ polymerization, the peaks at 1430 and 1325 cm^{-1} were assigned to the C = C vibration deformation of the quinoid ring and benzenoid ring, respectively [30]. For TOCNF-CNT@PANI-2 complexes, the N–H bonding vibration caused the major peaks at 3330 cm^{-1} shifted to lower wavenumbers, indicating that the TOCNF-CNT nanohybrids were coated with PANI [31]. Further, the vanishing of the sharp band at 1600 cm^{-1} from the carbonyl functional groups of cellulose revealed that PANI coated onto TOCNF-CNT nanohybrids successfully. For TOCNF-CNT@PANI/PVA-2, the distinct peaks at 1430 and 1325 cm^{-1} were attributed to the asymmetric B–O–C bonding, 840 cm^{-1} was assigned as B–O bonding of free $B(OH)_4{}^-$, and 660 cm^{-1} was ascribed to B–O–B bonding in the borate molecule networks, suggesting the presence of borax and borate [32], which further confirmed the existence of borate cross-linking network between the PVA molecule chains, TOCNF–CNT@PANI nanohybrids, and borate within the hydrogels [33].

Figure 2f shows the XRD diffraction patterns of CNTs, TOCNFs, TOCNF-CNT nanohybrids, TOCNF-CNT@PANI-2 nanocomposite, and TOCNF-CNT/PVA-2 hydrogel samples. A diffraction peak of CNTs at $2\theta = 26°$ arose from interlayer spacing (002), reflecting the characteristic of graphite. The diffraction peak at $2\theta = 43°$ arose from in-plane crystal lattice (100) [34]. In the XRD diffraction patterns of TOCNFs, a sharp peak and a broad peak at $2\theta = 22.2°$ and 15.0°, attributed to (002) and (101) planes, suggested the crystallization from cellulose I [22]. Comparing with the XRD profile of pure CNTs, the TOCNF-CNT nanohybrids showed two additional peaks at $2\theta = 15.0°$ and $22.2°$, reflecting the characteristic of cellulose I. These observations suggested that CNTs and TOCNFs were combined and remained integrality [35].

For the XRD spectra of TOCNF-CNT@PANI-2 nanocomposite, due to in situ polymerization of ANI monomers, the peaks located at around $2\theta = 15.0°$ and $22.2°$ were wider than that of TOCNF-CNT nanohybrids, which could be attributed to the overlapping of diffraction peaks at $2\theta = 19.4°$ and $14.9°$ from (020) and (011) crystal planes of PANI [36]. The peak intensity at $2\theta = 25.8°$ was enhanced, which was due to the overlapping of diffraction peaks of π-π stacking corresponding to the co-facially stacked conjugated backbones from the polymer chains of PANI [37]. After the incorporation of PVA hydrogel, the peaks of CNFs at $2\theta = 15°$ disappeared, and a broad new peak emerged at $2\theta = 22.2°$, which contained the diffraction peaks at $2\theta = 19.4°$ corresponding to the orthogonal lattice from PVA with semi-crystalline structure. All these revealed the strong interactions between PVA, borax, and TOCNF-CNT@PANI-2 nanocomposite and built a 3D network in the composite hydrogels [38].

3.3. Compression Test and Microstructures of Hydrogels

Figure 3a shows the stress-strain curves of these hydrogels under compression. The measured stresses at the 90% strain level were 52.3 ± 0.3, 86.1 ± 3.9, 108.4 ± 4.3, and 152.3 ± 5.1 kPa for TOCNF-CNT/PVA, TOCNF-CNT@PANI/PVA-1, TOCNF-CNT@PANI/PVA-3, and TOCNF-CNT@PANI/PVA-2, respectively. Thus, the stress of TOCNF-CNT@PANI/PVA-1 hydrogel with PANI at the 90% strain level was almost 1.6-fold than that of TOCNF-CNT/PVA. PANI nanoparticles combined with TOCNF-CNT nanofiber to form TOCNF-CNT@PANI composite fibers with a "core-shell" structure. The composite fibers based on good dispersibility and interfacial adhesion inside the hydrogel effectively transferred the load, thereby improving the mechanical strength of PVA hydrogel [39]. With the increase of PANI content, the stress of TOCNF-CNT@PANI/PVA increased first and then decreased. The stress of TOCNF-CNT@PANI/PVA-3 was 108.4 ± 4.3 kPa, which was lower than that of TOCNF-CNT@PANI/PVA-2 with 152.3 ± 5.1 kPa. This phenomenon could be attributed to the TOCNF-CNT biological template being insufficient to carry and disperse these excess PANI. Aggregated PANI prevented effective cross-linking between PVA and borax and disrupted the integrity of the network in the hydrogel. Under external force, the stress concentration caused by agglomeration would weaken the mechanical strength [40].

The TOCNF-CNT@PANI/PVA-2 possessed the highest mechanical strength in all the hydrogels. Its σ value (152.3 ± 5.1 kPa) at $\varepsilon = 90\%$ and E_e value (61.0 ± 0.8 kPa) in the σ-ε curve were 2.9-fold and 4.2-fold more than those ($\sigma = 52.3 \pm 0.3$ kPa, $E_e = 14.4 \pm 0.3$ kPa) of TOCNF-CNT/PVA hydrogel. The specific compressive stress (σ_s) value of TOCNF-CNT@PANI/PVA-2 was 128 kPa cm^3 g^{-1}, which was 2.8-fold larger than that of TOCNF-CNT/PVA with 45.9 kPa cm^3 g^{-1}. In Figure 3b, TOCNF-CNT@PANI/PVA-2 had the largest energy absorption (E_a) value. In the E_a-ε curves of hydrogels, the E_a with $\varepsilon = 90\%$ was selected to compare the mechanical properties of hydrogels. In particular, the E_a value of TOCNF-CNT@PANI/PVA-2 at $\varepsilon = 90\%$ was 3.2 ± 0.5 kJ m^{-3}, which was approximately 4 times larger than TOCNF-CNT/PVA with 0.8 ± 0.4 kJ m^{-3}. All the values of strength and physical properties are collected in Table 1.

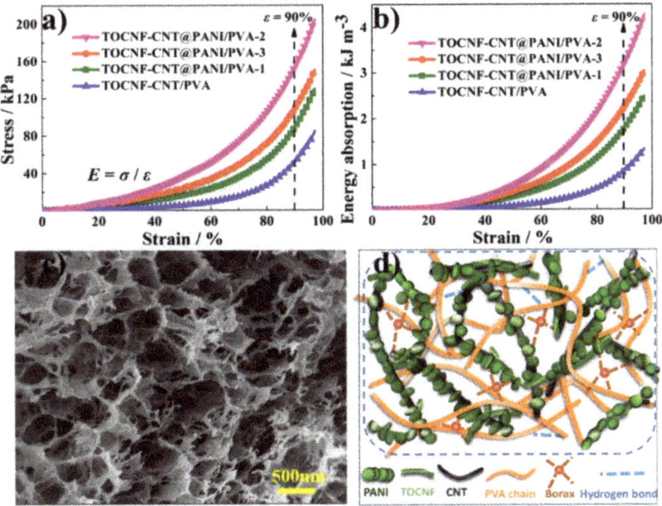

Figure 3. (**a**) Stress-stain curves under compression; (**b**) energy absorption-strain curves of hydrogels; (**c**) SEM image of TOCNF-CNT@PANI/PVA-2 composite hydrogel; (**d**) idealized 3D cross-linking network of TOCNF-CNT@PANI/PVA-2 composite hydrogel.

Table 1. Physical-mechanical characteristics of various hydrogels.

Sample	σ at $\varepsilon = 90\%$ [kPa]	ρ [g cm^{-3}]	σ_s [kPa cm^3 g^{-1}]	E_a at $\varepsilon = 90\%$ [kJ m^{-3}]	E_e [kPa]	W_c [wt%]
TOCNF-CNT/PVA	52.3 ± 0.3	1.14 ± 0.07	~45.9	0.8 ± 0.4	14.4 ± 0.3	95.4 ± 0.2
TOCNF-CNT@PANI/PVA-1	86.1 ± 3.9	1.17 ± 0.12	~73.5	1.7 ± 0.6	38.2 ± 0.6	94.3 ± 0.1
TOCNF-CNT@PANI/PVA-2	152.3 ± 5.1	1.19 ± 0.10	~128.0	3.2 ± 0.5	61.0 ± 0.8	95.15 ± 0.18
TOCNF-CNT@PANI/PVA-3	108.4 ± 4.3	1.23 ± 0.14	~88.1	2.2 ± 0.5	51.7 ± 0.7	94.90 ± 0.14

Note: (1) σ means stress; (2) ε means strain; (3) ρ means density; (4) σ_s means specific stress; (5) E_a means energy absorption; (6) E_e means compressive elastic modulus; (7) W_c means water content.

The improvement of mechanical properties was due to the effective enhancement of TOCNF-CNT@PANI composite fibers with a "core-shell" structure. In addition, the CNTs were entangled with each other to form a lot of contact junctions in the interstitial space between the PVA molecular chain. These contact junctions built the continuous conductive network in the hydrogel matrix. Figure 3c shows the microstructure of TOCNF-CNT@PANI/PVA-2 composite hydrogel, and Figure 3d presents the schematic diagram of the 3D network structure. The composite hydrogel possessed an interconnected porous structure, and each pore had a diameter of 200–500 nm. The wall of the pore was formed by a hydrogel matrix with a thickness of 10–30 nm. The entangled TOCNF-CNT@PANI composite fibers penetrated through the hole wall and built a hierarchical network. The specific framework could effectively promote the transport of electrons, improving the electrical conductivity of the hydrogel. Among them, TOCNFs were important to promote the formation of hierarchical microstructure, which profited from their excellent natural characteristics of hydrophilicity, high aspect-ratio, mechanical strength, and flexibility. TOCNFs combined with CNTs through hydrogen bonding and chain entanglement and served as nanocarriers to disperse CNTs in aqueous media [13]. The well-dispersed TOCNF-CNT nanohybrids were coated by PANI to form TOCNF-CNT@PANI composite fibers with "core-shell" structure. The composite fibers could further improve the mechanical strength and electrical properties of hydrogel [41]. The CNTs in composite fiber could effectively transfer the force from the PVA molecule chains. Moreover, an efficient and stable CNTs electric network could improve the electrical conductivity of TOCNF-CNT@PANI/PVA composite hydrogel. The results showed that the hierarchical network microstructure inside the composite hydrogel increased the interface between the electrolyte and the electroactive material, demonstrating a broad application prospect in flexible electrodes.

3.4. Dynamic Viscoelastic Performance of Hydrogels

Figure 4a shows the G' curves of hydrogel samples based on strain at $\omega = 1$ Hz. Within the LVR, the G'' and G' of the hydrogel were independent of strain, as determined by dynamic strain scanning tests. The critical strain (γ_c) of hydrogel was a strain point, where the G' value decreased from the platform value by 5%, indicating deviation from LVR [38]. The G' value corresponding to the strain higher than γ_c would gradually decrease, indicating that the quasi-solid hydrogel had changed to a quasi-liquid state. The γ_c values of TOCNF-CNT/PVA, TOCNF-CNT@PANI/PVA-1, TOCNF-CNT@PANI/PVA-2, and TOCNF-CNT@PANI/PVA-2 were 2.5%, 2.1%, 1.2%, and 1.5%, respectively. Therefore, in the following dynamic oscillation measurement, the γ_c value was selected as $\gamma = 1\%$, which could ensure that deformations of the hydrogel samples were within the LVR. For all the hydrogel samples in the LVR, the G' values were independent of strain, and the corresponding G'_{max} was 2.8, 4.1, 7.5, and 5.1 kPa, respectively (Figure 4a). The G'_{max} of TOCNF-CNT@PANI/PVA-1 was 1.5 times that of TOCNF-CNT/PVA. TOCNF-CNT@PANI/PVA-2 possessed the largest G'_{max} (7.5 kPa), which was nearly 1.8-fold larger than TOCNF-CNT@PANI/PVA-1 (4.1 kPa) and 1.5-fold greater than TOCNF-CNT@PANI/PVA-3 (5.1 kPa). Incorporation of an appropriate amount of PANI could improve remarkably the stiffness of hydrogel. The shorter the LVR, the closer the sample was to the solid-state. Compared with TOCNF-CNT@PANI/PVA-1 and TOCNF-CNT@PANI/PVA-3, it could be known that the TOCNF-CNT@PANI/PVA-2 possessed a higher G'_{max} and shorter LVR, indicating that TOCNF-CNT@PANI/PVA-2 was the strongest hydrogel. The result was consistent with the mechanical strength test.

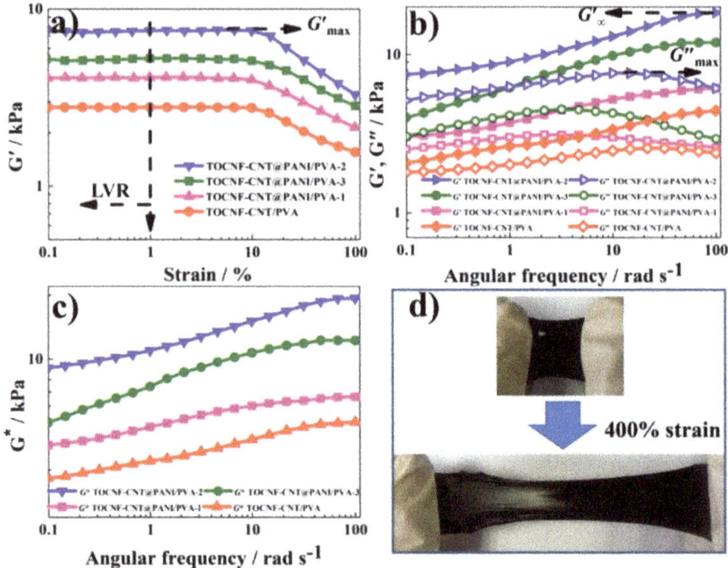

Figure 4. Dynamic viscoelastic properties of hydrogels at 25 °C. (**a**) storage modulus (G') curves based on strain (γ) at angular frequency (ω) = 1 Hz; (**b**) G' and loss modulus (G'') curves based on ω at $\gamma = 1\%$; (**c**) complex modulus (G^*) curves based on ω at $\gamma = 1\%$; (**d**) stretching demonstration of TOCNF-CNT@PANI/PVA-2 hydrogel.

In order to study the effect of TOCNF-CNT@PANI composite fiber on the viscoelasticity of hydrogels, the G' (elasticity) and G'' (viscosity) of hydrogels versus ω at $\gamma = 1\%$ in the LVR are shown in Figure 4b. As shown, the G' and G'' curves of all the hydrogels followed similar trends. With the increase of ω, the G' increased monotonically and arrived at plateau value (G'_∞), indicating the formation of neighboring polymer chains entanglements; the G'' increased preliminarily to reach the maximum value (G''_{max}), then decreased gradually. For all the hydrogels, the G' values were always higher than the G'' values throughout the ω range, suggesting hydrogels showed typical solid-like characteristics, indicating that a dynamic cross-linked network was established inside hydrogel [6,42]. The G'_∞ and G''_{max} values of TOCNF-CNT/PVA were 4.3 and 2.6 kPa, respectively. After the introduction of PANI, the G'_∞ (6.1 kPa) and G''_{max} (3.1 kPa) values of TOCNF-CNT@PANI/PVA-1 were 1.4 and 1.2 times those of TOCNF-CNT/PVA, respectively. It was shown that the combination of PANI and TOCNF-CNT to form a TOCNF-CNT@PANI composite fiber with a "core-shell" structure could significantly improve the viscoelasticity of hydrogel. Comparing between TOCNF-CNT@PANI/PVA-1, TOCNF-CNT@PANI/PVA-2, and TOCNF-CNT@PANI/PVA-3, the TOCNF-CNT@PANI/PVA-2 showed the highest G'_∞ (18.2 kPa) and G''_{max} (7.6 kPa). These data of dynamic viscoelastic properties are summarized in Table 2. It showed that an appropriate proportion of PANI could develop a hierarchical network structure and increase viscoelasticity together with TOCNF-CNT. However, excessive PANI would form aggregation due to insufficient TOCNF-CNT to disperse and load. A large amount of PANI blocked the cross-linking between PVA and borax, reducing the dynamic viscoelasticity of the hydrogel [33]. Figure 4c shows the curves of complex modulus (G^*) versus ω, which provided a clear contrast of viscoelasticity. The trend was TOCNF-CNT@PANI/PVA-2 > TOCNF-CNT@PANI/PVA-3 > TOCNF-CNT@PANI/PVA-1 > TOCNF-CNT/PVA within the entire range of ω. The TOCNF-CNT@PANI/PVA-2 showed the highest G^*, further proving that TOCNF-CNT@PANI/PVA-2 was the most quasi-solid hydrogel among these hydrogels. In Figure 4d, a piece of rubbery TOCNF-CNT@PANI/PVA-2 hydrogel could be stretched to 400% strain without damage, exhibiting excellent flexibility, viscoelasticity, and efficient energy dissipation capability. Inside hydrogel, flexible PVA chains and long TOCNF-CNT@PANI composite fiber were physically entangled or hydrogen-bonded to build a 3D network, which could unravel and reconstruct the energy dissipating ability of the hydrogel.

Table 2. Rheological characteristics from viscoelasticity curves.

Parameter	TOCNF-CNT/PVA	TOCNF-CNT@PANI/PVA-1	TOCNF-CNT@PANI/PVA-2	TOCNF-CNT@PANI/PVA-3
γ_c (%)	2.5	2.1	1.2	1.5
G'_{max} (kPa)	2.8	4.0	7.5	5.1
G'_∞ (kPa)	4.3	6.1	18.2	11.8
G''_{max} (kPa)	2.6	3.1	7.6	4.5

Note: (1) γ_c means critical strain; (2) G'_{max} means the maximum value of storage modulus based on strain; (3) G'_∞ means the plateau value of storage modulus based on angular frequency; (4) G''_{max} means the maximum value of loss modulus based on angular frequency.

3.5. Self-Healing Performance of the Hydrogels

The composite hydrogel possessed a dynamic self-healing PVA-borate network, and the TOCNF-CNT@PANI composite fibers network provided an additional platform to strengthen the structure [33]. In Figure 5a, the G' and G'' of TOCNF-CNT@PANI/PVA-2 were 8.1 and 4.2 kPa at $\gamma = 1\%$, respectively. The value of G' was larger than that of G'', indicating that the elastic character of hydrogel became the dominant factor. When the strain increased to $\gamma = 80\%$, the corresponding G' and G'' of TOCNF-CNT@PANI/PVA-2 were 0.2 and 0.6 kPa, respectively. The value of G' was less than that of G'', indicating hydrogel turned to the quasi-liquid state. Interestingly, when the strain dropped to 1% again, the corresponding G' and G'' immediately restored the original values, indicating that the hydrogel recovered the quasi-solid state. The rapid and repeatable phase transition between the quasi-solid state and quasi-liquid state demonstrated the intrinsic preeminent self-healing capability of the hydrogel.

To visualize the self-healing property of the hydrogel, two blocks of TOCNF-CNT@PANI/PVA-2 were pushed together for 20 s, and their contact surfaces would fuse. In addition, after the self-healed hydrogel was stretched to 300% strain, there was no crack at the healing interface (Figure 5b).

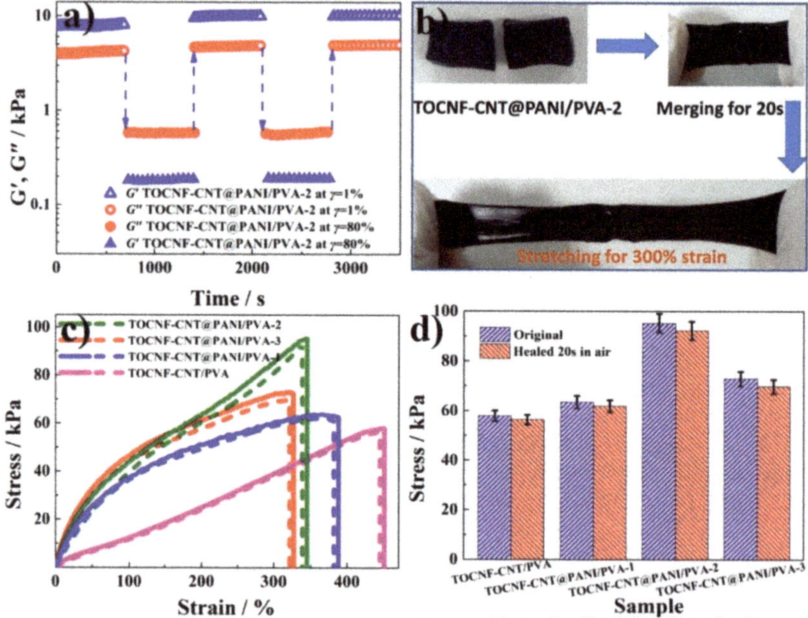

Figure 5. (**a**) The storage modulus (G') and loss modulus (G'') dependence of time in continuous step strain measurements. (**b**) Illustration of self-healing property for TOCNF-CNT@PANI/PVA-2 hydrogel. (**c**) Hydrogel stretching curve before and after 20 s self-healing. The solid lines represent the original hydrogels, and the dashed lines represent the self-healed hydrogels. (**d**) Histogram of tensile stress of the original hydrogels and the self-healed hydrogels after healing in the air for 20 s.

The stress-strain curves of hydrogels are shown in Figure 5c. The maximum break strain values of original TOCNF-CNT/PVA, TOCNF-CNT@PANI/PVA-1, TOCNF-CNT@PANI/PVA-2, and TOCNF-CNT@PANI/PVA-3 were 450.5 ± 23.2%, 387.8 ± 19.0%, 345.1 ± 15.1%, and 326.7 ± 12.5%, respectively. The highest tensile stress values of original TOCNF-CNT/PVA, TOCNF-CNT@PANI/PVA-1, TOCNF-CNT@PANI/PVA-2, and TOCNF-CNT@PANI/PVA-3 were 57.9 ± 2.1, 63.5 ± 2.5, 95.3 ± 3.2, and 72.9 ± 2.9 kPa, respectively. The TOCNF-CNT@PANI/PVA-2 composite hydrogel possessed the highest tensile stress, which was 1.6 times that of TOCNF-CNT/PVA hydrogel without PANI, indicating that the proper incorporation of PANI could effectively improve the tensile stress of composite hydrogels. This was consistent with previous measurements of mechanical strength and dynamic viscoelasticity.

To calculate the self-healing efficiency of hydrogels, these self-healed hydrogels after 20 s healing in the air were measured by a tensile test. The tensile curves of the self-healed hydrogels were basically the same as that of the original hydrogels. The maximum break strain values of healed TOCNF-CNT/PVA, TOCNF-CNT@PANI/PVA-1, TOCNF-CNT@PANI/PVA-2, and TOCNF-CNT@PANI/PVA-3 were 443.8 ± 22.1%, 381.1 ± 18.0%, 338.5 ± 13.3%, and 321.0 ± 10.2%, respectively. The corresponding η_k values were 98.5%, 98.3%, 98.1%, and 98.2%, respectively. The maximum tensile stress of healed TOCNF-CNT/PVA, TOCNF-CNT@PANI/PVA-1, TOCNF-CNT@PANI/PVA-2, and TOCNF-CNT@PANI/PVA-3 were 56.3 ± 2.0, 61.9 ± 2.4, 92.3 ± 3.7, and 69.7 ± 2.8 kPa, respectively. The corresponding η_F values were 97.2%, 97.5%, 96.8%, and 96.7%, respectively. These data of dynamic viscoelasticity properties are summarized in Table 3, which demonstrated the outstanding self-healing property of as-prepared hydrogels with the dynamic borate-assisted cross-linking network. The complexation between borate and hydroxyl was extremely fast (0.33 s), and TOCNFs and PVA chains contained large number of hydroxyl groups so that the hydrogels could recover quickly [43].

Table 3. Tensile strength and tensile rate of hydrogels before and after a 20 s self-healing.

Sample	$F_{original}$ (kPa)	$K_{original}$ (%)	F_{healed} (kPa)	K_{healed} (%)	η_F (%)	η_K (%)
TOCNF–CNT/PVA	57.9 ± 2.1	450.5 ± 23.2	56.3 ± 2.0	443.8 ± 22.1	97.2	98.5
TOCNF–CNT@PANI/PVA-1	63.5 ± 2.5	387.8 ± 19.0	61.9 ± 2.4	381.1 ± 18.0	97.5	98.3
TOCNF–CNT@PANI/PVA-2	95.3 ± 3.2	345.1 ± 15.1	92.3 ± 3.7	338.5 ± 13.3	96.8	98.1
TOCNF–CNT@PANI/PVA-3	72.9 ± 2.9	326.7 ± 12.5	69.7 ± 2.8	321.0 ± 10.2	96.7	98.2

Note: (1) $F_{original}$ means the break stress of initial hydrogel; (2) $K_{original}$ means the break strain of initial hydrogel; (3) F_{healed} means the break stress of healed hydrogel; (4) K_{healed} means the break strain of healed hydrogel; (5) η_F means the healing efficiencies of break stress; (6) η_K means the healing efficiencies of break strain.

3.6. Conductivity Analysis of Hydrogels

The composite hydrogel not only possessed excellent mechanical properties and self-healing ability but also possessed outstanding electrical conductivity due to the existence of CNTs and PANI in the hydrogels. The conductivity was quantitatively characterized by the *I–V* measurement at potential ranging from −4 to 4 V. In Figure 6a, the *I–V* curves of composite hydrogels are all linear and non-hysteretic, indicating the excellent electro-conductive character. The conductivity of the TOCNF-CNT@PANI/PVA-3, TOCNF-CNT@PANI/PVA-2, TOCNF-CNT@PANI/PVA-1, and TOCNF-CNT/PVA composite hydrogels were 15.3, 12.8, 8.2, and 6.4 S m^{-1}, respectively. The value was superior to phytic acid cross-linked polyaniline/poly(N-isopropylacrylamide) (PANI/PNIPAM) conductive hydrogels (~0.8 S m^{-1}) [15]. polyaniline-poly(styrene sulfonate) (PANI-PSS) hydrogels (~10^{-2} S m^{-1}) strengthened by sorbitol derivatives (DBS) supramolecular nanofibers [40]. Theoretically, CNTs and PANI were the main active material of conductive network within the composite hydrogels. With PANI as the shell and CNTs as the core, a composite fiber with a "core-shell" structure was formed. The electrical conductivity of the composite fiber was higher than that of bare CNTs fibers [44]. The conductivity of hydrogel increased rapidly when the mass ratio of ANI to TOCNF-CNTs increased from 1:1 to 2:1. However, the conductivity of hydrogel increased slowly, when the mass ratio of ANI to TOCNF-CNTs changed from 2:1 to 3:1. It could be concluded that a 2:1 ratio of ANI and TOCNF-CNTs could form the most perfect conductive network. As the ratio of ANI to TOCNF-CNT increased to 3:1, the TOCNF-CNT skeleton framework was insufficient to load excess PANI, which resulted in a slow increase in conductivity. Consequently, the well-integrating and stability of the TOCNF-CNT@PANI conductive network with the "core-shell" structure offered an effective electron-transfer pathway in the hydrogel. TOCNF-CNT@PANI/PVA-2 was selected for the next experiment based on the previous mechanical test results.

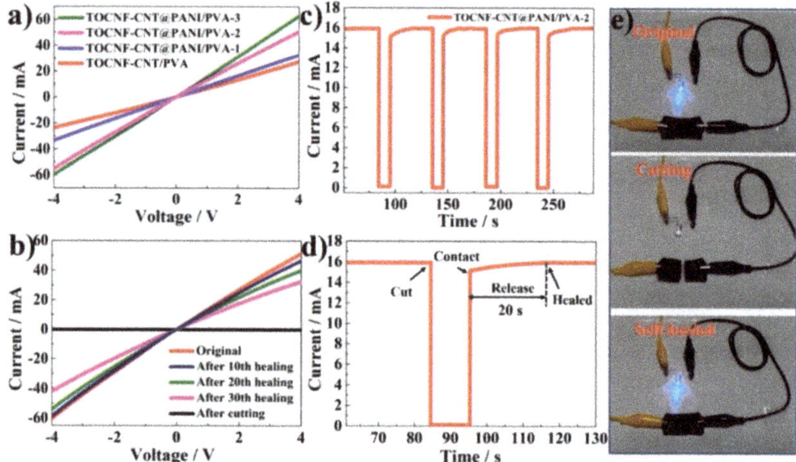

Figure 6. (**a**) Current-voltage (*I–V*) curves of TOCNF-CNT@PANI/PVA with different doping concentrations of PANI. (**b**) *I–V* curves of TOCNF-CNT@PANI/PVA-2 hydrogel after multiple cutting/healing cycles. (**c**) Cycling of the cutting/healing processes for TOCNF-CNT@PANI/PVA-2 at the same location under ambient conditions. (**d**) Time dependence of the electrical healing process by *I-V* measurements under ambient conditions. (**e**) Optical images of TOCNF-CNT@PANI/PVA-2 hydrogel under a cutting/healing cycle in a circuit with a light-emitting diode (LED) bulb.

The electrical conductivity's self-healing efficiency of composite hydrogels was further investigated. The conductivity of original, cutting, and self-healed TOCNF-CNT@PANI/PVA-2 hydrogel was characterized by the *I–V* measurement in Figure 6b. After 10th, 20th, and 30th self-healing, the conductivity of TOCNF-CNT@PANI/PVA-2 hydrogel was 12.8, 11.6, 10.0, and 8.0 S m^{-1}, respectively. The self-healing efficiency was calculated by σ_r/σ_i (σ_r is the healing conductivity, and σ_i is the original conductivity) [45]. After 10th, 20th, and 30th self-healing, the self-healing efficiency of TOCNF-CNT@PANI/PVA-2 hydrogel was 90.6%, 78.1%, and 62.5%, respectively. The average efficiency was 99.1% for each self-healing cycle, indicating the composite hydrogel possessed significant and repeatable electrical restoration performance.

By repeating the complete cutting/self-healing process, without any external force at room temperature, the conductivity of TOCNF-CNT@PANI/PVA-2 hydrogel was tested through *I-V* measurement. Figure 6c shows the time-current flow of TOCNF-CNT@PANI/PVA-2 hydrogel at the same location during repeated cutting/healing processes. In Figure 6d, when the hydrogel was completely cut in half to form an open circuit, the current dropped to zero. Then, the two fractured parts contacted each other, and the current quickly recovered to the initial value through a 20 s in situ self-healing. The conductivity of the hydrogel sample remained stable during the cycle, indicating that the conductivity had a high self-healing efficiency during the cutting-healing process.

As shown in Figure 6e, the self-healing conductive performance of the composite hydrogel was visually displayed through a closed-loop composed of light-emitting diode (LED), TOCNF-CNT/PVA-2 hydrogel, and power components. The LED indicator was lighted with a voltage of 5 V. The LED indicator was extinguished when the TOCNF-CNT/PVA-2 hydrogel was completely separated. However, the LED indicator lit up again, after pushing the two separated parts together for self-healing, illustrating the excellent self-healing conductive property of the composite hydrogel. The hierarchical 3D network consisting of PANI, CNTs, and TOCNFs formed a continuous conducting pathway for electron transport. The dynamically reversible cross-linking points from different borate-induced complexes provided inherent and repeatable self-healing capabilities for hydrogels, exhibiting promise for the self-healing electrode materials [46].

3.7. Electrochemical Properties of Composite Hydrogels

In order to evaluate the effect of the incorporated PANI on the electrochemical behavior of the composite hydrogel electrode, a CV test was performed, as shown in Figure 7a. In the CV test, the potential range was −0.2 to 0.8 V at a scan rate of 40 mV s^{-1}, in 6 M KOH electrolyte with platinum sheet counter electrode and Hg/HgO reference electrode. In Figure 7a, the CV curve of TOCNF-CNT/PVA exhibited regular rectangular and symmetric shapes, which reflected the typical characteristics of the electric double layer charge (EDLC) storage. Moreover, the CV curves of composite hydrogel containing PANI possessed a larger current density and different shape. The increase in current density indicated greater capacitance, which was due to the pseudo-capacitance effect of PANI. The deformation of the CV curve was attributed to the diffusion and migration of limited ions in the polymer block and the ohmic resistance due to the thick polymer layer [47]. However, the voltammograms of PANI-based hydrogel possessed clear faradaic oxidation and reduction peaks. Three pairs characteristic peaks arose at 0, 0.4, and 0.6 V; the peaks arose at 0 and 0.6 V were related to the redox behavior of PANI through the leucoemeraldine and pernigraniline states; the peaks at 0.4 V were assigned to the electron transition from the protonation/deprotonation of PANI [14,19]. Among these voltammograms, TOCNF-CNT@PANI/PVA-2 possessed the largest loop area, corresponding to the highest specific capacitance. Furthermore, the G-CD behaviors of these composite hydrogel electrodes were measured at 0.4 A g^{-1} current density from −0.2 to 0.8 V with 6 M KOH electrolyte (Figure 7b). The G-CD curves of the TOCNF-CNT/PVA hydrogel-based electrode exhibited a symmetrical triangle, indicating that it was an electric double-layer capacitor with reversible capacitance characteristics. For all the samples, the G-CD profiles were nearly triangular, demonstrating their excellent capacitive performances. Based on Equation (3), the C_s was calculated from the G-CD curves data. The C_s values of TOCNF-CNT/PVA, TOCNF-CNT@PANI/PVA-1, TOCNF-CNT@PANI/PVA-2, and TOCNF-CNT@PANI/PVA-3 were 84.9, 127.3, 226.8, and 184.4 F g^{-1} at 0.4 A g^{-1} current density, respectively. It was observed that composite hydrogel containing PANI possessed a higher specific capacitance than TOCNF-CNT/PVA hydrogel. The 3D network structure of TOCNF-CNT could load PANI and enable greater contact with electrolytes, thereby forming more active sites inside the hydrogel. Moreover, when CNTs were used as the filler for PANI to build a "core-shell" structure composite, the porous structure could further improve the capacitance performance. The high specific capacitance originated from two different charge storage methods: (1) the EDLC storage in CNTs nano-core and (2) the oxidation and reduction chemistry (pseudo-capacitance) of the PANI nano-shell [48].

The specific capacitances of composite hydrogels remained approximately 80%. For all samples, the relationships between the specific capacitance and the current density are shown in Figure 7c. At the same current density, these composite hydrogels containing PANI possessed higher specific capacitances than TOCNF-CNT/PVA hydrogels, indicating that PANI significantly increased the specific capacitance of the composite hydrogel. The sp^2-hybridized carbon atoms of CNTs formed π-π stacking interactions with the quinoid ring of the PANI without destroying the graphitized plane of CNTs [49,50]. Among these hydrogels, the TOCNF-CNT@PANI/PVA-2 possessed the largest specific capacitance. It could be attributed to the appropriate ratio of PANI to CNTs, which allowed the PANI to better combine with the CNT networks. The developed pore structure and large specific surface area were beneficial to the charge accumulation and enhanced the specific capacitance.

Nyquist plots from EIS of the composite hydrogel electrodes are shown in Figure 7d. The Nyquist plots of hydrogel electrodes showed a typical semicircle in the high-frequency region. The intercept of semicircle represented the equivalent series resistance (ESR), and the diameter of semicircle represented the charge-transfer resistance (R_{ct}) of the interface. Correspondingly, the diameter of a semicircle of TOCNF-CNT@PANI/PVA-2 hydrogel was the smallest in all samples, indicating that TOCNF-CNT@PANI/PVA-2 hydrogel possessed the lowest resistance. It was because the cross-linked 3D network structure in composite hydrogels provided an ideal charge-transfer path. Nyquist plots showed a straight line at the low-frequency region, and nearly vertical shape reflected the ideal capacitance characteristics [51,52].

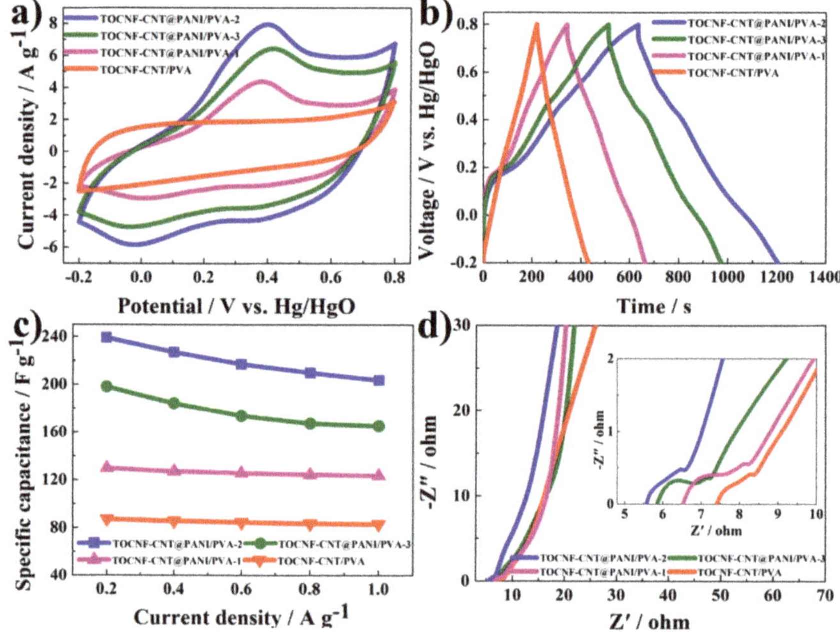

Figure 7. (**a**) CV (cyclic voltammetry) curves at 40 mV s^{-1} scan rate; (**b**) G-CD (galvanostatic charge-discharge) curves at 0.4 A g^{-1} current density; (**c**) specific capacitance of the hydrogel electrodes at different current densities; (**d**) Nyquist plot of hydrogel electrode and enlarged illustration of high-frequency region.

3.8. Self-Healable and Flexible Performance of the Supercapacitor

The self-healable and flexible solid-state supercapacitor was fabricated based on TOCNF-CNT@PANI/PVA-2 hydrogel electrode and TOCNF/PVA hydrogel electrolyte in a sandwich structure. The detailed fabrication process was described in the experimental part. Due to the inherent flexibility and self-healing ability of the PVA hydrogel, the interfaces between hydrogel electrode and electrolyte could be completely combined, thereby manufacturing an integrated solid supercapacitor device. The assembled supercapacitor could withstand cutting, bending, and another mechanical damage, but the electrochemical performance was not obviously affected. As demonstrated above, the dynamically reversible PVA-borate cross-linking network provided the inherent, repeatable, and effective self-healable ability for the composite hydrogel (Figure 8a).

Figure 8b,c shows the electrochemical performances of the self-healing supercapacitor after multiple cutting/healing cycles. The CV and G-CD curves of supercapacitor had no obvious deformation after multiple cutting/healing cycles, indicating that the capacitance had not been significantly reduced. As calculated by the G-CD curves at 0.6 A g^{-1} current density, the initial specific capacitance of the supercapacitor was 138 F g^{-1}, and the specific capacitance after 1, 5, 10 cutting/healing cycles was 137.3, 134.7, and 124.1 F g^{-1}, respectively. The corresponding capacitance retention was 99.5%, 97.6%, and 90.0%, respectively. For the as-prepared supercapacitor device, the self-healing capability was more outstanding than those reports. For example, a CNT film was spread on a self-healable substrate to manufacture electrode, combining the self-healable electrodes and polyvinylpyrrolidone-sulfuric acid (PVP-H$_2$SO$_4$) gel electrolyte to fabricate a supercapacitor. Its capacitance retention reached 85.7% after the 5th cutting/healing cycles [53]. By coating PANI and CNT nanomaterials on the surface of polymer fibers with self-healing ability to develop a novel filamentous self-healable supercapacitor, its capacitance retention was 92% after one cutting/healing cycle [54]. In Figure 8c, the voltage drop in the G-CD curves was due to the electron transfer resistance of the solid-state hydrogel-based electrolyte, which resulted in the specific capacitance of the supercapacitor being smaller than that of the electrode [3].

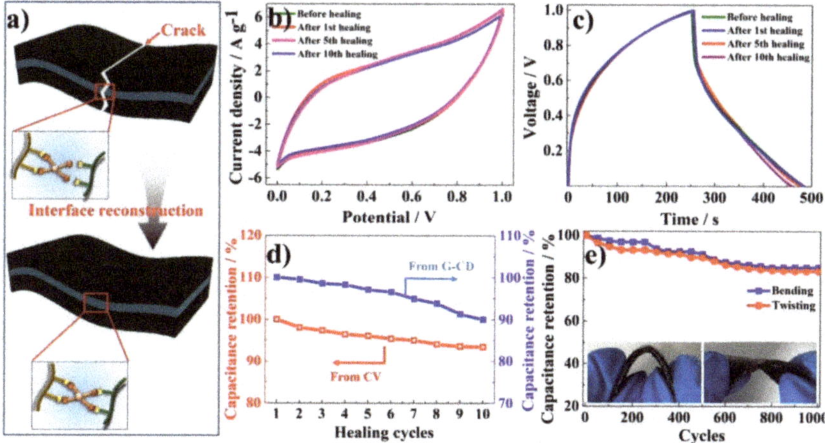

Figure 8. (**a**) Schematic and structural illustration of the self-healing process due to the dynamic borate bond of the supercapacitor. (**b**) CV curves at initial and after the 1st, 5th, and 10th self-healing at a scan rate of 40 mV s^{-1}. (**c**) G-CD curves at initial and after the 1st, 5th, and 10th self-healing at a current density of 0.6 A g^{-1}. (**d**) Self-healing efficiency derived from both G-CD and CV curves at self-healing cycles from 1st to 10th. (**e**) Capacitance retention over 1000 bending and twisting cycles at an angle of 180°.

The in-situ measurement method was used to evaluate the electrochemical performance of the assembled supercapacitor device under the bending or twisting state. Figure 8e shows the variation of capacitance according to the number of cycles, which were calculated from their G-CD curves at a constant current density of 1 A g^{-1} after each bending cycle. A bending cycle started from flat-state, passed through a 180° bending-state, and then returned to flat-state. One twisting cycle was similar. In Figure 8e, the capacitance retention of the supercapacitor device was 85.0% and 82.3% after 1000 bending cycles and twisting cycles, respectively. The performance was comparable supercapacitor to be tested in a flat state. Such a flexible solid-state supercapacitor with PANI hydrogel electrode possessed capacitance retention of 86% after 1000 consecutive charge-discharge cycles [55]. It even was superior to the polyaniline-sodium alginate (PANI-SA) hydrogel supercapacitor reported previously (typically 71% retention for over 1000 cycles) [51]. The improved cycling stability could be due to the quasi-solid hydrogel, further protecting the active PANI and avoiding the delamination of the CNT fibers as conductive pathways. The delamination from continuous expansion and shrinkage of the PANI molecular chain during the charge-discharge cycles could cause performance degradation. The superior capacitance retention under bending/twisting cycles also suggested that the contact between different layers of the supercapacitor device was excellent, which could benefit from the inherent self-healing property of PVA-based hydrogels [43]. The superior flexibility and self-healing solid-state supercapacitor had promising potential applications in a flexible electronic device.

4. Conclusions

A unique, flexible, and self-healing ECHs were synthesized through introducing TOCNF-CNT@PANI nanohybrid with a "core-shell" structure into viscoelastic PVA hydrogel matrix. The nanohybrid built a 3D hierarchical framework in the hydrogel matrix, which not only improved the viscoelasticity but also enhanced the electrochemical performance of the ECHs. The ECH possessed a high mechanical toughness ($\sigma_s \approx 128$ kPa cm^3 g^{-1} and $E_e \approx 61$ kPa), excellent viscoelastic characteristics ($G'_\infty \approx 18.2$ kPa and $G''_{max} \approx 7.6$ kPa), and ideal electroconductivity (up to 15.3 S m^{-1}). The specific capacitance of the TOCNF-CNT@PANI/PVA-2 hydrogel electrode was 226.8 F g^{-1} at a current density of 0.4 A g^{-1}. The ECHs exhibited fast self-healing character within 20 s at room temperature and superior flexile performance due to the reversible and dynamic borate-associated network. The symmetric solid-state supercapacitor was fabricated by the TOCNF-CNT@PANI/PVA-2 hydrogel electrodes and TOCNF/PVA electrolyte; the capacitance retention was 90% after 10 cutting/healing cycles; the capacitance retention was 85.0% and 82.3% after 1000 bending and twisting cycles, respectively. Consequently, the novel ECHs provided an alternative platform for personal wearable electronic devices.

Author Contributions: Conceptualization, J.H. and X.X.; methodology, J.H. and X.X.; validation, Y.L., Y.Y., J.H. and X.X.; formal analysis, J.H. and X.X.; investigation, Q.W.; resources, J.H. and X.X.; writing—original draft preparation, H.W. and S.K.B. and S.Z.; writing—review and editing, Y.L., Y.Y., J.H., X.X., Q.W. and H.X.; supervision, J.H. and X.X. All authors have read and agreed to the published version of the manuscript.

Funding: The authors are thankful for the financial support from National Natural Science Foundation of China (31770609), Natural Science Foundation of Jiangsu Province for Outstanding Young Scholars (BK20180090), Qing Lan Project of Jiangsu Province (2019), 333 Project Foundation of Jiangsu Province (BRA2018337), Priority Academic Program Development (PAPD), and Analysis and Test Center of Nanjing Forestry University.

Conflicts of Interest: The authors declare no conflicts of interest.

References

1. Wang, K.; Zhang, X.; Li, C.; Sun, X.; Meng, Q.; Ma, Y.; Wei, Z. Chemically crosslinked hydrogel film leads to integrated flexible supercapacitors with superior performance. *Adv. Mater.* **2015**, *27*, 7451–7457. [CrossRef]
2. Wang, Z.; Li, H.; Tang, Z.; Liu, Z.; Ruan, Z.; Ma, L.; Yang, Q.; Wang, D.; Zhi, C. Hydrogel electrolytes for flexible aqueous energy storage devices. *Adv. Funct. Mater.* **2018**, *28*. [CrossRef]
3. Chen, C.R.; Qin, H.; Cong, H.P.; Yu, S.H. A highly stretchable and real-time healable supercapacitor. *Adv. Mater.* **2019**, *31*. [CrossRef]
4. Sun, X.; Cai, M.; Chen, L.; Qiu, Z.; Liu, Z. Electrodes of carbonized MWCNT-cellulose paper for supercapacitor. *J. Nanopart. Res.* **2017**, *19*. [CrossRef]
5. Wang, Z.H.; Carlsson, D.O.; Petter, T.; Kai, H.; Zhang, P.; Leif, N.; Maria, S. Surface modified nanocellulose fibers yield conducting polymer-based flexible supercapacitors with enhanced capacitances. *ACS Nano* **2015**, *9*, 7563–7571. [CrossRef]
6. Ding, Q.; Xu, X.; Yue, Y.; Mei, C.; Huang, C.; Jiang, S.; Wu, Q.; Han, J. Nanocellulose-mediated electroconductive self-Healing hydrogels with high strength, plasticity, viscoelasticity, stretchability, and biocompatibility toward multifunctional applications. *ACS Appl. Mater. Interfaces* **2018**, *10*, 27987–28002. [CrossRef]
7. Huang, Y.; Zhong, M.; Shi, F.; Liu, X.; Tang, Z.; Wang, Y.; Huang, Y.; Hou, H.; Xie, X.; Zhi, C. An intrinsically stretchable and compressible supercapacitor containing a polyacrylamide hydrogel electrolyte. *Angew. Chem.* **2017**, *56*, 9141–9145. [CrossRef]
8. Shi, Y.; Zhang, J.; Bruck, A.M.; Zhang, Y.; Li, J.; Stach, E.A.; Takeuchi, K.J.; Marschilok, A.C.; Takeuchi, E.S.; Yu, G. A Tunable 3D nanostructured conductive gel framework electrode for high-performance lithium ion batteries. *Adv. Mater.* **2017**, *29*. [CrossRef]
9. Das, S.; Chakraborty, P.; Mondal, S.; Shit, A.; Nandi, A.K. Enhancement of energy storage and photo response properties of folic acid-polyaniline hybrid hydrogel by in situ growth of Ag nanoparticles. *ACS Appl. Mater. Interfaces* **2016**, *8*, 28055–28067. [CrossRef]
10. Luo, W.; Hayden, J.; Jang, S.-H.; Wang, Y.; Zhang, Y.; Kuang, Y.; Wang, Y.; Zhou, Y.; Rubloff, G.W.; Lin, C.F.; et al. Highly conductive, light weight, robust, corrosion-resistant, scalable, all-fiber based current collectors for aqueous acidic batteries. *Adv. Energy Mater.* **2017**, *8*. [CrossRef]
11. Zhu, H.; Parvinian, S.; Preston, C.; Vaaland, O.; Ruan, Z.; Hu, L. Transparent nanopaper with tailored optical properties. *Nanoscale* **2013**, *5*, 3787–3792. [CrossRef]
12. Meng, C.; Liu, C.; Chen, L.; Hu, C.; Fan, S. Highly flexible and all-solid-state paperlike polymer supercapacitors. *Nano Lett.* **2010**, *10*, 4025–4031. [CrossRef]

13. Hamedi, M.M.; Hajian, A.; Fall, A.B. Highly conducting, strong nanocomposites based on nanocellulose-assisted aqueous dispersions of single-wall carbon nanotubes. *ACS Nano* **2014**, *8*, 2467–2476. [CrossRef]

14. Chen, P.Y.; Dorval, C.; Hyder, M.N.; Qi, J.; Belcher, A.M.; Hammond, P.T. Carbon nanotube–polyaniline core–shell nanostructured hydrogel for electrochemical energy storage. *RSC Adv.* **2015**, *5*, 37970–37977. [CrossRef]

15. Shi, Y.; Ma, C.; Peng, L.; Yu, G. Conductive "smart" hybrid hydrogels with PNIPAM and nanostructured conductive polymers. *Adv. Funct. Mater.* **2015**, *25*, 1219–1225. [CrossRef]

16. Wu, X.; Lu, C.; Zhang, X.; Zhou, Z. Conductive natural rubber/carbon black nanocomposites via cellulose nanowhisker templated assembly: Tailored hierarchical structure leading to synergistic property enhancements. *J. Mater. Chem. A* **2015**, *3*, 13317–13323. [CrossRef]

17. Hao, G.P.; Hippauf, F.; Oschatz, M. Stretchable and semitransparent conductive hybrid hydrogels for flexible supercapacitors. *ACS Nano* **2014**, *8*, 7138–7146. [CrossRef]

18. Huan, S.; Bai, L.; Cheng, W.; Han, G. Manufacture of electrospun all-aqueous poly(vinyl alcohol)/cellulose nanocrystal composite nanofibrous mats with enhanced properties through controlling fibers arrangement and microstructure. *Polymer* **2016**, *92*, 25–35. [CrossRef]

19. Zhao, Z.Y.; Miao, Y.F.; Yang, Z.Q.; Wang, H.; Sang, R.J.; Fu, Y.C.; Huang, C.X.; Wu, Z.H.; Zhang, M.; Sun, S.J.; et al. Effects of sulfuric acid on the curing behavior and bonding performance of tannin-sucrose adhesive. *Polymers* **2018**, *10*, 651. [CrossRef]

20. Zhao, Z.Y.; Sun, S.J.; Wu, D.; Zhang, M.; Huang, C.X.; Umemura, K.; Yong, Q. Synthesis and characterization of sucrose and ammonium dihydrogen phosphate (SADP) adhesive for plywood. *Polymers* **2019**, *11*, 1909. [CrossRef]

21. Zhao, Z.Y.; Shin, H.; Xu, W.; Wu, Z.H.; Tanaka, S.; Sun, S.J.; Zhang, M.; Kozo, K.Y.; Umemura, K. A Novel eco-friendly wood adhesive composed by sucrose and ammonium dihydrogen phosphate. *Polymers* **2018**, *10*, 1251. [CrossRef] [PubMed]

22. Wang, H.; Zhu, E.; Yang, J.; Zhou, P.; Sun, D.; Tang, W. Bacterial cellulose nanofiber-supported polyaniline nanocomposites with flake-shaped morphology as supercapacitor electrodes. *J. Phys. Chem. C* **2012**, *116*, 13013–13019. [CrossRef]

23. Zhao, Z.Y.; Sakai, S.; Wu, D.; Chen, Z.; Zhu, N.; Huang, C.X.; Sun, S.J.; Zhang, M.; Umemura, K.; Yong, Q. Exploration of sucrose–citric acid adhesive: Investigation of optimal hot-pressing conditions for plywood and curing behavior. *Polymers* **2019**, *11*, 1996. [CrossRef]

24. Sun, S.J.; Zhao, Z.Y.; Umemura, K. Further Exploration of sucrose-citric acid adhesive: Synthesis and application on plywood. *Polymers* **2019**, *11*, 1875. [CrossRef]

25. Popenomo, G.; Guan, C.; Song, X. Hollow fiber membrane decorated with Ag/MWNTs: Toward effective water disinfection and biofouling control. *ACS Nano* **2011**, *5*, 10033–10040.

26. Bai, L.; Bossa, N.; Qu, F.; Winglee, J.; Li, G.; Sun, K.; Liang, H.; Wiesner, M.R. Comparison of hydrophilicity and mechanical properties of nanocomposite membranes with cellulose nanocrystals and carbon nanotubes. *Environ. Sci. Technol.* **2017**, *51*, 253–262. [CrossRef]

27. Han, J.; Zhou, C.; Wu, Y.; Liu, F.; Wu, Q. Self-assembling behavior of cellulose nanoparticles during freeze-drying: Effect of suspension concentration, particle size, crystal structure, and surface charge. *Biomacromolecules* **2013**, *14*, 1529–1540. [CrossRef]

28. Mougel, J.-B.; Adda, C.; Bertoncini, P.; Capron, I.; Cathala, B.; Chauvet, O. Highly efficient and predictable noncovalent dispersion of single-walled and multi-walled carbon nanotubes by cellulose nanocrystals. *J. Phys. Chem. C* **2016**, *120*, 22694–22701. [CrossRef]

29. Tang, Y.; He, Z.; Mosseler, J.A.; Ni, Y. Production of highly electro-conductive cellulosic paper via surface coating of carbon nanotube/graphene oxide nanocomposites using nanocrystalline cellulose as a binder. *Cellulose* **2014**, *21*, 4569–4581. [CrossRef]

30. Feng, J.; Li, J.; Lv, W.; Xu, H.; Yang, H.; Yan, W. Synthesis of polypyrrole nano-fibers with hierarchical structure and its adsorption property of Acid Red G from aqueous solution. *Synth. Met.* **2014**, *191*, 66–73. [CrossRef]

31. Simotwo, S.K.; DelRe, C.; Kalra, V. Supercapacitor electrodes based on high-purity electrospun polyaniline and polyaniline-carbon nanotube nanofibers. *ACS Appl. Mater. Interfaces* **2016**, *8*, 21261–21269. [CrossRef]

32. Koysuren, O.; Karaman, M.; Dinc, H. Preparation and characterization of polyvinyl borate/polyvinyl alcohol (PVB/PVA) blend nanofibers. *J. Appl. Polym. Sci.* **2012**, *124*, 2736–2741. [CrossRef]

33. Spoljaric, S.; Salminen, A.; Luong, N.D.; Seppälä, J. Stable, self-healing hydrogels from nanofibrillated cellulose, poly(vinyl alcohol) and borax via reversible crosslinking. *Eur. Polym. J.* **2014**, *56*, 105–117. [CrossRef]

34. Cardoso, R.M.; Montes, R.H.O.; Lima, A.P.; Dornellas, R.M.; Nossol, E.; Richter, E.M.; Munoz, R.A.A. Multi-walled carbon nanotubes: Size-dependent electrochemistry of phenolic compounds. *Electrochim. Acta* **2015**, *176*, 36–43. [CrossRef]

35. Kuzmenko, V.; Saleem, A.M.; Staaf, H.; Haque, M.; Bhaskar, A.; Flygare, M.; Svensson, K.; Desmaris, V.; Enoksson, P. Hierarchical cellulose-derived CNF/CNT composites for electrostatic energy storage. *J. Micromech. Microeng.* **2016**, *26*, 12401–12407. [CrossRef]

36. Liu, S.; Yu, T.; Wu, Y.; Li, W.; Li, B. Evolution of cellulose into flexible conductive green electronics: A smart strategy to fabricate sustainable electrodes for supercapacitors. *RSC Adv.* **2014**, *4*, 34134–34143. [CrossRef]

37. Jasim, A.; Ullah, M.W.; Shi, Z.; Lin, X.; Yang, G. Fabrication of bacterial cellulose/polyaniline/single-walled carbon nanotubes membrane for potential application as biosensor. *Carbohydr. Polym.* **2017**, *163*, 62–69. [CrossRef]

38. Han, J.; Lei, T.; Wu, Q. High-water-content mouldable polyvinyl alcohol-borax hydrogels reinforced by well-dispersed cellulose nanoparticles: Dynamic rheological properties and hydrogel formation mechanism. *Carbohydr. Polym.* **2014**, *102*, 306–316. [CrossRef]

39. Zhou, C.; Wu, Q. A novel polyacrylamide nanocomposite hydrogel reinforced with natural chitosan nanofibers. *Colloids Surf. B Biointerfaces* **2011**, *84*, 155–162. [CrossRef]

40. Li, W.; Zeng, X.; Wang, H.; Wang, Q.; Yang, Y. Polyaniline-poly(styrene sulfonate) conducting hydrogels reinforced by supramolecular nanofibers and used as drug carriers with electric-driven release. *Eur. Polym. J.* **2015**, *66*, 513–519. [CrossRef]

41. Shi, X.; Hu, Y.; Tu, K.; Zhang, L.; Wang, H.; Xu, J.; Zhang, H.; Li, J.; Wang, X.; Xu, M. Electromechanical polyaniline–cellulose hydrogels with high compressive strength. *Soft Matter* **2013**, *9*, 10129–10134. [CrossRef]

42. Wang, Q.; Mynar, J.L.; Yoshida, M.; Lee, E.; Lee, M.; Okuro, K.; Kinbara, K.; Aida, T. High-water-content mouldable hydrogels by mixing clay and a dendritic molecular binder. *Nature* **2010**, *463*, 339–343. [CrossRef] [PubMed]

43. Chen, W.P.; Hao, D.Z.; Hao, W.J.; Guo, X.L.; Jiang, L. Hydrogel with ultrafast self-healing property both in air and underwater. *ACS Appl. Mater. Interfaces* **2018**, *10*, 1258–1265. [CrossRef]

44. Chen, T.; Cai, Z.; Qiu, L.; Li, H.; Ren, J.; Lin, H.; Yang, Z.; Sun, X.; Peng, H. Synthesis of aligned carbon nanotube composite fibers with high performances by electrochemical deposition. *J. Mater. Chem. A* **2013**, *1*, 2211–2216. [CrossRef]

45. Cai, G.; Wang, J.; Qian, K.; Chen, J.; Li, S.; Lee, P.S. Extremely stretchable strain sensors based on conductive self-healing dynamic cross-links hydrogels for human-motion detection. *Adv. Sci.* **2017**, *4*, 160–169. [CrossRef] [PubMed]

46. Shi, Y.; Yu, G. Designing hierarchically nanostructured conductive polymer gels for electrochemical energy storage and conversion. *Chem. Mater.* **2016**, *28*, 2466–2477. [CrossRef]

47. Lee, S.-Y.; Kim, J.-I.; Park, S.-J. Activated carbon nanotubes/polyaniline composites as supercapacitor electrodes. *Energy* **2014**, *78*, 298–303. [CrossRef]

48. Liu, R.; Ma, L.; Huang, S.; Mei, J.; Xu, J.; Yuan, G. A flexible polyaniline/graphene/bacterial cellulose supercapacitor electrode. *New J. Chem.* **2017**, *41*, 857–864. [CrossRef]

49. Kim, J.; Ju, H.; Inamdar, A.I.; Jo, Y.; Han, J.; Kim, H.; Im, H. Synthesis and enhanced electrochemical supercapacitor properties of Ag–MnO_2–polyaniline nanocomposite electrodes. *Energy* **2014**, *70*, 473–477. [CrossRef]

50. Giri, S.; Das, C.K. Preparation and electrochemical characterization of polyaniline functionalized copper bridges carbon nanotube for supercapacitor applications. *J. Nanosci. Nanotechnol.* **2014**, *14*, 6373–6381. [CrossRef] [PubMed]

51. Huang, H.; Zeng, X.; Li, W.; Wang, H.; Wang, Q.; Yang, Y. Reinforced conducting hydrogels prepared from the in situ polymerization of aniline in an aqueous solution of sodium alginate. *J. Mater. Chem. A* **2014**, *2*, 16516–16522. [CrossRef]

52. Pan, L.; Yu, G.; Zhai, D.; Lee, H.R.; Zhao, W.; Liu, N.; Wang, H.; Tee, B.C.; Shi, Y.; Cui, Y.; et al. Hierarchical nanostructured conducting polymer hydrogel with high electrochemical activity. *Proc. Natl. Acad. Sci. USA* **2012**, *109*, 9287–9292. [CrossRef] [PubMed]

53. Wang, H.; Zhu, B.; Jiang, W.; Yang, Y.; Leow, W.R.; Wang, H.; Chen, X. A mechanically and electrically self-healing supercapacitor. *Adv. Mater.* **2014**, *26*, 3638–3643. [CrossRef] [PubMed]

54. Sun, H.; You, X.; Jiang, Y.; Guan, G.; Fang, X.; Deng, J.; Chen, P.; Luo, Y.; Peng, H. Self-healable electrically conducting wires for wearable microelectronics. *Angew. Chem.* **2014**, *53*, 9526–9531. [CrossRef] [PubMed]

55. Wang, K.; Zhang, X.; Li, C.; Zhang, H.; Sun, X.; Xu, N.; Ma, Y. Flexible solid-state supercapacitors based on a conducting polymer hydrogel with enhanced electrochemical performance. *J. Mater. Chem. A* **2014**, *2*, 19726–19732. [CrossRef]

MDPI

St. Alban-Anlage 66

4052 Basel

Switzerland

Tel. +41 61 683 77 34

Fax +41 61 302 89 18

www.mdpi.com

Nanomaterials Editorial Office

E-mail: nanomaterials@mdpi.com

www.mdpi.com/journal/nanomaterials